Sustainability in Plant and Crop Protection

Series Editor

Aurelio Ciancio, Sezione di Bari, Consiglio Nazionale delle Ricerche Istituto per la Protezione delle Piante, Bari, Italy

More information about this series at http://www.springer.com/series/13031

Doris Zúñiga-Dávila
Fernando González-Andrés
Ernesto Ormeño-Orrillo
Editors

Microbial Probiotics for Agricultural Systems

Advances in Agronomic Use

Springer

Editors
Doris Zúñiga-Dávila
Laboratorio de Ecología Microbiana y Biotecnología, Departamento de Biología-Facultad de Ciencias
Universidad Nacional Agraria La Molina
Lima, Peru

Fernando González-Andrés
Research Group IQUIMAB, Institute of Environment Natural Resources and Biodiversity
University of León
León, Spain

Ernesto Ormeño-Orrillo
Laboratorio de Ecología Microbiana y Biotecnología, Departamento de Biología-Facultad de Ciencias
Universidad Nacional Agraria La Molina
Lima, Peru

All the chapters included in this book have been peer-reviewed with a positive result. They were selected among the works presented at the III IBEMPA (Ibero-American Conference on Beneficial Plant-Microorganisms-Environment Interactions), held on November 6–10, 2017, in Lima, Peru.

ISSN 2567-9805 ISSN 2567-9821 (electronic)
Sustainability in Plant and Crop Protection
ISBN 978-3-030-17599-3 ISBN 978-3-030-17597-9 (eBook)
https://doi.org/10.1007/978-3-030-17597-9

This Springer imprint is published by the registered company Springer Nature Switzerland AG.
The registered company address is: Gewerbestrasse 11, 6330 Cham, Switzerland

To Prof. Dr. José-Miguel Barea in memoriam

Preface

This volume, entitled *Microbial Probiotics for Agricultural Systems: Advances in Agronomic Use* includes 17 chapters derived from selected contributions to the III IBEMPA (Ibero-American Conference on Beneficial Plant-Microorganisms-Environment Interactions) held at the "Universidad Agraria La Molina" in Lima (Peru) during November 6–10, 2017. The international event gathered more than 150 scientists from Spain, Portugal, Mexico, Cuba, Colombia, Venezuela, Ecuador, Peru, Brasil, Bolivia, Chile, Argentina, and Uruguay. IBEMPA conferences stand out for bringing together the most outstanding scientists in the agronomical aspects of microbial plant biostimulants. For this reason, this book aims at providing a reference in the agronomic application of microbial probiotics.

The book consists of two main parts, the first devoted to review articles (6 chapters), and the second is composed of original research articles (11 chapters). Five chapters cover the aspects related to the "economy" of microbes and the interaction of microbes-plant-environment (Chaps. 1, 2, 3, 8, and 9). They aim at understanding the agronomic performance of microorganisms. The other chapters gather recent evidences of agricultural uses of inoculants, focusing on new applications and on key points required to overcome known inconsistences in microbial-based products performance, on the field scale. Three chapters (Chaps. 6, 7, and 15) focus on the characterization, conservation, and exploitation of microbial biodiversity as a key strategy to sustain the agronomic performance of related products.

The book attempts to bridge the gap between the extensive literature on the –omics of the symbiotic relationships of soil microorganisms with plants and the environment, and the increasing demand of knowledge needed for the development of commercial products.

The editors gratefully acknowledge the ALAR and SEFIN executive board and the staff at the Universidad Nacional Agraria, La Molina, for their participation in the organization of the event, as well as all the sponsors of the Congress.

This book is dedicated *in memoriam* to Prof. Dr. José Miguel Barea, an exceptional scientist and man, for all his teachings and scientific contributions.

Lima, Peru — Doris Zúñiga-Dávila
León, Spain — Fernando González-Andrés
Lima, Peru — Ernesto Ormeño-Orrillo

Obituary

In memory of José Miguel Barea Navarro, pioneer of mycorrhizal research in Spain.

Last April 3rd, 2018, our beloved colleague and friend Prof. Dr. José Miguel Barea Navarro died in Granada. José Miguel was born in Granada in 1942. He graduated in Pharmacy, at the University of Granada, in 1965. In 1968, he defended his PhD thesis entitled "Studies on soil microorganisms capable of mineralizing organic phosphates". The thesis was evaluated as "Excellent *cum laude*" and he was awarded the highest honor from the University of Granada, known as the "Extraordinary PhD Prize". Between 1968 and 1972 he was Professor of Microbiology at the University of Granada, later becoming part of the Spanish Council for Scientific Research (CSIC) as Tenured Scientist (1972–1976), Research Scientist (1976–1984), Research Professor (1984–2012), and Associate Professor *ad honorem* since 2012.

The whole of his research activity was developed at the Estación Experimental del Zaidín (EEZ-CSIC, Granada, Spain), where he was Deputy Director (1983–1987) and Director (1989–1997). He was a very dynamic person, and until the end he remained involved in all the activities of his research group. Although in the end his health deteriorated, he remained active and taught master classes and postgraduate courses practically until his last days.

José Miguel developed his entire professional career in the field of beneficial plant–microbe interactions and, more specifically, in the study of arbuscular mycorrhizas and their interactions with rhizospheric microorganisms. In 1972, he moved to Rothamsted Experimental Station (UK) to work with Dr. Margaret Brown on phosphate solubilizing bacteria. There he met Dr. Barbara Mosse, considered the "mother" of mycorrhizas worldwide, who instilled interest in this plant–fungus symbiosis. Upon his return to Granada, in collaboration with his wife Charo Azcón, also a scientist, he started this line of research at EEZ. José Miguel was a leader in Spain in this research line, and among the pioneers in Europe. Most researchers who currently work in this discipline in Spain and many other Ibero-American countries have been trained under his supervision.

Throughout his long scientific career, José Miguel supervised 32 PhD theses, published more than 170 scientific papers in journals with a high impact index, and about 100 book chapters. He led multiple research projects, funded mainly by the Spanish government and the EU. His way of understanding science, always open to sharing his knowledge and exploring new possibilities, led José Miguel to collaborate with researchers from practically the entire world. His links with other European researchers were particularly strong and led to a time of great development of mycorrhiza research in Europe. He contributed notably to the development of several COST Actions (European Cooperation in Science and Technology), promoted by the EU, on arbuscular mycorrhizas. He also organized in Granada some of the most relevant congresses on mycorrhizas, such as the 4th European Symposium on Mycorrhiza (4ESM), in 1994 and the 5th International Conference on Mycorrhiza (ICOM5), in 2006.

Especially intense were his ties with Ibero-America. Of the 32 PhD theses supervised by José Miguel, 10 corresponded to Ibero-American students. His open and enthusiastic personality and the novelty of his proposals contributed to turn his research group into a magnet for visiting researchers from different parts of the world. Most of the trained postdocs returned to their countries of origin and began a line of research following the ideas developed under the guidelines of José Miguel.

His close relationship with Ibero-America is reflected in his participation in the Ibero-American Conferences on Beneficial Interactions between Microorganism-Plant-Environment (IBEMPA). José Miguel participated in all the conferences developed to date and held in La Habana (2009), Seville (2013), and Lima (2017). In fact, the IBEMPA developed in Lima in November 2017 in which he gave the inaugural conference, was almost his last participation at an international congress.

José Miguel had a great teaching vocation, as he enjoyed teaching classes, showing his ability to talk clearly and enjoyably. From the beginning he became involved in teaching at the University of Granada. Later, once he joined CSIC, although his work focused mainly on research, he never left teaching. Thus, since 1968 he was a professor of the International Course on Plant Biology, an annual course held uninterrupted since 1963 in the EEZ, being its director since 1998. In addition, he participated as a professor in Masters from the Universities of Granada and Murcia, and in multiple international postgraduate courses. José Miguel was one of the founding members of the Doctorate in Natural Resources Sciences and the Scientific Nucleus in Natural Resources of the La Frontera University (BIOREN-UFRO, Temuco, Chile). As a result of these collaborations, José Miguel was awarded the Rectoral Medal by the La Frontera University and was appointed Honorary Professor of the Faculty of Agronomy at the University of Buenos Aires (Argentina).

Throughout his scientific career, José Miguel was recognized with different awards. He received the research prize of the General Savings Bank of Granada in 1980 and the prize for scientific research of the Academy of Mathematical, Physical-Chemical, and Natural Sciences of Granada in 1984. In 2012, he received recognition from the Mexican Society of Mycorrhizal Symbiosis "for his contributions to the knowledge on mycorrhizae" and he was appointed as member of the Academy of Mathematical, Physical-Chemical, and Natural Sciences of Granada, where he delivered the conference "Mycorrhizas and Climate Change." At the First Research Conference on Mycorrhiza in Spain, held in Granada during 2012, he was honored in recognition of his scientific contributions and pioneering work in research on the subject.

One of the main reasons for its success was its extraordinary intelligence and its enormous capacity for work. He was always one of the first to arrive at the institute, even during the weekends. But that enthusiasm for his work was compatible with a great vitality and ability to enjoy life. He was very fond of sports, especially cycling and soccer. He was able to organize a trip to the Pyrenees with the whole family to see his idol Miguel Indurain race a stage of the *Tour de France* or go to Lisbon with his sons to see the UEFA Champions League final.

I would like to highlight his great human quality. José Miguel was a cheerful person, friendly and always willing to make jokes. We all remember his great memory and his incredible ability to tell jokes for hours. He loved to travel. He organized his trips up to the last detail. His travel experience, together with his great memory, made him a "living guide" to which all those who planned a trip in Spain or Europe asked for his advice. His recommendations went beyond the monuments or natural areas to visit.

José Miguel was an extremely generous person and always found reasons to propose a celebration that he organized or paid with pleasure. He was also generous in terms of time and work. The doors of his office were always open for any consultation and their laboratories and equipment available to researchers from any part of the world, who wanted to work or expand knowledge about the research lines developed there.

His absence has been felt by the entire scientific community that knew him, and the innumerable condolences we have received are a good proof of this. José Miguel did not leave anyone indifferent. Wherever he went, he left friends and disciples who remember fondly the excellent teacher and friend that he was. His loss as a friend and colleague is irreparable, but fortunately we will always keep his teachings and his happy and affectionate memory. RIP.

Estación Experimental del Zaidín Conchi Azcón
Granada, Spain

Contents

Part II Original Research Papers

Part I
Review Articles

Chapter 1
Cyclic di-GMP Regulation in Beneficial Plant-Microbe Interactions

Daniel Pérez-Mendoza and Juan Sanjuán

Abstract Cyclic diguanilate, c-di-GMP, has emerged as an ubiquitous secondary messenger in bacteria. It influences a wide range of cellular processes, including cell-cell signalling, cell cycle progression and virulence. However, it is best known for controlling the transition from a planktonic/motile lifestyle to a sessile biofilm mode of growth. Plant-interacting bacteria can live as free motile single cells or in association with their plant hosts, which provide them with numerous adaptive advantages. Also, the permanent exchange of multiple signals between plant and bacteria during the establishment of mutualistic interactions need to be integrated by the bacteria in order to modify their lifestyle to the new ecological niche. Most of the studied systems evidenced a temporal and spatial requirement for opposite c-di-GMP-metabolizing processes at different stages, culminating with an effective interaction with the plant host. In this review we highlight the importance of c-di-GMP signaling in beneficial plant-bacteria interactions, gathering the yet limited but significant knowledge acquired from different rhizobial and other plant growth-promoting rhizobacteria (PGPR).

Keywords Plant-bacteria interactions · C-di-GMP · Bacterial signalling · Rhizobia · PGPRs · Symbiosis

1.1 Sensing the Environment and Acting Accordingly

The ability of bacteria to adopt a particular lifestyle under specific environmental conditions is crucial for their survival. In order to cope with this ambitious task, microbes have developed signal transductions systems connecting the sense of specific environmental cues to appropriate changes in their physiology and gene

---

D. Pérez-Mendoza (✉) · J. Sanjuán
Dpto. Microbiología del Suelo y Sistemas Simbióticos, Estación Experimental del Zaidín, CSIC, Granada, Spain
e-mail: dpmendoza@eez.csic.es

D. Zúñiga-Dávila et al. (eds.), *Microbial Probiotics for Agricultural Systems*, Sustainability in Plant and Crop Protection,
https://doi.org/10.1007/978-3-030-17597-9_1

expression. However, the link between primary signals, also known as first messengers, with the appropriate cellular responses is not direct in most bacterial transduction systems (Galperin 2004). Instead, the perception of first messengers normally alters the intracellular levels of a second group of molecules, known as a bacterial second messengers (Gomelsky and Galperin 2013). Second messengers of different chemical nature can operate in bacterial transduction pathways, but without doubt one of the most important classes are those based on nucleotides (Pesavento and Hengge 2009). The presence of Nucleotide Second Messengers (NSMs) in bacterial signalling provides important advantages including: (i) the ability to easily diffuse within the cell serving as (ii) a single integrated output from multiple sensory inputs, which often allows to (iii) amplify the environmental signals, providing a (iv) quick and accurate cellular response to the changing milieu.

Much of the knowledge on NSM regulation has been acquired through the study of bacteria interacting with animal hosts, resulting in a scarcity of knowledge on bacteria associated with plants, either mutualistic or pathogenic. Despite this, the pace of discoveries lately accelerated, with new studies emphasizing the importance of NSM economy for plant-interacting bacteria. Associations between bacteria and plants vary from extracellular to intracellular, extending from the rhizosphere to the phyllosphere, and can range from mutualism to pathogenesis. However, in all cases the competence to colonize plants' habitats is important for the success of the interaction. Also, the permanent exchange of multiple signals between plants and bacteria during the establishment of the interaction must be integrated by the bacteria, in order to modify their lifestyle (from free-living to intimately interactive with a plant host) and to coordinate, in time and space, the expression of essential determinants for host invasion and colonization (Fig. 1.1; Soto et al. 2011).

1.2 Cyclic-di-GMP, a Universal Bacterial NSM

Cyclic mono- (cAMP, cGMP) and di-nucleotides (c-di-GMP, c-di-AMP and c-AMP-GMP), as well as linear nucleotides (ppGpp, pppGpp) have been described as important bacterial NSMs, regulating a multitude of processes including motility, cell division, substrate utilization, virulence factor production or biofilm formation (Hengge et al. 2016). Among them, c-di-GMP has recently experienced the most rapid advances and is indeed the most comprehensively studied NSM in bacteria. The bis-(3′,5′)-cyclic diguanosine monophosphate (cyclic diguanylate, c-di-GMP, cdG) was discovered 30 years ago as an allosteric activator of bacterial cellulose synthase (Ross et al. 1987; Römling and Galperin 2017). It is currently considered an ubiquitous second messenger in bacteria. Best known for its role in controlling the transition from a planktonic/motile lifestyle to a sessile biofilm mode of growth, this cyclic dinucleotide influences a wide range of cellular processes (see review in Römling et al. 2013). Although our understanding of c-di-GMP regulation in bacteria has exponentially grown during the last 10–15 years, knowledge on the

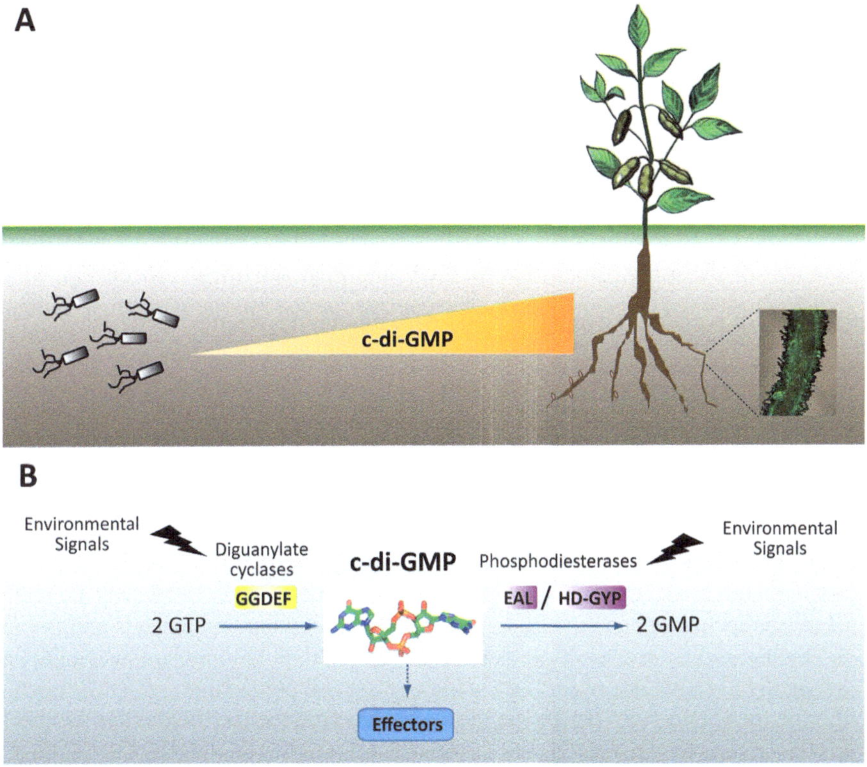

Fig. 1.1 Transition to a plant-associated lifestyle. (**a**) Cyclic-di-GMP is a key bacterial molecule controlling the transition from a motile saprophytic lifestyle in soil to a sessile community in intimate association with the plant hosts. A detail of a root colonization by a GFP-labeled Rhizobacteria with high intracellular levels of c-di-GMP is shown. (**b**) Intracellular c-di-GMP levels depend on the opposite activities of diguanylate cyclases, with GGDEF domains (in yellow), and phosphodiesterases, with EAL or HD-GYP domains (in purple), in accordance with environmental and physiological cues

molecular mechanisms of c-di-GMP regulation and the role of this second messenger in the life cycles of bacteria is still fragmented and far from complete. It is considered a universal bacterial second messenger because c-di-GMP metabolic enzymes can be found in strains of almost all eubacterial taxa (Römling et al. 2013). Numerous species, particularly those with complex life-cycles, including plant-interacting bacteria, contain dozens of c-di-GMP metabolic enzymes and display diverse c-di-GMP-triggered phenotypes. On the other hand, bacteria living in stable environments, e.g. obligate intracellular pathogens or extremophiles that occupy stable and unique ecological niches, usually have less-developed transduction systems and encode few or even none c-di-GMP related enzymes (Galperin et al. 2010).

1.3 A Versatile Regulatory Molecule

Cyclic-di-GMP signalling systems are generally composed of four major constituents: (i) diguanylate cyclases (DGCs, synthesize c-di-GMP from two GTP molecules), (ii) phosphodiesterases (PDEs, degrade c-di-GMP), (iii) c-di-GMP binding effectors, which interact with (iv) target components to produce a molecular output (Fig. 1.1b; Hengge 2009). The synthesis and degradation of c-di-GMP is relatively well understood. In general, the GGDEF domain of DGCs and the EAL or HD-GYP domains of PDEs are responsible for DGC and PDE activities, respectively, which can be modulated through various mechanisms, in accordance with environmental and physiological cues. Balanced control of these opposite activities determines c-di-GMP homeostasis within the cell (Sondermann et al. 2012). On the other hand, much remains to be discovered about the effector molecules that bind the dinucleotide and elicit downstream responses in the cell. To start with, the diversity of effector molecules is extraordinary and includes proteins with characteristic PilZ, GIL, or degenerate GGDEF and EAL domains, c-di-GMP specific binding domains in transcription factors, and even RNA motifs (riboswitches) (Amikam and Galperin 2006; Fang et al. 2014; Mills et al. 2011; Römling et al. 2013; Ryjenkov et al. 2006; Sondermann et al. 2012). The diversity of c-di-GMP effectors evidences the ability of this molecule to interact with a large number of cellular components and to exert its regulatory action at multiple tiers: transcriptional, post-transcriptional and post-translational levels (Hengge 2009). Although several predictable c-di-GMP binding motifs are known, i.e., the PilZ and degenerate GGDEF and EAL domains, c-di-GMP can bind to a diverse range of protein folds which cannot be bioinformatically predicted in advance (Chou and Galperin 2016; Pérez-Mendoza et al. 2017). Thus, the growing number of functions regulated by c-di-GMP and the multiplicity of DGCs and PDEs, combined with the complexity of c-di-GMP signalling, contrast with the comparatively few c-di-GMP receptors/effectors identified so far. This situation has been compared to a dysfunctional army with far more officers than soldiers (Römling et al. 2013), suggesting that many c-di-GMP binding proteins (the soldiers) still await discovery.

1.4 Transitioning for a Better Life

Plant-interacting bacteria can live as free motile single cells or as part of a sessile community and/or in association with their hosts (Fig. 1.1a). Each lifestyle provides different adaptive advantages: motility enables bacteria to seek and colonize new ecological niches, while in biofilms or in association with plants they are more protected from adverse conditions (Hall-Stoodley et al. 2004). The detection of and the response to plant-released signals is one of the first events in plant-bacteria interactions. This involves bacterial chemotaxis toward plant root exudates or wound saps (Scharf et al. 2016). Bacterial signalling systems transduce chemical stimuli into different forms of flagellar-mediated motility (swarming and swimming), directing their movements toward the plant host. Bacterial motility is almost universally repressed by c-di-GMP (Wolfe and Visick 2008). This is part of the requirement for the transition from single motile cells to a surface-attached sessile state. Furthermore, this NSM is able to adjust the velocity providing an accurate control of this process (Boehm et al. 2010). This regulation relies on the motility control at both transcriptional and post-translational levels. Transcription of different motility gene clusters are under control of master regulators responsive to c-di-GMP, such as FleQ in Pseudomonads. Cyclic-di-GMP binds to FleQ to dampen its activity, causing down-regulation of flagella gene expression (Baraquet and Harwood 2013). On the other hand, flagella rotation requires a quick and fine-tuned regulation on the scale of seconds. Thus, its control must also operate at the post-translational level. Different effectors (YcgRs and related proteins) have been described, which after binding c-di-GMP, are able to act as a 'molecular brake' controlling the direction and/or the velocity of the flagellar rotation (Paul et al. 2010; Boehm et al. 2010). This dual regulation of motility at both transcriptional and posttranslational levels by c-di-GMP can be referred to as so-called 'sustained sensing'. This involves multiple receptors to regulate different steps, in the same biological process and provides an additional level of regulation complexity, being biologically meaningful (Orr et al. 2016).

As mentioned above, c-di-GMP plays also a key role in attachment to surfaces, and the formation of biofilm communities. The formation of a biofilm associated to the plant host allows for a lifestyle that is entirely different from the planktonic state. The biofilm is a dynamic three-dimensional structure where bacteria live in a self-produced matrix of extracellular polymeric substances. The fundamental roles of the biofilms, such as social cooperation, resource capture and enhanced survival, rely on the structural and functional properties of the matrix (Flemming et al. 2016). The composition of biofilm matrixes greatly differ among different species and growth conditions, but generally consists of polysaccharides, nucleic acids, lipids and proteins, all of them essential for biofilm structure and ecological function. The production of many matrix components is directly or indirectly controlled by c-di-GMP. The production and/or the secretion of exopolysaccharides (EPS) seems to be universally activated by c-di-GMP (Fuqua 2010; Liang 2015). EPS are a major fraction of the biofilm matrix, and are very often essential for biofilm formation. To date, nearly a dozen bacterial EPS (i.e. cellulose, alginate, curdlan, etc.) are known to be regulated by c-di-GMP (recently reviewed in Pérez-Mendoza and Sanjuán 2016). Proteins are also macromolecules recognized as crucial components of bacterial biofilm matrixes (Fong and Yildiz 2015), including proteinaceus adhesins and motility organelles (flagella, pili). Similarly to EPSs, c-di-GMP directly or indirectly activates the function of some adhesins and the production of amyloid curli fibres (Lindenberg et al. 2013; Fuqua 2010). Thus, c-di-GMP can be considered not just an intracellular modulator of cell adaptive responses to the ever changing environment, but also a driver of cellular processes leading to changing and adapting the extracellular environment, every single moment, to the needs of a bacterial community.

Different studies evidence that c-di-GMP also influences further steps in host-microbe interactions. For obvious reasons, most of the genetic screens have been focused on how c-di-GMP signaling pathways affect the infection and virulence in different animal and plant pathogens. Initial studies suggested that constitutively elevated c-di-GMP levels were detrimental for acute infections in animal pathogens (Tischler and Camilli 2005). However, subsequent genetic surveys in other bacteria have uncovered a more complex situation. A systematic screen for virulence in the plant pathogen *Xanthomonas campestris*, for example, did not find a direct correlation between the predicted changes in c-di-GMP concentrations and virulence. Proteins with either a role in c-di-GMP synthesis (containing GGDEF domains) or in degradation (containing EAL and HD-GYP domains) contribute to the virulence of *X. campestris* on Chinese radish (Ryan et al. 2007). In the same vein, infection studies in animal pathogens with *Vibrio* and *Salmonella* spp. revealed a temporal and spatial requirement for opposite c-di-GMP-metabolizing processes at different stages, in the course of infection (Ahmad et al. 2011; Tamayo et al. 2008). On the other hand, some diseases completely depend on the perpetuation of the bacterial infection in the host. The majority of chronic infections involve biofilm formation where c-di-GMP plays important roles (Tamayo et al. 2007). Furthermore, for some pathogens this NSM seems to be crucial "to decide" the infection strategy that the microbe is going to pursue. Cyclic-di-GMP together with other regulators (e.g. GacA/GacS and quorum sensing systems) inversely control functions involved in acute *versus* chronic infections in *Pseudomonas aeruginosa* (Moscoso et al. 2011). Some c-di-GMP signaling proteins involved in virulence are conserved between animal and plant pathogens. For example, a deletion on the DGC *dgcP* gene in *P. savastanoi* pv. savastanoi NCPPB 3335 and *P. aeruginosa* PAK reduced their virulence in olive plants and in a mouse acute lung injury model, respectively (Aragón et al. 2015). Nevertheless, what seems to be clear after numerous studies with different pathogenic models is that finely tuned c-di-GMP signaling is required for an adequate infection of the eukaryotic host.

Mechanisms by which c-di-GMP signaling affects host invasion include: cell adherence, cytotoxicity, intracellular infection, secretion of virulence factors, resistance to oxidative stress, and modulation of immune responses (reviewed in Römling et al. 2013). Different types of protein secretion systems (SS) have been shown to be subject to c-di-GMP regulation. Secretion of adhesins dependent on type I secretion systems (SS) (Pérez-Mendoza et al. 2011), Plant Cell Wall-degrading Enzymes (PCWDE) reliant on type II-SS (Ryan et al. 2007; Yi et al. 2010), and the expression of several proteins involving type III-SS (Yi et al. 2010), are regulated by c-di-GMP in different phytopathogenic bacteria. Furthermore, it has been reported that c-di-GMP can also allosterically regulate various secretion systems (i.e. T2SS, T3SS, T6SS) by direct binding to ATPases associated with these macromolecular complexes (Trampari et al. 2015; Roelofs et al. 2015; Wang et al. 2016). Other studies with different bacterial phytopathogens have revealed that c-di-GMP is a key molecule to elude plant immunity (recently reviewed in Martínez-Gil and Ramos 2018).

1.5 Cyclic-di-GMP in Beneficial Plant-Microbe Interactions

Plant-interacting bacteria generally encode a high number of proteins with c-di-GMP-related domains (up to 1% of the total functions in some strains; https://www.ncbi.nlm.nih.gov/Complete_Genomes/c-di-GMP.html). In spite of this the information available on c-di-GMP signaling in plant-interacting bacteria is limited. This scarcity of knowledge is even more pronounced for plant beneficial bacteria. Nonetheless, several c-di-GMP regulated processes (e.g. motility, EPS production, biofilm formation, transcriptional regulators, etc.…) reported in well-studied animal pathogenic strains (e.g. *Pseudomonas aeruginosa* PAO1), display some degree of conservation and can be extrapolated to different Plant Growth Promoting Rhizobacteria (PGPRs) belonging to this genus, e.g. *Pseudomonas fluorescens* and *P. putida* (Xiao et al. 2016a; Fazli et al. 2014; Martínez-Granero et al. 2014b). In contrast, other c-di-GMP regulated traits, such as different multi-repeat adhesins (*lap* system), are conserved among environmental Pseudomonads such as *P. putida* KT2440, *P. fluorescens* PfO1 and *P. fluorescens* WCS365, but are absent from pathogenic species such as *P. aeruginosa* and *P. syringae* (Hinsa et al. 2003). Very recent studies in other PGPR strains, e.g. *Azospirillum brasilense*, have recently evidenced the importance of c-di-GMP for biofilm formation, motility and the aerotaxis of this α-proteobacterium, commonly used as a biofertilizer (Mata et al. 2018; O'Neal et al. 2017; Ramírez-Mata et al. 2016). Different studies on members of the bacteria collectively referred to as rhizobia, foresee an important role of this and other NSMs on the establishment of rhizobia-legume symbiotic interactions (Schäper et al. 2015, 2017; Pérez-Mendoza et al. 2014, 2015; Gao et al. 2014; Soto et al. 2011; Wang et al. 2010; Krol et al. 2016; Romero-Jiménez et al. 2015).

1.5.1 Motility

As for most bacteria, c-di-GMP has a generally inhibitory role on swarming and swimming motility in beneficial plant-interacting bacteria. Two DGC (WspR and SadC) and one PDE (BifA) are implicated in the regulation of swimming motility and biofilm formation in *P. fluorescens* F113 (Martínez-Granero et al. 2014a). Furthermore, Martínez-Granero and co-workers identified FglZ as a c-di-GMP effector whose subcellular localization depends on c-di-GMP intracellular levels. In this strain, FlgZ acts as a YcgR-like protein involved in the c-di-GMP-mediated repression of swimming motility (Martínez-Granero et al. 2014a). As an exception to the rule, Tlp1 protein binds c-di-GMP via its C-terminal PilZ domain and promotes persistent motility by increasing swimming velocity and decreasing swimming reversal frequency in *A. brasilense*. Changes in O_2 concentration result in rapid changes in the intracellular c-di-GMP levels of this microaerophilic bacterium. Using this c-di-GMP regulated motility *A. brasilense* quickly navigates to zones of low O_2 concentration (Russell et al. 2013).

An artificial increment of the intracellular levels of c-di-GMP caused inhibition of both surface and swimming motilities in various rhizobial strains: *Rhizobium leguminosarum* bv. viciae UPM791, *R. etli* strains CFN42 and CNPAF512, and *Sinorhizobium meliloti* 8530 and Rm2011 (Pérez-Mendoza et al. 2014; Romero-Jiménez et al. 2015; Gao et al. 2014; Schäper et al. 2015). However, a direct correlation between specific DGC and/or PDE with a motility phenotype is not so evident. For example, different screening surveys carried out with mutants in c-di-GMP-metabolizing proteins in *S. meliloti* showed contradictory results. In a preliminary study with strain 1021, 14 mutants in different GGDEF and/or EAL containing proteins showed swimming differences compared to the wild type (Wang et al. 2010). In contrast, a more recent and detailed study on Rm2011 showed no swimming differences between mutants in 21 out of 22 c-di-GMP-related genes and the wild type, and only overexpression of seven of these genes resulted in phenotypic changes (Schäper et al. 2015). Nevertheless, a motility-associated c-di-GMP receptor (McrA, SMc00507) containing a PilZ domain was identified in this study. Overexpression of *mcrA* in Rm2011 wild type resulted in slightly reduced motility and its combined overexpression with a DGC led to a nonmotile phenotype. Furthermore, this motility repression was completely abolished when one of the c-di-GMP binding conserved motifs of the PilZ domain, RXXXR, was mutated (Amikam and Galperin 2006), strongly suggesting that c-di-GMP-bound McrA negatively controls motility in *S. meliloti*.

1.5.2 EPS Production and Biofilm Formation

Plant-interacting bacteria produce a complex array of EPS which provide not only protection against environmental biotic and abiotic stresses, but also play important roles in cell aggregation and biofilm formation. These are critical processes during the interaction of bacteria with their plant hosts. Rhizobia produce several c-di-GMP-activated EPS including: cellulose, curdlan, unipolar glucomannan polysaccharide (UPP) and the mixed-linkage β-glucan (MLG), recently reviewed in Pérez-Mendoza and Sanjuán (2016). However, once again, the assignment of specific DGC and/or PDE, involved in the c-di-GMP pool required for the activation of most of these EPS, remains elusive. In addition, c-di-GMP directly or indirectly participates also in the regulation of the symbiotically active EPS produced by *S. meliloti*. Galactoglucan (EPS II) production is negatively affected by c-di-GMP, at least in part by a transcriptional repression of biosynthetic genes (*wgeA*). In contrast, succinoglycan (EPS I) production seems activated by c-di-GMP. This EPS I activation is not produced at the transcription of *exoY*. Enhanced levels of EPS I may hence be mediated by transcriptional regulation of other biosynthetic gene/s or at post-translational level (Schäper et al. 2015). Strikingly, a single rhizobial strain usually produces more than one of these EPS in response to high levels of c-di-GMP. This opens the interesting question of how these plant-interacting bacteria manage to control the production of each of these energy costing EPS, at the required place and time. The overproduction of these EPS induced by c-di-GMP usually promotes different forms of biofilm formation, flocculation and surface attachment to abiotic and biotic surfaces (Schäper et al. 2015; Romero-Jiménez et al. 2015; Pérez-Mendoza et al. 2015, 2014; Gao et al. 2014). Indeed, mutants in the biosynthetic genes of some of these EPS display impaired colonization of plant roots (see below).

PGPR strains belonging to Pseudomonads also produce c-di-GMP regulated EPS important for biofilm formation, generating a characteristic wrinkly spreader colony morphotype under high intracellular levels of c-di-GMP. Cyclic di-GMP-regulated EPS include two specific EPS (Pea and Peb) in *P. putida* and alginate and cellulose in *P. putida* and *P. fluorescens* (Fazli et al. 2014). However, biofilm formation by these Pseudomonads environmental strains also depends on other important c-di-GMP-regulated adhesins i.e. large cell-surface adhesive proteins known as Lap proteins. Indeed, LapA seems to be the most important biofilm matrix component for *in vitro* biofilm formation by *P. putida* (Gjermansen et al. 2010) and *P. fluorescens* (Hinsa et al. 2003). The presence of LapA on the bacterial cell surface depends on a sophisticated phosphate-dependent c-di-GMP modulation system, which implicates a periplasmic protease, LapG, whose activity is mediated by the inner membrane protein LapD. LapD possesses degenerate GGDEF and EAL domains and binds c-di-GMP. Decreasing the intracellular level of c-di-GMP liberates LapG from LapD, enabling cleavage and release of LapA from the outer membrane. This leads to a collapse of the structural integrity of the biofilm (Fuqua 2010; Boyd and O'Toole 2012; Fazli et al. 2014). Furthermore, c-di-GMP also regulates *lapA* and *lapF* (a second large surface adhesin present in *P. aeruginosa* but not in *P. fluorescens*) at the transcriptional level. Transcriptional regulator FleQ is required for the modulation of *lapA* and cellulose expression by c-di GMP, but does not seem to be necessary for that of *lapF* and Pea exopolysaccharide (Xiao et al. 2016b).

1.5.3 Plant-Related and Other Relevant Phenotypes

Adhesins LapA and LapF are involved in efficient corn seed colonization by *P. putida* (Martínez-Gil et al. 2010; Espinosa-Urgel et al. 2000). Furthermore, the overexpression of a specific DGC, which is induced in the rhizosphere by corn root exudates (PP4959), leads to increased LapA-dependent biofilm formation in this bacterium (Matilla et al. 2011). In *P. fluorescens* F113, a master regulator conserved in Pseudomonads named AmrZ, transcriptionally regulates most DGC and PDE of this strain. An *amrZ* mutant displays reduced c-di-GMP levels and presents a pleiotropic phenotype, showing increased swimming motility, decreased biofilm formation and very limited ability for competitive colonization of rhizosphere (Muriel et al. 2018).

The role of cyclic-di-GMP in plant-related phenotypes in rhizobia is more controversial. On one hand, competitive nodulation tests performed with c-di-GMP-related mutants in *S. meliloti* 1021 *versus* wild type indicated that the nodule occupation ratio of 11 single mutants on putative DGC and/or PDE was 20% lower than in wild type (Gao et al. 2014). On the contrary, a similar systematic analysis with single knockout mutants in 19 out of the 20 putative DGC and/or PDE-encoding genes in *S. meliloti* Rm2011 showed that none of them were impaired in symbiosis, and formed pink nitrogen-fixing root nodules indistinguishable from the wild type. Furthermore, *Medicago sativa* plants inoculated with a *S. meliloti* Rm2011 strain that did not produce detectable levels of c-di-GMP (cdG^0, mutant in all putative DGC and/or PDE), yet showed nodulation efficiency, nodule occupancy and competitiveness similar to wild type (Schäper et al. 2015). Differences between strains and assay conditions may account for this discrepancy. However, the latter study indicates that c-di-GMP does not significantly affect the symbiotic ability of *S. meliloti*, at least under laboratory conditions. On the other hand, mutants on biosynthetic genes of the c-di-GMP-regulated exopolysaccharide MLG and UPP were impaired in attachment and root colonization by *S. meliloti* 8530 and *R. leguminosarum* bv. viciae 3841, respectively (Williams et al. 2008; Pérez-Mendoza et al. 2015). However, a *S. meliloti* mutant in MLG or UPP production was as competitive for nodulation as the wild type (Pérez-Mendoza et al. 2015; Schäper et al. 2015). Furthermore, even when a *R. leguminosarum* mutant in UPP production was outcompeted by the wild type in mixed inoculations, it was still able to form functional nodules on pea (Williams et al. 2008). Other experiments with *R. etli* and *R. leguminosarum* showed that increasing c-di-GMP levels favor the early stages of interaction, as it enhanced adhesion to plant roots. However it decreased the symbiotic efficiency, as plant growth and nitrogen contents were reduced (Pérez-Mendoza et al. 2014).

In *S. meliloti*, c-di-GMP content strongly dropped in the stationary phase (c-di-GMP content of exponentially growing cells was 10- to 30-fold higher than that of cells in stationary phase), and high levels of c-di-GMP directly or indirectly provoked *S. meliloti* cell elongation (Schäper et al. 2015). Furthermore, rhizobia and other α-proteobacteria present orthologues of *Caulobacter crescentus* proteins (e.g. DivK, CcKA, PleD) involved in c-di-GMP regulation of cell cycle (Thanbichler 2009).

1.6 Conclusion

It is now widely accepted that there are striking analogies between the molecular mechanisms governing the establishment of pathogenic and symbiotic bacteria-plant associations. Both types of bacteria must exert a strict control of the functions involved in the interaction with their hosts. Accordingly, Cyclic di-GMP is now known to be a crucial second messenger in host-bacteria interactions. Knowledge on bacteria that associate with plants is now accumulating, with several studies emphasizing the importance of c-di-GMP signaling in plant-interacting bacteria. Different c-di-GMP-regulated traits such as movement, attachment, colonization and settlement in beneficial bacteria, result in a positive impact on the fitness of the host plant. On the bacterial side, accurate transition from the saprophytic to a plant-associated lifestyle provides countless advantages. Our current knowledge about the c-di-GMP signaling governing this transition clearly indicates that important and complex regulatory pathways, with numerous players, are involved.

Acknowledgements Ana Matia is acknowledged for helping in the preparation of the figure. DPM was supported by Andalucía Talent Hub Program launched by the Andalusian Knowledge Agency, co-funded by the European Union's Seventh Framework Program, Marie Skłodowska-Curie actions (COFUND – Grant Agreement n° 291780) and the Ministry of Economy, Innovation, Science and Employment of the Junta de Andalucía. Support from grants BIO2014-55075-P and BIO2017-83533-P (Spanish Ministry of Economy) are also acknowledged.

References

Ahmad, I., Lamprokostopoulou, A., Le Guyon, S., Streck, E., Barthel, M., Peters, V., Hardt, W. D., & Romling, U. (2011). Complex c-di-GMP signaling networks mediate transition between virulence properties and biofilm formation in Salmonella enterica serovar Typhimurium. *PLoS One, 6*(12), e28351. https://doi.org/10.1371/journal.pone.0028351.

Amikam, D., & Galperin, M. Y. (2006). PilZ domain is part of the bacterial c-di-GMP binding protein. *Bioinformatics, 22*(1), 3–6.

Aragón, I. M., Pérez-Mendoza, D., Moscoso, J. A., Faure, E., Guery, B., Gallegos, M. T., Filloux, A., & Ramos, C. (2015). Diguanylate cyclase DgcP is involved in plant and human *Pseudomonas* spp. infections. *Environmental Microbiology, 17*(11), 4332–4351. https://doi.org/10.1111/1462-2920.12856.

Baraquet, C., & Harwood, C. S. (2013). Cyclic diguanosine monophosphate represses bacterial flagella synthesis by interacting with the Walker A motif of the enhancer-binding protein FleQ. *Proceedings of the National Academy of Sciences, 110*(46), 18478–18483. https://doi.org/10.1073/pnas.1318972110.

Boehm, A., Kaiser, M., Li, H., Spangler, C., Kasper, C. A., Ackermann, M., Kaever, V., Sourjik, V., Roth, V., & Jenal, U. (2010). Second messenger-mediated adjustment of bacterial swimming velocity. *Cell, 141*(1), 107–116. https://doi.org/10.1016/j.cell.2010.01.018.

Boyd, C. D., & O'Toole, G. A. (2012). Second messenger regulation of biofilm formation: Breakthroughs in understanding c-di-GMP effector systems. *Annual Review of Cell and Developmental Biology, 28*, 439–462. https://doi.org/10.1146/annurev-cellbio-101011-155705.

Chou, S. H., & Galperin, M. Y. (2016). Diversity of cyclic Di-GMP-binding proteins and mechanisms. *Journal of Bacteriology, 198*(1), 32–46. https://doi.org/10.1128/JB.00333-15.

Espinosa-Urgel, M., Salido, A., & Ramos, J. L. (2000). Genetic analysis of functions involved in adhesion of Pseudomonas putida to seeds. *Journal of Bacteriology, 182*(9), 2363–2369.

Fang, X., Ahmad, I., Blanka, A., Schottkowski, M., Cimdins, A., Galperin, M. Y., Römling, U., & Gomelsky, M. (2014). GIL, a new c-di-GMP-binding protein domain involved in regulation of cellulose synthesis in enterobacteria. *Molecular Microbiology, 93*(3), 439–452. https://doi.org/10.1111/mmi.12672.

Fazli, M., Almblad, H., Rybtke, M. L., Givskov, M., Eberl, L., & Tolker-Nielsen, T. (2014). Regulation of biofilm formation in *Pseudomonas* and *Burkholderia* species. *Environmental Microbiology, 16*(7), 1961–1981. https://doi.org/10.1111/1462-2920.12448.

Flemming, H. C., Wingender, J., Szewzyk, U., Steinberg, P., Rice, S. A., & Kjelleberg, S. (2016). Biofilms: An emergent form of bacterial life. *Nature Reviews Microbiology, 14*(9), 563–575. https://doi.org/10.1038/nrmicro.2016.94.

Fong, J. N., & Yildiz, F. H. (2015). Biofilm matrix proteins. *Microbiology Spectrum, 3*(2). https://doi.org/10.1128/microbiolspec.MB-0004-2014.

Fuqua, C. (2010). Passing the baton between laps: Adhesion and cohesion in *Pseudomonas putida* biofilms. *Molecular Microbiology, 77*(3), 533–536.

Galperin, M. Y. (2004). Bacterial signal transduction network in a genomic perspective. *Environmental Microbiology, 6*(6), 552–567. https://doi.org/10.1111/j.1462-2920.2004.00633.x.

Galperin, M. Y., Higdon, R., & Kolker, E. (2010). Interplay of heritage and habitat in the distribution of bacterial signal transduction systems. *Molecular BioSystems, 6*(4), 721–728. https://doi.org/10.1039/b908047c.

Gao, S., Romdhane, S. B., Beullens, S., Kaever, V., Lambrichts, I., Fauvart, M., & Michiels, J. (2014). Genomic analysis of cyclic-di-GMP-related genes in rhizobial type strains and functional analysis in *Rhizobium etli. Applied Microbiology and Biotechnology, 98*(10), 4589–4602. https://doi.org/10.1007/s00253-014-5722-7.

Gjermansen, M., Nilsson, M., Yang, L., & Tolker-Nielsen, T. (2010). Characterization of starvation-induced dispersion in *Pseudomonas* putida biofilms: genetic elements and molecular mechanisms. *Molecular Microbiology, 75*(4), 815–826. https://doi.org/10.1111/j.1365-2958.2009.06793.x.

Gomelsky, M., & Galperin, M. Y. (2013). Bacterial second messengers, cGMP and c-di-GMP, in a quest for regulatory dominance. *The EMBO Journal, 32*(18), 2421–2423. https://doi.org/10.1038/emboj.2013.193.

Hall-Stoodley, L., Costerton, J. W., & Stoodley, P. (2004). Bacterial biofilms: From the natural environment to infectious diseases. *Nature Reviews Microbiology, 2*(2), 95–108. https://doi.org/10.1038/nrmicro821.

Hengge, R. (2009). Principles of c-di-GMP signalling in bacteria. *Nature Reviews Microbiology, 7*(4), 263–273. https://doi.org/10.1038/nrmicro2109.

Hengge, R., Grundling, A., Jenal, U., Ryan, R., & Yildiz, F. (2016). Bacterial signal transduction by cyclic Di-GMP and other nucleotide second messengers. *Journal of Bacteriology, 198*(1), 15–26. https://doi.org/10.1128/JB.00331-15.

Hinsa, S. M., Espinosa-Urgel, M., Ramos, J. L., & O'Toole, G. A. (2003). Transition from reversible to irreversible attachment during biofilm formation by *Pseudomonas fluorescens* WCS365 requires an ABC transporter and a large secreted protein. *Molecular Microbiology, 49*(4), 905–918.

Krol, E., Klaner, C., Gnau, P., Kaever, V., Essen, L.-O., & Becker, A. (2016). Cyclic mononucleotide- and Clr-dependent gene regulation in *Sinorhizobium meliloti. Microbiology, 162*(10), 1840–1856. https://doi.org/10.1099/mic.0.000356.

Liang, Z. X. (2015). The expanding roles of c-di-GMP in the biosynthesis of exopolysaccharides and secondary metabolites. *Natural Product Reports, 32*(5), 663–683. https://doi.org/10.1039/c4np00086b.

Lindenberg, S., Klauck, G., Pesavento, C., Klauck, E., & Hengge, R. (2013). The EAL domain protein YciR acts as a trigger enzyme in a c-di-GMP signalling cascade in *E. coli* biofilm control. *The EMBO Journal, 32*(14), 2001–2014.

Martínez-Gil, M., & Ramos, C. (2018). Role of cyclic di-GMP in the bacterial virulence and evasion of the plant immunity. *Current Issues in Molecular Biology, 25*, 199–222. https://doi.org/10.21775/cimb.025.199.

Martínez-Gil, M., Yousef-Coronado, F., & Espinosa-Urgel, M. (2010). LapF, the second largest *Pseudomonas putida* protein, contributes to plant root colonization and determines biofilm architecture. *Molecular Microbiology, 77*(3), 549–561. https://doi.org/10.1111/j.1365-2958.2010.07249.x.

Martínez-Granero, F., Navazo, A., Barahona, E., Redondo-Nieto, M., Gonzalez de Heredia, E., Baena, I., Martin-Martin, I., Rivilla, R., & Martin, M. (2014a). Identification of flgZ as a flagellar gene encoding a PilZ domain protein that regulates swimming motility and biofilm formation in Pseudomonas. *PLoS One, 9*(2), e87608. https://doi.org/10.1371/journal.pone.0087608.

Martínez-Granero, F., Redondo-Nieto, M., Vesga, P., Martín, M., & Rivilla, R. (2014b). AmrZ is a global transcriptional regulator implicated in iron uptake and environmental adaption in P. fluorescens F113. *BMC Genomics, 15*(1), 237. https://doi.org/10.1186/1471-2164-15-237.

Mata, A. R., Pacheco, C. M., Cruz Pérez, J. F., Sáenz, M. M., & Baca, B. E. (2018). In silico comparative analysis of GGDEF and EAL domain signaling proteins from the Azospirillum genomes. *BMC Microbiology, 18*(1), 20. https://doi.org/10.1186/s12866-018-1157-0.

Matilla, M. A., Travieso, M. L., Ramos, J. L., & Ramos-Gonzalez, M. I. (2011). Cyclic diguanylate turnover mediated by the sole GGDEF/EAL response regulator in *Pseudomonas putida*: Its role in the rhizosphere and an analysis of its target processes. *Environmental Microbiology, 13*(7), 1745–1766. https://doi.org/10.1111/j.1462-2920.2011.02499.x.
Mills, E., Pultz, I. S., Kulasekara, H. D., & Miller, S. I. (2011). The bacterial second messenger c-di-GMP: Mechanisms of signalling. *Cellular Microbiology, 13*(8), 1122–1129. https://doi.org/10.1111/j.1462-5822.2011.01619.x.
Moscoso, J. A., Mikkelsen, H., Heeb, S., Williams, P., & Filloux, A. (2011). The *Pseudomonas aeruginosa* sensor RetS switches Type III and Type VI secretion via c-di-GMP signalling. *Environmental Microbiology, 13*(12), 3128–3138. https://doi.org/10.1111/j.1462-2920.2011.02595.x.
Muriel, C., Arrebola, E., Redondo-Nieto, M., Martínez-Granero, F., Jalvo, B., Pfeilmeier, S., Blanco-Romero, E., Baena, I., Malone, J. G., Rivilla, R., & Martín, M. (2018). AmrZ is a major determinant of c-di-GMP levels in *Pseudomonas fluorescens* F113. *Scientific Reports, 8*(1), 1979. https://doi.org/10.1038/s41598-018-20419-9.
O'Neal, L., Ryu, M. H., Gomelsky, M., & Alexandre, G. (2017). Optogenetic manipulation of cyclic Di-GMP (c-di-GMP) levels reveals the role of c-di-GMP in regulating aerotaxis receptor activity in Azospirillum brasilense. *Journal of Bacteriology, 199*(18). https://doi.org/10.1128/JB.00020-17.
Orr, M. W., Galperin, M. Y., & Lee, V. T. (2016). Sustained sensing as an emerging principle in second messenger signaling systems. *Current Opinion in Microbiology, 34*, 119–126. https://doi.org/10.1016/j.mib.2016.08.010.
Paul, K., Nieto, V., Carlquist, W. C., Blair, D. F., & Harshey, R. M. (2010). The c-di-GMP binding protein YcgR controls flagellar motor direction and speed to affect chemotaxis by a "backstop brake" mechanism. *Molecular Cell, 38*(1), 128–139. https://doi.org/10.1016/j.molcel.2010.03.001.
Pérez-Mendoza, D., & Sanjuán, J. (2016). Exploiting the commons: Cyclic diguanylate regulation of bacterial exopolysaccharide production. *Current Opinion in Microbiology, 30*, 36–43. https://doi.org/10.1016/j.mib.2015.12.004.
Pérez-Mendoza, D., Coulthurst, S. J., Humphris, S., Campbell, E., Welch, M., Toth, I. K., & Salmond, G. P. (2011). A multi-repeat adhesin of the phytopathogen, *Pectobacterium atrosepticum*, is secreted by a type I pathway and is subject to complex regulation involving a non-canonical diguanylate cyclase. *Molecular Microbiology, 82*(3), 719–733.
Pérez-Mendoza, D., Aragón, I. M., Prada-Ramírez, H. A., Romero-Jiménez, L., Ramos, C., Gallegos, M. T., & Sanjuán, J. (2014). Responses to elevated c-di-GMP levels in mutualistic and pathogenic plant-interacting bacteria. *PLoS One, 9*(3), e91645. https://doi.org/10.1371/journal.pone.0091645.
Pérez-Mendoza, D., Rodríguez-Carvajal, M. A., Romero-Jiménez, L., Farias, G. A., Lloret, J., Gallegos, M. T., & Sanjuan, J. (2015). Novel mixed-linkage beta-glucan activated by c-di-GMP in *Sinorhizobium meliloti*. *Proceedings of the National Academy of Sciences of the United States of America, 112*(7), E757–E765. https://doi.org/10.1073/pnas.1421748112.
Pérez-Mendoza, D., Bertinetti, D., Lorenz, R., Gallegos, M. T., Herberg, F. W., & Sanjuan, J. (2017). A novel c-di-GMP binding domain in glycosyltransferase BgsA is responsible for the synthesis of a mixed-linkage beta-glucan. *Scientific Reports, 7*(1), 8997. https://doi.org/10.1038/s41598-017-09290-2.
Pesavento, C., & Hengge, R. (2009). Bacterial nucleotide-based second messengers. *Current Opinion in Microbiology, 12*(2), 170–176. https://doi.org/10.1016/j.mib.2009.01.007.
Ramírez-Mata, A., López-Lara, L. I., Xiqui-Vázquez, M. L., Jijón-Moreno, S., Romero-Osorio, A., & Baca, B. E. (2016). The cyclic-di-GMP diguanylate cyclase CdgA has a role in biofilm formation and exopolysaccharide production in *Azospirillum brasilense*. *Research in Microbiology, 167*(3), 190–201. https://doi.org/10.1016/j.resmic.2015.12.004.
Roelofs, K. G., Jones, C. J., Helman, S. R., Shang, X., Orr, M. W., Goodson, J. R., Galperin, M. Y., Yildiz, F. H., & Lee, V. T. (2015). Systematic identification of cyclic-di-GMP binding proteins

in vibrio cholerae reveals a novel class of cyclic-di-GMP-binding ATPases associated with type II secretion systems. *PLoS Pathogens, 11*(10), e1005232. https://doi.org/10.1371/journal.ppat.1005232.

Romero-Jiménez, L., Rodríguez-Carbonell, D., Gallegos, M. T., Sanjuán, J., & Pérez-Mendoza, D. (2015). Mini-Tn7 vectors for stable expression of diguanylate cyclase PleD* in Gram-negative bacteria. *BMC Microbiology, 15*(1), 190. https://doi.org/10.1186/s12866-015-0521-6.

Römling, U., & Galperin, M. Y. (2017). Discovery of the Second Messenger Cyclic di-GMP. In K. Sauer (Ed.), *c-di-GMP signaling: Methods and protocols* (pp. 1–8). New York: Springer. https://doi.org/10.1007/978-1-4939-7240-1_1.

Römling, U., Galperin, M. Y., & Gomelsky, M. (2013). Cyclic di-GMP: The first 25 years of a universal bacterial second messenger. *Microbiology and Molecular Biology Reviews, 77*(1), 1–52. https://doi.org/10.1128/MMBR.00043-12.

Ross, P., Weinhouse, H., Aloni, Y., Michaeli, D., Weinbergerohana, P., Mayer, R., Braun, S., Devroom, E., Vandermarel, G. A., Vanboom, J. H., & Benziman, M. (1987). Regulation of cellulose synthesis in *Acetobacter xylinum* by cyclic diguanylic acid. *Nature, 325*(6101), 279–281.

Russell, M. H., Bible, A. N., Fang, X., Gooding, J. R., Campagna, S. R., Gomelsky, M., & Alexandre, G. (2013). Integration of the second messenger c-di-GMP into the chemotactic signaling pathway. *mBio, 4*(2), e00001–e00013. https://doi.org/10.1128/mBio.00001-13.

Ryan, R. P., Fouhy, Y., Lucey, J. F., Jiang, B. L., He, Y. Q., Feng, J. X., Tang, J. L., & Dow, J. M. (2007). Cyclic di-GMP signalling in the virulence and environmental adaptation of *Xanthomonas campestris*. *Molecular Microbiology, 63*(2), 429–442. https://doi.org/10.1111/j.1365-2958.2006.05531.x.

Ryjenkov, D. A., Simm, R., Römling, U., & Gomelsky, M. (2006). The PilZ domain is a receptor for the second messenger c-di-GMP: The PilZ domain protein YcgR controls motility in enterobacteria. *The Journal of Biological Chemistry, 281*(41), 30310–30314.

Schäper, S., Krol, E., Skotnicka, D., Kaever, V., Hilker, R., Sogaard-Andersen, L., & Becker, A. (2015). Cyclic di-GMP regulates multiple cellular functions in the symbiotic alphaproteobacterium *Sinorhizobium meliloti*. *Journal of Bacteriology, 198*(3), 521–535. https://doi.org/10.1128/JB.00795-15.

Schäper, S., Steinchen, W., Krol, E., Altegoer, F., Skotnicka, D., Søgaard-Andersen, L., Bange, G., & Becker, A. (2017). AraC-like transcriptional activator CuxR binds c-di-GMP by a PilZ-like mechanism to regulate extracellular polysaccharide production. *Proceedings of the National Academy of Sciences, 114*(24), E4822–E4831. https://doi.org/10.1073/pnas.1702435114.

Scharf, B. E., Hynes, M. F., & Alexandre, G. M. (2016). Chemotaxis signaling systems in model beneficial plant–bacteria associations. *Plant Molecular Biology, 90*(6), 549–559. https://doi.org/10.1007/s11103-016-0432-4.

Sondermann, H., Shikuma, N. J., & Yildiz, F. H. (2012). You've come a long way: C-di-GMP signaling. *Current Opinion in Microbiology, 15*(2), 140–146. https://doi.org/10.1016/j.mib.2011.12.008.

Soto, M. J., Nogales, J., Pérez-Mendoza, D., Gallegos, M. T., Olivares, J., & Sanjuán, J. (2011). Pathogenic and mutualistic plant-bacteria interactions: Ever increasing similarities. *Central European Journal of Biology, 6*(6), 911–917.

Tamayo, R., Pratt, J. T., & Camilli, A. (2007). Roles of cyclic diguanylate in the regulation of bacterial pathogenesis. *Annual Review of Microbiology, 61*, 131–148. https://doi.org/10.1146/annurev.micro.61.080706.093426.

Tamayo, R., Schild, S., Pratt, J. T., & Camilli, A. (2008). Role of cyclic Di-GMP during el tor biotype vibrio cholerae infection: Characterization of the in vivo-induced cyclic Di-GMP phosphodiesterase CdpA. *Infection and Immunity, 76*(4), 1617–1627. https://doi.org/10.1128/IAI.01337-07.

Thanbichler, M. (2009). Spatial regulation in *Caulobacter crescentus*. *Current Opinion in Microbiology, 12*(6), 715–721. https://doi.org/10.1016/j.mib.2009.09.013.

Tischler, A. D., & Camilli, A. (2005). Cyclic diguanylate regulates *Vibrio cholerae* virulence gene expression. *Infection and Immunity, 73*(9), 5873–5882. https://doi.org/10.1128/IAI.73.9.5873-5882.2005.

Trampari, E., Stevenson, C. E. M., Little, R. H., Wilhelm, T., Lawson, D. M., & Malone, J. G. (2015). Bacterial rotary export ATPases are allosterically regulated by the nucleotide second messenger cyclic-di-GMP. *Journal of Biological Chemistry, 290*(40), 24470–24483. https://doi.org/10.1074/jbc.M115.661439.

Wang, Y., Xu, J., Chen, A., Wang, Y., Zhu, J., Yu, G., Xu, L., & Luo, L. (2010). GGDEF and EAL proteins play different roles in the control of *Sinorhizobium meliloti* growth, motility, exopolysaccharide production, and competitive nodulation on host alfalfa. *Acta Biochimica et Biophysica Sinica (Shanghai), 42*(6), 410–417.

Wang, Y.-C., Chin, K.-H., Tu, Z.-L., He, J., Jones, C. J., Sanchez, D. Z., Yildiz, F. H., Galperin, M. Y., & Chou, S.-H. (2016). Nucleotide binding by the widespread high-affinity cyclic di-GMP receptor MshEN domain. *Nature Communications, 7*, 12481. https://doi.org/10.1038/ncomms12481.

Williams, A., Wilkinson, A., Krehenbrink, M., Russo, D. M., Zorreguieta, A., & Downie, J. A. (2008). Glucomannan-mediated attachment of *Rhizobium leguminosarum* to pea root hairs is required for competitive nodule infection. *Journal of Bacteriology, 190*(13), 4706–4715.

Wolfe, A. J., & Visick, K. L. (2008). Get the message out: Cyclic-di-GMP regulates multiple levels of flagellum-based motility. *Journal of Bacteriology, 190*(2), 463–475. https://doi.org/10.1128/JB.01418-07.

Xiao, Y., Nie, H., Liu, H., Chen, W., & Huang, Q. (2016a). Expression of the diguanylate cyclase GcbA is regulated by FleQ in response to cyclic di-GMP in *Pseudomonas putida* KT2440. *Environmental Microbiology Reports, 8*(6), 993–1002. https://doi.org/10.1111/1758-2229.12478.

Xiao, Y., Nie, H., Liu, H., Luo, X., Chen, W., & Huang, Q. (2016b). C-di-GMP regulates the expression of lapA and bcs operons via FleQ in Pseudomonas putida KT2440. *Environmental Microbiology Reports, 8*(5), 659–666. https://doi.org/10.1111/1758-2229.12419.

Yi, X., Yamazaki, A., Biddle, E., Zeng, Q., & Yang, C. H. (2010). Genetic analysis of two phosphodiesterases reveals cyclic diguanylate regulation of virulence factors in *Dickeya dadantii*. *Molecular Microbiology, 77*(3), 787–800. https://doi.org/10.1111/j.1365-2958.2010.07246.x.

Chapter 2
NADPH Oxidases Have Key Roles in Mutulistic Associations with Rhizobia or with Mycorrhizal Fungi in Root Legumes

Jesús Montiel and Carmen Quinto

Abstract Symbiotic associations between plants and microorganisms appeared early in evolution, conferring great advantages to both partners, under limited nutrient conditions. Arbuscular mycorrhizal symbiosis (AMS) is one of the oldest known mutualistic relationship between roots and microorganisms, enhancing nitrogen and phosphorus acquisition into the plant. Legumes form a specialized organ called nodule, in presence of soil diazotroph bacteria (rhizobia), providing assimilable nitrogen to the plant host, a process known as root nodule symbiosis (RNS). Several reports indicate that RNS recruited molecular components from the signaling pathway of the AMS. This latter has been confirmed by several evidences, including legume mutants impaired in establishing both symbiotic associations. Recent reports indicate that members of the plant NADPH oxidase gene family, known as respiratory burst oxidase homologs (*Rbohs*) participate in both symbioses. RBOHs are membrane proteins specialized in reactive oxygen species (ROS) generation, and are produced at different stages of AMS and RNS. RBOH-mediated ROS have positive or negative roles in these interactions, depending on the oxidase isoform implicated in a particular process, and the type of symbiosis established. Herein, the participation of *Rboh* genes in both interactions is analyzed and discussed, proposing working models that explain the dynamics of such oxidases.

Keywords NADPH oxidase · ROS · Symbiosis · Nodule · Legume · Symbiosis

---

J. Montiel
Department of Molecular Biology and Genetics, Aarhus University, Aarhus, Denmark

Departamento de Biología Molecular de Plantas, Instituto de Biotecnología, Universidad Nacional Autónoma de México, UNAM, Cuernavaca, Morelos, Mexico

C. Quinto (✉)
Departamento de Biología Molecular de Plantas, Instituto de Biotecnología, Universidad Nacional Autónoma de México, UNAM, Cuernavaca, Morelos, Mexico
e-mail: quinto@ibt.unam.mx

D. Zúñiga-Dávila et al. (eds.), *Microbial Probiotics for Agricultural Systems*, Sustainability in Plant and Crop Protection,
https://doi.org/10.1007/978-3-030-17597-9_2

2.1 Plant NADPH Oxidases

NADPH oxidases are plasma membrane enzymes that catalyze the reduction of oxygen to generate superoxide anions using the reducing power of NADPH. They have been identified and studied in species ranging from fungi to mammals, and generate reactive oxygen species (ROS) (Aguirre and Lambeth 2010). In plants, these oxidases are commonly referred to as respiratory burst oxidase homologs (RBOHs) and display a distinctive primary structure (Torres and Dangl 2005). The N-terminal region is highly variable, but contains two conserved EF-hand motifs and regulatory sequences. Six transmembrane regions can be predicted, where a pair of conserved histidine residues in the third and fifth constitute the heme group of the enzyme. Finally, the C-terminal is characterized by NADPH and FAD binding sites (Kaur et al. 2018).

Several *Rbohs* constitute gene families in diverse plant species, showing specific and overlapping spatio-temporal expression profiles in different plant tissues and under diverse experimental conditions (Torres et al. 2002; Zhang et al. 2009; Marino et al. 2011; Montiel et al. 2012; Cheng et al. 2013; Kaur and Pati 2016; Montiel et al. 2016; Morales et al. 2016). In the last decades, a growing body of evidence has demonstrated the key role of RBOHs in a wide range of biological processes, including developmental programs, responses to environmental stimuli and microbe interactions (Suzuki et al. 2011; Marino et al. 2012; Montiel et al. 2016; Wang et al. 2016). In this chapter, we will focus on the findings made by different research groups, revealing the interplay of certain *Rboh* gene members during arbuscular mycorrhizal (AM) and root nodule symbiosis (RNS).

2.2 RBOHs Are Positive and Negative Regulators in the AMS

It has been estimated that arbuscular mycorrhizal symbiosis (AMS) evolved 400 My ago with the first bryophyte-like plants, facilitating land colonization and the acquisition of limiting growth elements present in the soil, such as nitrogen and phosphorus (Brundrett 2002). Arbuscular mycorrhizal (AM) fungi of the monophyletic fungal lineage Glomeromycota establish mutualistic symbiotic associations with the majority of vascular plants (Smith and Read 2008). AM fungi make an important contribution to the global phosphate and carbon cycling budget of terrestrial ecosystems, and also play a key role in providing disease resistance to the host plants. The AMS is initiated in the rhizosphere, where strigolactones are exudated from the roots, functioning as chemoattractants and promoting fungal germination and growth (Akiyama et al. 2005). In response, lipochito-oligosaccharides known as Myc-factors are produced by AM fungi, establishing a molecular crosstalk with the plant partner (Maillet et al. 2011). Upon recognition of a compatible host, the fungal hyphae attach to the root epidermal surface forming a structure known as appressorium. Simultaneously, the cytoplasm of the inner root cells is reorganized, tracing the route for hyphal invasion and the development of the pre-penetration apparatus (PPA) (Genre et al. 2005, 2008). The hypha penetrates intracellularly and intercellularly, surrounded by the plant plasma membrane and the cell wall. In the root cortex, the fungal hyphae differentiate to ramified arbuscules where nutrient compounds are exchanged between both the macro and the micro symbionts (Schmitz and Harrison 2014).

Legume roots exposed to AM spores exhibit signaling symbiotic responses such as intracellular calcium oscillations and ROS production (Navazio et al. 2007; Chabaud et al. 2011; Kiirika et al. 2012). The oxidative burst of root sections from *Medicago truncatula* challenged with AM spores of *Glomus intraradices* is different from that triggered by spores of the pathogenic oomycete *Aphanomyces euteiches* (Kiirika et al. 2012). This suggests that a ROS-specific symbiotic signal is induced in the plant partner by the AM factors. Roots of *M. truncatula*, *Lotus japonicus*, *Zea mays* and *Nicotiana tabacum* show high accumulation of ROS in arbuscular containing cells and in neighboring cells (Fester and Hause 2005; Lanfranco et al. 2005). The relevance of ROS regulation during AM has been further confirmed using reverse genetics in plants of *M. truncatula* and *Phaseolus vulgaris* (Kiirika et al. 2012; Arthikala et al. 2013; Belmondo et al. 2016). Mycorrhizal hyphal colonization is exacerbated in *M. truncatula* roots silenced by RNA interference (RNAi) of ROP9 (a plant-specific small G protein), involved in the regulation of RBOH. The oxidative burst triggered by AM spores is prevented in the root fragments of MtROP9-RNAi plants, supporting the close connection of ROP9 with ROS metabolism (Kiirika et al. 2012). This latter report suggests a negative regulating role of RBOH-mediated ROS production in the AM-*M. truncatula* symbiosis. This notion was strengthened by the characterization of *RbohB* in *P. vulgaris*, which is upregulated in AM roots. Down-regulation of *PvRbohB* by RNAi induced early colonization by AM with a greater number of appressoria and intraradical hyphae (Arthikala et al. 2013). An opposite effect was observed in *P. vulgaris* transgenic roots over-expressing *PvRbohB* (Arthikala et al. 2014). In this direction, a recent study indicates that the involvement of RBOH in the AMS, is more complex than that originally contemplated. Specific silencing of *MtRbohE* generated abundant intercellular hyphae with a few number of arbuscules in *M. truncatula* transgenic roots. This suggested a positive role of MtRBOHE-promoted ROS production in the root cells, invaded by the fungi arbuscules (Belmondo et al. 2016). As mentioned above, ROS are generated in arbuscule containing cells and this response is certainly perceived by the fungal symbiont. For instance, the *Gigaspora margarita* CuZn superoxide dismutase is upregulated in AM roots of *M. truncatula* and *L. japonicus* (Lanfranco et al. 2005). ROS are versatile molecules, capable to modify the cell wall structure in plant cells (Ros Barceló and Gómez Ros 2009), and this remodeling is a prerequisite to allow hyphal growth and penetration. On the other hand, lignification of the cell wall can be achieved through cross-linking of flavonoid compounds, in a peroxidase-dependent process mediated by hydrogen peroxide (Liszkay et al. 2003; Shigeto and Tsutsumi 2016). In contrast, the hydroxyl radical is able to cleave cell wall polymers, promoting cell wall loosening (Liszkay et al. 2004; Fig. 2.1).

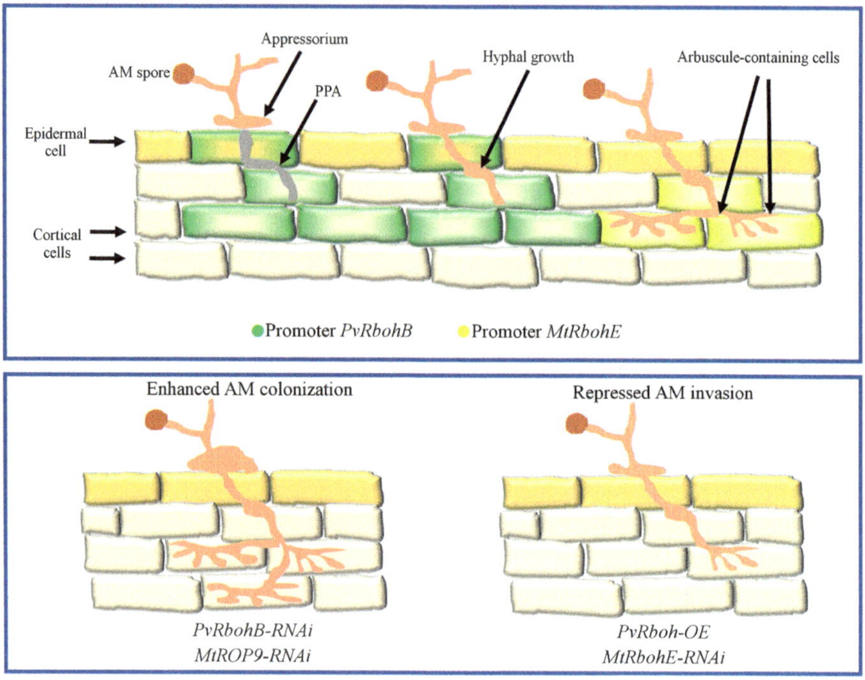

Fig. 2.1 RBOH-mediated ROS production in the AMS. *PvRbohB* promoter is activated during hyphal invasion, in the PPA and at the appressorium formation in *P. vulgaris* composite plants. In *M. truncatula* plants, *MtRbohE* promoter is mainly expressed in the arbuscular-containing cells. AM colonization is enhanced in *PvRbohB*-RNAi roots and *MtROP9*-RNAi lines but repressed in *MtRbohE*-RNAi transgenic roots and when *PvRbohB* is over-expressed (*PvRbohB*-OE)

In *P. vulgaris,* PvRBOHB seems to regulate hyphal invasion, restricting its growth within the root, as *PvRbohB*-RNAi lines developed abnormally wider appressorium (Arthikala et al. 2013). In *M. truncatula, MtRbohE* is preferably facilitating arbuscules development within the host root cells (Belmondo et al. 2016). These differences between *PvRbohB* and *MtRbohE* during the AMS, are most likely related to their expression profiles and subcellular localizations, in each of these two legumes. In addition, *PvRbohB* promoter is strongly activated in the pre-penetration apparatus, at the site of appressorium and along the hyphal invasion (Arthikala et al. 2013). In contrast, the *MtRbohE* promoter is particularly induced in arbuscule-containing root cells (Belmondo et al. 2016). All together these data suggest that during AM invasion, the development of hyphal structures and infection, are regulated by one RBOH isoform, whereas arbuscule colonization is facilitated by a different RBOH. Both in *M. truncatula* and *P. vulgaris* only one *Rboh* gene has been characterized during the AMS. Nonetheless, MtROP9 influences the expression, and possibly the activity, of a different *MtRboh* member, which appears to be negatively regulated during AM colonization in the roots (Kiirika et al. 2012).

2.3 RBOHs Are Required at Different Steps of Root Nodule Symbiosis (RNS)

Unlike the AMS, which is widespread between numerous plant families, the RNS is mainly associated with legumes (Coba de la Peña et al. 2017). The legume–rhizobia interaction initiates in the rhizosphere with a molecular cross-talk between the legume root hairs and Gram-negative soil bacteria called rhizobia. Flavonoids and nodulation factors (NFs) produced by legume roots and bacteria, respectively, are critical players in the species-specific recognition during this symbiosis (Oldroyd and Downie 2008). Flavonoid and isoflavonoid compounds secreted by the legume roots are perceived specifically by rhizobia in the soil. The chemical recognition of these molecules by the bacteria leads to the specific synthesis and release of N-acetyl glucosamine oligomers called NFs, which act as phytohormones (Perret et al. 2000). After this molecular species-specific dialogue, the bacteria invade the root hair of the legume via an infection thread (IT), a tubular structure designed by the legume host to allow and guide rhizobial infection. Meanwhile, mitotic divisions in the cortex cells originate the nodule primordium formation, which eventually develops into the matured nitrogen fixing nodule. The bacterial content within the IT is released into the nodule cells, where rhizobia differentiate to bacteroids that accomplish biological nitrogen fixation in exchange of carbon compounds (Oldroyd and Downie 2008).

Throughout the RNS, different types of ROS are produced from the early signaling pathway to nodule senescence (Santos et al. 2001; Ramu et al. 2002; Alesandrini et al. 2003; D'Haeze et al. 2003; Shaw and Long 2003; Rubio et al. 2004; Cárdenas et al. 2008; Montiel et al. 2016; Robert et al. 2018). A fast, transient and specific oxidative burst is induced in *P. vulgaris* root hair cells, after minutes of NF exposure, which is prevented by the NADPH oxidase inhibitor, dyphenilene iodonium (DPI) (Cárdenas et al. 2008). This response is most likely promoted by the activity of at least two oxidases, PvRBOHA and PvRBOHB. The specific silencing of each of these genes in fact affects the early symbiotic signaling pathway of *P. vulgaris* transgenic roots inoculated with rhizobia (Montiel et al. 2012; Arthikala et al. 2017).

These two oxidases also participate in subsequent stages of the RNS. *PvRbohA* mediates rhizobial infection in the root hairs, as *PvRbohA*-RNAi roots displayed reduced infection events after rhizobial inoculation. The superoxide anion accumulated in the infection pocket (Santos et al. 2001) likely facilitates structural changes of the plant cell wall, allowing rhizobial entrance. This type of ROS is also abundant in the IT of the rhizobial infected root hair cells (Santos et al. 2001). The progression of the IT is aborted at the base of the root hairs in *PvRbohA*-RNAi and *PvRbohB*-RNAi transgenic roots (Montiel et al. 2012; Arthikala et al. 2017) (Fig. 2.2). Both proteins are remarkably abundant in the infection pocket, however, their distribution is different in other zones of the infected root hair cell and surrounding tissues. A weak YFP-PvRBOHA signal can be observed throughout the IT membrane, the base of the infected cell and the neighboring cortical cells

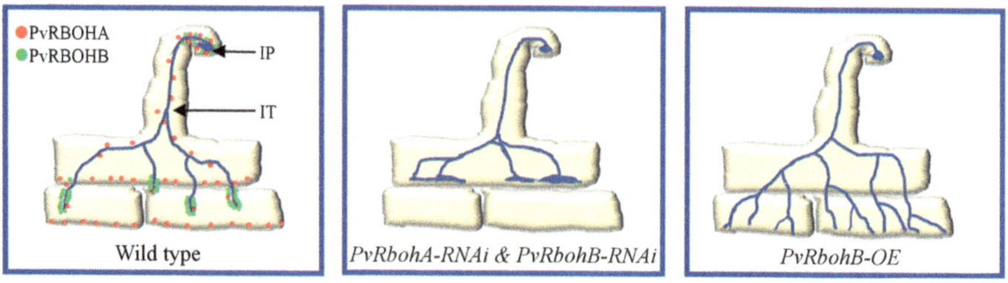

Fig. 2.2 Interplay of *Rbohs* during rhizobial infection and nodule organogenesis. In the rhizobial-infected root hair, YFP-PvRBOHA and YFP-PvRBOHB are mainly located in the infection pocket (IP). YFP-PvRBOHA shows a weak distribution throughout the plasma membrane of the infection threads IT, while YFP-PvRBOHB is abundantly localized at the sites of IT progression. However, the migration of the IT is blocked in *P. vulgaris* transgenic roots silenced in *PvRbohA* or *PvRbohB* but the infection is enhanced in lines over-expressing *PvRbohB* (*PvRbohB*-OE)

(Arthikala et al. 2017), while YFP-PvRBOHB is specially localized at the sites of IT progression (Montiel et al. 2016) (Fig. 2.2). ROS production in these regions of the IT is evidently required to sustain its progression, through a coordinated action of these NADPH oxidases. The expression profile of these genes is also different during nodule primordium formation. The dividing cortical cells show promoter activity of *PvRbohB* (Montiel et al. 2012), while the *PvRbohA* promoter expression is restricted to the emerging vascular bundles (Arthikala et al. 2017).

Recent studies confirm the strong connection of RBOHs and auxins during the developmental processes in plants (Orman-Ligeza et al. 2016; Mangano et al. 2017). In *P. vulgaris*, the AUX1-like gene is slightly upregulated after rhizobial infection, but this response is abolished in *PvRbohA*-RNAi transgenic roots (Arthikala et al. 2017).

The reports described above illustrate the interplay of RBOHs during rhizobial infection and nodule organogenesis. In spite of these, these oxidases also participate in nodule functioning and senescence. In *M. truncatula*, the early steps of the RNS are not affected in *MtRbohA*-RNAi or *MtRbohE*-RNAi composite plants (Marino et al. 2011; Belmondo et al. 2016). Nonetheless, both nitrogen fixation and the expression of the bacterial genes *nifD* and *nifH*, that encode the Mo-Fe and Fe subunits of the nitrogenase enzymatic complex, are significantly reduced (Marino et al. 2011). Likewise, nitrogen fixation is severely decreased in *PvRbohA*-RNAi and *PvRbohB*-RNAi nodules (Montiel et al. 2012; Arthikala et al. 2017). On the contrary, nodule numbers and nitrogen fixation rates are enhanced when *PvRbohB* is over-expressed (Arthikala et al. 2014). Interestingly, the ectopic over-expression of *PvRbohB* also delayed nodule senescence in *P. vulgaris* (Arthikala et al. 2014). An imbalance in redox status is the hallmark of senescence during nodule aging, characterized by a reduced content of antioxidant compounds, oxidation of biomolecules and a drastic decrease of nitrogen fixation (Matamoros et al. 2003). Senescent nodules of *Glycine max* and *Pisum sativum* manifest an altered structure of the symbiosome, accompanied by hydrogen peroxide accumulation (Alesandrini et al. 2003; Rubio et al. 2004; Puppo et al. 2005). None of the *MtRbohs* tested by Marino et al. (2011) were up regulated during nodule senescence in *M. truncatula* plants, nonetheless other unidentified Rboh isoforms could be implicated in this process.

2.4 Conclusions

Studies accomplished in the roots of two legumes, *P. vulgaris* and *M. truncatula*, indicate that certain *Rbohs* genes have a key role during AMS and RNS. The ROS produced by these oxidases seem to be involved in two processes, with a signaling role when the symbiotic signals produced by the microsymbiont are perceived, and in promoting cell wall remodeling, to facilitate and limit both rhizobial and fungal colonization. Intriguingly, the NADPH oxidases characterized in these two legumes are not putative orthologs, based on their phylogenetic proximity. This suggests that during evolution a different set of *Rbohs* genes were recruited by distinct legumes, as confirmed by several differences observed in the expression profiles of *MtRbohs* and *PvRbohs* during AMS and RNS. The characterization of other *Rbohs* genes would unveil their complex dynamics and their presumed interplay with other components of the symbiotic signaling pathway, such as reactive nitrogen species and phytohormones.

Acknowledgements This work was partially supported by Dirección General de Asuntos del Personal Académico–Universidad Nacional Autónoma de México, UNAM-DGAPA IN204115, to C.Q.

References

Aguirre, J., & Lambeth, J. D. (2010). Nox enzymes from fungus to fly to fish and what they tell us about Nox function in mammals. *Free Radical Biology & Medicine, 49*(9), 1342–1353. https://doi.org/10.1016/j.freeradbiomed.2010.07.027.

Akiyama, K., Matsuzaki, K., & Hayashi, H. (2005). Plant sesquiterpenes induce hyphal branching in arbuscular mycorrhizal fungi. *Nature, 435*(7043), 824–827. https://doi.org/10.1038/nature03608.

Alesandrini, F., Mathis, R., van de Sype, G., Herouart, D., & Puppo, A. (2003). Possible roles for a cysteine protease and hydrogen peroxide in soybean nodule development and senescence. *The New Phytologist, 158*, 131–138.

Arthikala, M. K., Montiel, J., Nava, N., Santana, O., Sanchez-Lopez, R., Cardenas, L., & Quinto, C. (2013). PvRbohB negatively regulates Rhizophagus irregularis colonization in Phaseolus vulgaris. *Plant & Cell Physiology, 54*(8), 1391–1402. https://doi.org/10.1093/pcp/pct089.

Arthikala, M. K., Sanchez-Lopez, R., Nava, N., Santana, O., Cardenas, L., & Quinto, C. (2014). RbohB, a Phaseolus vulgaris NADPH oxidase gene, enhances symbiosome number, bacteroid size, and nitrogen fixation in nodules and impairs mycorrhizal colonization. *The New Phytologist, 202*(3), 886–900. https://doi.org/10.1111/nph.12714.

Arthikala, M. K., Montiel, J., Sanchez-Lopez, R., Nava, N., Cardenas, L., & Quinto, C. (2017). Respiratory burst oxidase homolog gene a is crucial for rhizobium infection and nodule maturation and function in common bean. *Frontiers in Plant Science, 8*, 2003. https://doi.org/10.3389/fpls.2017.02003.

Belmondo, S., Calcagno, C., Genre, A., Puppo, A., Pauly, N., & Lanfranco, L. (2016). The Medicago truncatula MtRbohE gene is activated in arbusculated cells and is involved in root cortex colonization. *Planta, 243*(1), 251–262. https://doi.org/10.1007/s00425-015-2407-0.

Brundrett, M. C. (2002). Coevolution of roots and mycorrhizas of land plants. *The New Phytologist, 154*(2). https://doi.org/10.1046/j.1469-8137.2002.00397.x.

Cárdenas, L., Martinez, A., Sanchez, F., & Quinto, C. (2008). Fast, transient and specific intracellular ROS changes in living root hair cells responding to Nod factors (NFs). *The Plant Journal, 56*(5), 802–813. https://doi.org/10.1111/j.1365-313X.2008.03644.x.

Chabaud, M., Genre, A., Sieberer, B. J., Faccio, A., Fournier, J., Novero, M., Barker, D. G., & Bonfante, P. (2011). Arbuscular mycorrhizal hyphopodia and germinated spore exudates trigger Ca2+ spiking in the legume and nonlegume root epidermis. *The New Phytologist, 189*(1), 347–355. https://doi.org/10.1111/j.1469-8137.2010.03464.x.

Cheng, C., Xu, X., Gao, M., Li, J., Guo, C., Song, J., & Wang, X. (2013). Genome-wide analysis of respiratory burst oxidase homologs in grape (Vitis vinifera L.). *International Journal of Molecular Sciences, 14*(12), 24169–24186. https://doi.org/10.3390/ijms141224169.

Coba de la Peña, T., Fedorova, E., Pueyo, J. J., & Lucas, M. M. (2017). The symbiosome: Legume and rhizobia co-evolution toward a nitrogen-fixing organelle? *Frontiers in Plant Science, 8*, 2229. https://doi.org/10.3389/fpls.2017.02229.

D'Haeze, W., De Rycke, R., Mathis, R., Goormachtig, S., Pagnotta, S., Verplancke, C., Capoen, W., & Holsters, M. (2003). Reactive oxygen species and ethylene play a positive role in lateral root base nodulation of a semiaquatic legume. *Proceedings of the National Academy of Sciences of the United States of America, 100*(20), 11789–11794. https://doi.org/10.1073/pnas.1333899100.

Fester, T., & Hause, G. (2005). Accumulation of reactive oxygen species in arbuscular mycorrhizal roots. *Mycorrhiza, 15*(5), 373–379. https://doi.org/10.1007/s00572-005-0363-4.

Genre, A., Chabaud, M., Timmers, T., Bonfante, P., & Barker, D. G. (2005). Arbuscular mycorrhizal fungi elicit a novel intracellular apparatus in Medicago truncatula root epidermal cells before infection. *Plant Cell, 17*(12), 3489–3499. https://doi.org/10.1105/tpc.105.035410.

Genre, A., Chabaud, M., Faccio, A., Barker, D. G., & Bonfante, P. (2008). Prepenetration apparatus assembly precedes and predicts the colonization patterns of arbuscular mycorrhizal fungi within the root cortex of both Medicago truncatula and Daucus carota. *Plant Cell, 20*(5), 1407–1420. https://doi.org/10.1105/tpc.108.059014.

Kaur, G., & Pati, P. K. (2016). Analysis of cis-acting regulatory elements of respiratory burst oxidase homolog (Rboh) gene families in Arabidopsis and rice provides clues for their diverse functions. *Computational Biology and Chemistry, 62*, 104–118. https://doi.org/10.1016/j.compbiolchem.2016.04.002.

Kaur, G., Guruprasad, K., Temple, B. R. S., Shirvanyants, D. G., Dokholyan, N. V., & Pati, P. K. (2018). Structural complexity and functional diversity of plant NADPH oxidases. *Amino Acids, 50*(1), 79–94. https://doi.org/10.1007/s00726-017-2491-5.

Kiirika, L. M., Bergmann, H. F., Schikowsky, C., Wimmer, D., Korte, J., Schmitz, U., Niehaus, K., & Colditz, F. (2012). Silencing of the Rac1 GTPase MtROP9 in Medicago truncatula stimulates early mycorrhizal and oomycete root colonizations but negatively affects rhizobial infection. *Plant Physiology, 159*(1), 501–516. https://doi.org/10.1104/pp.112.193706.

Lanfranco, L., Novero, M., & Bonfante, P. (2005). The mycorrhizal fungus Gigaspora margarita possesses a CuZn superoxide dismutase that is up-regulated during symbiosis with legume hosts. *Plant Physiology, 137*(4), 1319–1330. https://doi.org/10.1104/pp.104.050435.

Liszkay, A., Kenk, B., & Schopfer, P. (2003). Evidence for the involvement of cell wall peroxidase in the generation of hydroxyl radicals mediating extension growth. *Planta, 217*(4), 658–667. https://doi.org/10.1007/s00425-003-1028-1.

Liszkay, A., van der Zalm, E., & Schopfer, P. (2004). Production of reactive oxygen intermediates (O(2)(.-), H(2)O(2), and (.)OH) by maize roots and their role in wall loosening and elongation growth. *Plant Physiology, 136*(2), 3114–3123. https://doi.org/10.1104/pp.104.044784; discussion 3001.

Maillet, F., Poinsot, V., Andre, O., Puech-Pages, V., Haouy, A., Gueunier, M., Cromer, L., Giraudet, D., Formey, D., Niebel, A., Martinez, E. A., Driguez, H., Becard, G., & Denarie, J. (2011). Fungal lipochitooligosaccharide symbiotic signals in arbuscular mycorrhiza. *Nature, 469*(7328), 58–63. https://doi.org/10.1038/nature09622.

Mangano, S., Denita-Juarez, S. P., Choi, H. S., Marzol, E., Hwang, Y., Ranocha, P., Velasquez, S. M., Borassi, C., Barberini, M. L., Aptekmann, A. A., Muschietti, J. P., Nadra, A. D., Dunand, C., Cho, H. T., & Estevez, J. M. (2017). Molecular link between auxin and ROS-mediated polar growth. *Proceedings of the National Academy of Sciences of the United States of America, 114*(20), 5289–5294. https://doi.org/10.1073/pnas.1701536114.

Marino, D., Andrio, E., Danchin, E. G., Oger, E., Gucciardo, S., Lambert, A., Puppo, A., & Pauly, N. (2011). A Medicago truncatula NADPH oxidase is involved in symbiotic nodule functioning. *The New Phytologist, 189*(2), 580–592. https://doi.org/10.1111/j.1469-8137.2010.03509.x.

Marino, D., Dunand, C., Puppo, A., & Pauly, N. (2012). A burst of plant NADPH oxidases. *Trends in Plant Science, 17*(1), 9–15. https://doi.org/10.1016/j.tplants.2011.10.001.

Matamoros, M. A., Dalton, D. A., Ramos, J., Clemente, M. R., Rubio, M. C., & Becana, M. (2003). Biochemistry and molecular biology of antioxidants in the rhizobia-legume symbiosis. *Plant Physiology, 133*(2), 499–509. https://doi.org/10.1104/pp.103.025619.

Montiel, J., Nava, N., Cardenas, L., Sanchez-Lopez, R., Arthikala, M. K., Santana, O., Sanchez, F., & Quinto, C. (2012). A Phaseolus vulgaris NADPH oxidase gene is required for root infection by rhizobia. *Plant & Cell Physiology, 53*(10), 1751–1767. https://doi.org/10.1093/pcp/pcs120.

Montiel, J., Arthikala, M. K., Cardenas, L., & Quinto, C. (2016). Legume NADPH oxidases have crucial roles at different stages of nodulation. *International Journal of Molecular Sciences, 17*(5). https://doi.org/10.3390/ijms17050680.

Morales, J., Kadota, Y., Zipfel, C., Molina, A., & Torres, M. A. (2016). The Arabidopsis NADPH oxidases RbohD and RbohF display differential expression patterns and contributions during plant immunity. *Journal of Experimental Botany, 67*(6), 1663–1676. https://doi.org/10.1093/jxb/erv558.

Navazio, L., Moscatiello, R., Genre, A., Novero, M., Baldan, B., Bonfante, P., & Mariani, P. (2007). A diffusible signal from arbuscular mycorrhizal fungi elicits a transient cytosolic calcium elevation in host plant cells. *Plant Physiology, 144*(2), 673–681. https://doi.org/10.1104/pp.106.086959.

Oldroyd, G. E., & Downie, J. A. (2008). Coordinating nodule morphogenesis with rhizobial infection in legumes. *Annual Review of Plant Biology, 59*, 519–546. https://doi.org/10.1146/annurev.arplant.59.032607.092839.

Orman-Ligeza, B., Parizot, B., de Rycke, R., Fernandez, A., Himschoot, E., Van Breusegem, F., Bennett, M. J., Perilleux, C., Beeckman, T., & Draye, X. (2016). RBOH-mediated ROS production facilitates lateral root emergence in Arabidopsis. *Development, 143*(18), 3328–3339. https://doi.org/10.1242/dev.136465.

Perret, X., Staehelin, C., & Broughton, W. J. (2000). Molecular basis of symbiotic promiscuity. *Microbiology and Molecular Biology Reviews, 64*(1), 180–201.

Puppo, A., Groten, K., Bastian, F., Carzaniga, R., Soussi, M., Lucas, M. M., de Felipe, M. R., Harrison, J., Vanacker, H., & Foyer, C. H. (2005). Legume nodule senescence: Roles for redox and hormone signalling in the orchestration of the natural aging process. *The New Phytologist, 165*(3), 683–701. https://doi.org/10.1111/j.1469-8137.2004.01285.x.

Ramu, S. K., Peng, H. M., & Cook, D. R. (2002). Nod factor induction of reactive oxygen species production is correlated with expression of the early nodulin gene rip1 in Medicago truncatula. *Molecular Plant-Microbe Interactions, 15*(6), 522–528. https://doi.org/10.1094/MPMI.2002.15.6.522.

Robert, G., Munoz, N., Alvarado-Affantranger, X., Saavedra, L., Davidenco, V., Rodriguez-Kessler, M., Estrada-Navarrete, G., Sanchez, F., & Lascano, R. (2018). Phosphatidylinositol 3-kinase at the very early symbiont perception: A local nodulation control under stress conditions? *Journal of Experimental Botany*. https://doi.org/10.1093/jxb/ery030.

Ros Barceló, A., & Gómez Ros, L. V. (2009). *Reactive oxygen species in plant cell walls*. Berlin: Springer.

Rubio, M. C., James, E. K., Clemente, M. R., Bucciarelli, B., Fedorova, M., Vance, C. P., & Becana, M. (2004). Localization of superoxide dismutases and hydrogen peroxide in legume root nodules. *Molecular Plant-Microbe Interactions, 17*(12), 1294–1305. https://doi.org/10.1094/MPMI.2004.17.12.1294.

Santos, R., Herouart, D., Sigaud, S., Touati, D., & Puppo, A. (2001). Oxidative burst in alfalfa-Sinorhizobium meliloti symbiotic interaction. *Molecular Plant-Microbe Interactions, 14*(1), 86–89. https://doi.org/10.1094/MPMI.2001.14.1.86.

Schmitz, A. M., & Harrison, M. J. (2014). Signaling events during initiation of arbuscular mycorrhizal symbiosis. *Journal of Integrative Plant Biology, 56*(3), 250–261. https://doi.org/10.1111/jipb.12155.

Shaw, S. L., & Long, S. R. (2003). Nod factor inhibition of reactive oxygen efflux in a host legume. *Plant Physiology, 132*(4), 2196–2204.

Shigeto, J., & Tsutsumi, Y. (2016). Diverse functions and reactions of class III peroxidases. *The New Phytologist, 209*(4), 1395–1402. https://doi.org/10.1111/nph.13738.

Smith, S. E., & Read, D. J. (2008). *Mycorrhizal symbiosis* (3rd ed.). London: Academic.

Suzuki, N., Miller, G., Morales, J., Shulaev, V., Torres, M. A., & Mittler, R. (2011). Respiratory burst oxidases: The engines of ROS signaling. *Current Opinion in Plant Biology, 14*(6), 691–699. https://doi.org/10.1016/j.pbi.2011.07.014.

Torres, M. A., & Dangl, J. L. (2005). Functions of the respiratory burst oxidase in biotic interactions, abiotic stress and development. *Current Opinion in Plant Biology, 8*(4), 397–403. https://doi.org/10.1016/j.pbi.2005.05.014.

Torres, M. A., Dangl, J. L., & Jones, J. D. (2002). Arabidopsis gp91phox homologues AtrbohD and AtrbohF are required for accumulation of reactive oxygen intermediates in the plant defense response. *Proceedings of the National Academy of Sciences of the United States of America, 99*(1), 517–522. https://doi.org/10.1073/pnas.012452499.

Wang, Y. J., Wei, X. Y., Jing, X. Q., Chang, Y. L., Hu, C. H., Wang, X., & Chen, K. M. (2016). The fundamental role of NOX family proteins in plant immunity and their regulation. *International Journal of Molecular Sciences, 17*(6). https://doi.org/10.3390/ijms17060805.

Zhang, H., Fang, Q., Zhang, Z., Wang, Y., & Zheng, X. (2009). The role of respiratory burst oxidase homologues in elicitor-induced stomatal closure and hypersensitive response in Nicotiana benthamiana. *Journal of Experimental Botany, 60*(11), 3109–3122. https://doi.org/10.1093/jxb/erp146.

Chapter 3
Linking Ammonia Volatilization with Moisture Content and Abundance of Nitrification and Denitrification Genes in N-Fertilized Soils

Antonio Castellano-Hinojosa, Jesús González-López, Antonio Vallejo, and Eulogio J. Bedmar

Abstract Application of N-fertilizers can lose nitrogen to the atmosphere via ammonia (NH_3) volatilization and also affect survival of microbial communities in the soil. Here, an agricultural soil was supplemented with urea, ammonium sulfate, potassium nitrate or urea together with the urease inhibitor N-(n-butyl) thiophosphoric triamide (NBPT) and maintained at 50 and 80% water-filled pore space (WFPS). Ammonia volatilization was monitored daily for 15 days. Abundance of total bacterial and archaeal populations and those of the nitrifier and denitrifier communities were determined by quantitative PCR of their corresponding 16S rRNA, *amoA* and *norB*, *nos*ZI and *nos*ZII genes. The highest NH_3 fluxes were detected at 50% WFPS after urea application and ammonium had intermediate volatilization potential regarding the control soil. Urea + NBPT reduced ammonia emissions at 50 and 80% WFPS. Addition of any of the fertilizers increased the total abundance of Bacteria and Archaea regardless of the moisture conditions, and urea + NBPT had no effect on soil microbial communities. Ammonium-based fertilizers increased the abundance of the PCR-amplified nitrification genes from soil since the beginning,

A. Castellano-Hinojosa
Department of Soil Microbiology and Symbiotic Systems, Estación Experimental del Zaidín, CSIC, Granada, Spain

Department of Microbiology, Faculty of Pharmacy, University of Granada, Granada, Spain

J. González-López
Department of Soil Microbiology and Symbiotic Systems, Estación Experimental del Zaidín, Granada, Spain

A. Vallejo
School of Agriculture, Technical University of Madrid, Madrid, Spain

E. J. Bedmar (✉)
Department of Microbiology, Faculty of Pharmacy, University of Granada, Granada, Spain
e-mail: eulogio.bedmar@eez.csic.es

D. Zúñiga-Dávila et al. (eds.), *Microbial Probiotics for Agricultural Systems*, Sustainability in Plant and Crop Protection,
https://doi.org/10.1007/978-3-030-17597-9_3

values that were delayed when urea + NBPT was added. Nitrogen fertilization increased the abundance of denitrifiers, particularly after nitrate application. Urea + NBPT did not affect the total abundance of the denitrification genes.

Keywords N-fertilizer · Ammonia volatilization · WFPS · Nitrification · Denitrification · qPCR

3.1 Introduction

The utilization of nitrogen fertilizers is essential to sustain the high rates of food production delivered by modern agriculture. Urea, ammonium and nitrate are used, among which urea is largely applied to agricultural soils due to its low cost, high N content and ease of transport (Heffer and Prud'homme 2011).

Nitrogen losses in soils occur mainly through nitrate (NO_3^-) leaching, ammonia (NH_3) volatilization and nitrogenous gas emissions (NO, N_2O and N_2), which diminish the N use efficiency (Erisman et al. 2015). The N leak due to NH_3 volatilization does not only cause environmental and human health problems, but also economic damage to farmers (Zaman and Blennerhassett 2010), with an estimated cost of around €50 billion per year in Europe (Sutton et al. 2011). In this sense, different strategies have been developed to mitigate N losses in soils. The transformation of urea into carbon dioxide (CO_2) and NH_3 is catalyzed by urease, and urease inhibitors were designed to depress urea hydrolysis with the concomitant decrease in ammonia volatilization (Abalos et al. 2012; Modolo et al. 2015). Among the urease inhibitors, N-(n-butyl) thiophosphoric triamide (NBPT) is widely used because it has been shown to be highly effective under laboratory (Gill et al. 1999) and field conditions (Sanz-Cobeña et al. 2008, 2011; Zaman et al. 2009; Rodríguez-Soares et al. 2012), thus improving the N use efficiency when a soil is treated with urea (Abalos et al. 2014; Yang et al. 2016). Since weather conditions, inhibitor concentration, soil characteristics, pH, etc., can affect NH_3 emissions, there is still some debate about the effectiveness of NBPT (Sanz-Cobeña et al. 2011).

The application to the soil of any chemical compound could alter the microbial communities. Using the 16S rRNA ribosomal gene as a molecular marker, the addition of NBPT to two pasture soils did not affect the total abundance of the analyzed bacterial community (Shi et al. 2017). Within the N cycle, nitrification and denitrification are main processes involved in redox transformations of compounds produced by N-fertilization (Butterbach-Bahl et al. 2013). Nitrification, the aerobic conversion of ammonia into nitrate, is catalyzed by the *amoA* gene-encoded ammonia monooxygenase (AMO) enzyme (Prosser and Nicol 2012). Different reports have shown that NBPT did not affect soil nitrification activity by this gene (Giovannini et al. 2009). Recently, Shi et al. (2017) have shown that NBPT also diminished the abundance of the bacterial and archaeal *amoA* genes. Denitrification is the biochemical pathway responsible for the reduction of nitrate to N_2 via the formation of nitrite, nitric oxide and nitrous oxide under oxygen-limiting conditions (Bedmar et al. 2013; Torres et al. 2016). Information on the effect of NBPT on denitrification is scarce, but its repeated application under field conditions decreased the abundance of the *narG* and *nirK* genes (Shi et al. 2017).

The application of NBPT to soils fertilized with urea or ammonium can result in the reduction of gaseous NH_3 as well as on NO_3^- leaching losses and N_2O release into the atmosphere. In this sense, the aim of this work was to study the effect of NBPT on ammonium volatilization by soils fertilized with urea, ammonium or nitrate at 50 and 80% water filled pore space (WFPS). Under those moisture conditions, we have also examined the effect of N-fertilization and the application of the urease inhibitor on the abundance of the nitrifying and denitrifying communities in soil.

3.2 Materials and Methods

3.2.1 Soil Sampling and Experimental Design

An Eutric Cambisol soil (30% clay, 12.5% silt, 57.5% sand, w/w; pH in water, 6.8; organic C, 4%; total N 0.2%; NO_3^- 6.8 mg kg^{-1}; exchangeable NH_4^+, not detected; HCO_3^- 244 mg kg^{-1}) of the FAO series (FAO 2017) was collected from an agricultural area (UTM coordinates 36° 43’ 53.5“ N, 3° 32’ 56.2” W) in the vicinity of Motril (Granada, Spain). Soil samples (30 × 30 cm, to a depth of 25 cm) were taken from 12 sites and pooled together to obtain a composite sample. Then, soil was further supplemented with either urea (CON_2H_4, UR), ammonium sulfate [$(NH_4)_2SO_4$, AS] or potassium nitrate (KNO_3, PN) to a final concentration of 260 kg equivalent N ha^{-1} as recommended for horticultural crops and leguminous plants by the Spanish Ministry of Agriculture, Food and Environment. A set of pots containing urea was supplemented with the urease inhibitor (UI) NBPT (0.25%) (UR + UI). Soil without fertilization (CS) was used as a control. Soil was used to fill 5 kg PVC pots (26 cm diam., 15 cm height) up to 8 cm from the rim, watered to reach 40% WFPS and kept for 4 days to avoid the pulse of respiration associated with wetting dry soils (Kieft et al. 1987). To initiate the experiments, each one of the fertilizers and the urease were diluted in 200 mL water and added to the pots (6 pots/treatment) from the top. The soil of half of the pots was adjusted to 50% WFPS and the other half to 80% WFPS. The pots were placed in a greenhouse at the farm of the Technical University of Madrid (Spain) that was maintained at 18 °C for 15 days. Their distribution was arranged in a factorial randomized complete block design. For each water content, half of the pots were used for NH_3 volatilization measurements (non-destructive assay; n = 3), and the other half was used to analyze the soil physicochemical properties (destructive assay; n = 3).

3.2.2 Soil Analyses

Sample corers (5 cm diam., 15 cm height) were manually inserted into the pots to sample the soil from the different treatments at 1, 5, 10 and 15 days after starting the experiment. The exchangeable ammonium and nitrate contents were assayed using ionic chromatography (Metrohm IC Compact 761) as indicated earlier (González-Martínez et al. 2016). Urease activity was determined as described by Nannipieri et al. (1980). The pH was measured after water extraction (1:5, w/v) for 2 h. Moisture content was calculated gravimetrically according to Danielson and Sutherland (1986).

3.2.3 NH_3 Emission Measurement

Ammonia volatilization was estimated using a gas flow-through system coupled to a chemiluminiscence ammonia analyzer (Thermo Scientific, model 17i analyzer) (Aneja et al. 2000; Walker et al. 2002). For this purpose, each container was closed with a Teflon®-covered lid provided with an inlet and an outlet holes. Air was pumped at a constant rate ($Q = 10\ L\ min^{-1}$) through the inlet hole. Gas samples were transported from the outlet hole to a T tube through 3-m Teflon® tubing, with one part venting to the atmosphere and the remainder entering into the analyzer at a flow rate of $0.5\ L\ min^{-1}$. An outlet hole was used to minimize pressure differences between the chamber and the atmosphere. Then, airflow samples were passed through a stainless steel converter, where NH_3 transforms to nitric oxide (NO) after reaction with ozone. The lower detectable limit was 1.0 ppb (v/v). The steady-state concentration of NH_3 in the chamber was reached after 20 min. A chamber coated with Teflon® was used as control. The air flow rate was checked daily with a flow-meter. Under steady-state conditions, the change of concentration with respect to time is zero, so that the NH_3 flux was calculated following the equation proposed by Kaplan et al. (1988) for reactive gases. Total NH_3-N losses were estimated by successive linear interpolations of the flux measurements per chamber. It was assumed that emissions followed a linear trend during the periods in which no samples were taken. Cumulative NH_3 fluxes were calculated as the sum of NH_3 volatilized at each measurement interval. Two samples were taken daily for the first 3 days and then one sample per day until the end of the experiment.

3.2.4 DNA Extraction and Quantification of Nitrification and Denitrification Genes

Soil DNA was extracted from 0.5 g soil samples (Correa-Galeote et al. 2014; Castellano-Hinojosa et al. 2018) taken 1, 5, 10 and 15 days after fertilization and purified using GeneClean (MP Bio) spin columns. DNA concentration was determined using the Qubit® ssDNA assay kit (Molecular Probes). Quantitative PCR (qPCR) was carried out using a Bio-Rad iCycler iQ5 Thermocycler (Bio-Rad Laboratories, USA) using SYBR Green as the fluorescent dye. The total bacterial and archaeal communities were determined using the corresponding 16S rRNA genes as molecular markers (Ochsenreiter et al. 2003; López-Gutiérrez et al. 2004). The *amoA* gene from ammonia-oxidizing Bacteria (AOB) and Archaea (AOA) served to estimate the size of the nitrifying communities (Rotthauwe et al. 1997; Tourna et al. 2008, respectively), while that of the denitrifying community was assessed via the *norB*, *nos*ZI and *nos*ZII genes (Braker and Tiedje 2003; Henry et al. 2006; Jones et al. 2013, respectively). Standard curves were obtained using serial dilutions of linearized plasmids containing the targeted genes amplified from bacterial strains. No template controls gave null values. The efficiency of PCR amplification for all target genes was between 90% and 100%. The quality of all qPCR amplifications was verified by electrophoresis in 1% agarose and by melting curve analysis.

3.2.5 Statistical Analyses

The Shapiro-Wilk test and the Bartlett's test were used to check normality and homoscedasticity assumptions, respectively. Since data did not fit the normal distribution, the Kruskal-Wallis test was selected to search for significant differences, and the Conover-Iman test was used for multiple comparisons among samples using the software package Statgraphics Centurion XVII (Statgraphics.net, Spain).

3.3 Results

3.3.1 Soil Physicochemical Properties

Regardless of the WFPS, the original soil pH of 6.8 remained largely unchanged during the 15 days experimental period for all treatments, except for urea, that varied from a calculated value of 8.0 at the beginning to a pH of 7.35 after 15 days. For each water content, treatment of soil with urea and UR + UI decreased the urease activity during the experimental time and the presence of NBPT delayed the urease hydrolysis. The urease activity was not found after treatment of soil with ammonium and nitrate, both at 50 and 80% WFPS. The NH_4^+-N concentration declined along the experimental time in soils treated with urea, ammonium and UR + UI. After 15 days, the amounts of ammonium detected were 49.2, 68.2 and 123.4 mg NH_4^+-N kg^{-1} at 50% WFPS, respectively. Ammonium was not detected either in the control or the nitrate-treated soils, during the experimental period. At 80% WFPS, the contents of NH_4^+-N were 57.2, 84.2 and 137.9 mg NH_4^+-N kg^{-1} when the soil was amended with urea, ammonium and UR + UI for 15 days, respectively. The content of NO_3^--N in soils treated with urea, ammonium and UR + UI increased with time. At the end of the experiment, the values at 50% WFPS were 48.4, 47.8 and 5.9 mg NO_3^--N kg^{-1}, respectively, and 28.2, 28.9 and 4.1 at 80% WFPS, respectively. The highest nitrate content was detected in the soil amended with nitrate, 187.8 and 124.9 mg NO_3^--N at 50 and 80%, respectively.

3.3.2 Ammonia Volatilization

The NH_3 fluxes by the soils treated with the different fertilizers are shown in Fig. 3.1. At 50% WFPS (Fig. 3.1a) the fluxes due to treatments with nitrate and UR + UI were similar to those of the control. The ammonia volatilization increased after

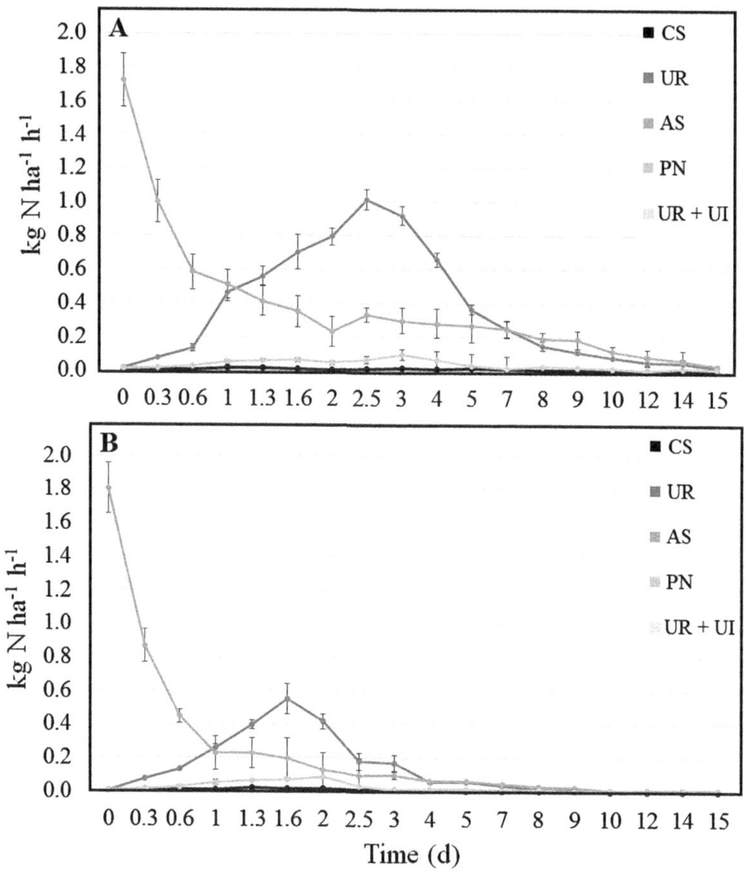

Fig. 3.1 Ammonia fluxes by soils treated with urea (UR), ammonium (AS), nitrate (PN) and urea + NBPT (UR + UI) at 50% (**A**) and 80% (**B**) WFPS during 15 days. Vertical lines indicate standard errors (n = 3)

Table 3.1 Ammonia cumulative emissions by soils treated with urea (UR), ammonium (AS), nitrate (PN) and urea + NBPT (UR + UI) at 50% and 80% WFPS

	NH_3 cumulative emission (kg N ha^{-1})	
Treatment	50% WFPS	80% WFPS
CS	2.0aA	1.9aA
UR	91.9 dB	38.2dA
AS	62.4cB	28.6cA
PN	8.8bA	6.6bA
UR + UI	2.3aA	1.8aA

Unfertilized soil was used as a control (CS). Ammonia was analyzed daily during 15 days. For each row and column, values followed by different lowercase or uppercase letters indicate statistical differences among treatments or WFPS, respectively. A Kruskal-Wallis and Conover-Iman tests were performed ($P < 0.05$; $n = 4$)

amendment with urea to reach a value of 1.0 kg N ha^{-1} h^{-1} after 2.5 days and decreased afterwards to a basal level of about 0.3 kg N ha^{-1} h^{-1} during the remaining experimental time. When the soil was treated with ammonium, the highest NH_3 fluxes (1.7 kg N ha^{-1} h^{-1}) were detected right after the fertilizer addition, and decreased gradually to the basal level. A similar pattern of NH_3 fluxes were observed at 80% WFPS with maximal values of 0.55 kg N ha^{-1} h^{-1} at 1.6 days for urea and 1.7 kg N ha^{-1} h^{-1} right after fertilization with ammonium (Fig. 3.1b). Regardless of the WFPS, the ammonia volatilization was negligible when the soil was fertilized with nitrate (Fig. 3.1a, b). Calculations of the NH_3 cumulative emissions during 15 days at 50% WFPS showed that soil treatments with either nitrate or UR + UI resulted in N losses of 8.8 and 2.3 kg N ha^{-1}, respectively, and that urea and ammonium fertilization released 91.9 kg and 62.4 kg N ha^{-1}, respectively (Table 3.1). At 80% WFPS, the emissions corresponding to the control, urea, ammonium, nitrate and UR + UI were 1.9, 38.2, 28.6, 6.6 and 1.8 kg N ha^{-1}, respectively (Table 3.1).

3.3.3 *Quantification of the 16S rRNA,* amoA, norB, nosZI *and* nosZII *Genes*

Under all conditions examined, the total abundance of members of the Bacteria domain at 50 (Fig. 3.2) and 80% WFPS (Fig. 3.3) was, on overage, 1.2 and 1.3 times higher than that of the Archaea, respectively. The addition of urea, ammonium or nitrate significantly increased the Bacteria and Archaea biomass along the experimental period as compared to the control soil by 0.6, 0.7 and 0.6 times after 15 days assay, respectively (Figs. 3.2 and 3.3). Increases were not found in the soil treated with UR + UI. At 50% WFPS, the total abundance of the *amoA* gene from Bacteria was, on average, 1.5-fold higher than that of the *amoA* from Archaea, and increased after treatment with urea and ammonium during the time of assay. Similar increases, however, were not observed in nitrate-and UR + UI-treated soils (Figs. 3.2 and 3.3).

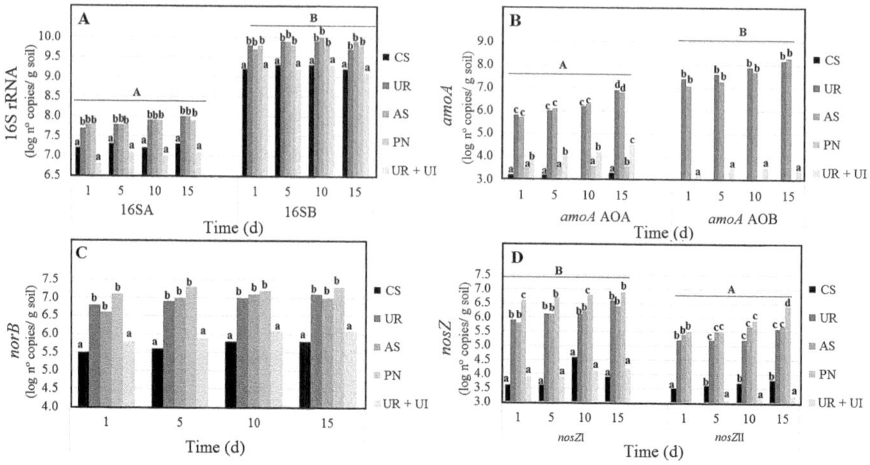

Fig. 3.2 Total abundance of the 16S rRNA gene from Bacteria and Archaea (**A**), *amoA* gene from Bacteria and Archaea (**B**), *norB* (**C**), and *nos*ZI and *nos*ZII denitrification genes (**D**) in soils treated with urea (UR), ammonium (AS), nitrate (PN) and urea + NBPT (UR + UI) at 50% WFPS during 15 d. Unfertilized soil was used as a control (CS). Samples were taken at 1, 5, 10 and 15 days after treatments. Data in columns or below horizontal lines with the same lowercase or uppercase letters, respectively, are not statistically different according to the Kruskal-Wallis and Conover-Iman tests ($P < 0.05$; $n = 4$)

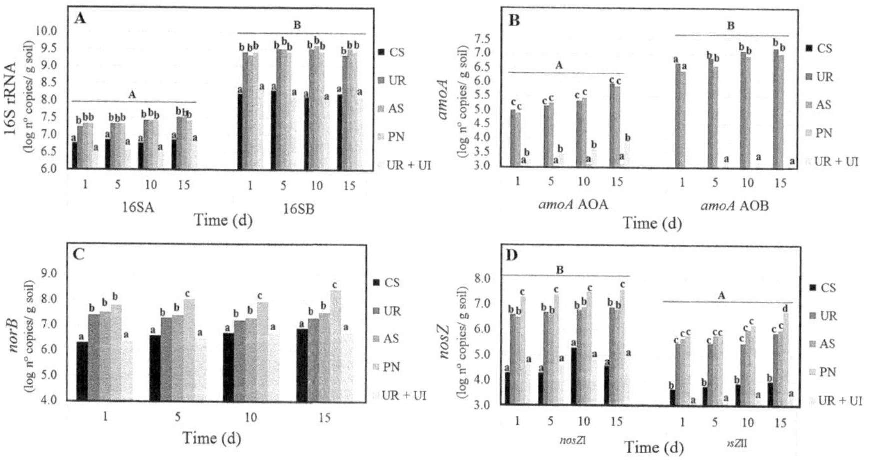

Fig. 3.3 Total abundance of the 16S rRNA gene from Bacteria and Archaea (**A**), *amoA* gene from Bacteria and Archaea (**B**), *norB* (**C**), and *nos*ZI and *nos*ZII denitrification genes (**D**) in soils treated with urea (UR), ammonium (AS), nitrate (PN) and urea + NBPT (UR + UI), at 80% WFPS during 15 d. Unfertilized soil was used as a control (CS). Samples were taken at 1, 5, 10 and 15 days after treatments. Data in columns or below horizontal lines with the same lowercase or uppercase letter, respectively, are not statistically different according to the Kruskal-Wallis and Conover-Iman tests ($P < 0.05$; $n = 4$)

The abundance of the bacterial *amoA* gene was also, on average, 1.2 times higher than that of the archaeal *amoA* at 80% WFPS, and their biomasses increased after treatment with urea and ammonium, but not when nitrate and UR + UI were applied (Fig. 3.3). The addition of urea, ammonia or nitrate at 80% WFPS increased, on average, 1.4 times the total abundance of the *norB*, *nos*ZI and *nos*ZII denitrification genes as compared to the 50% WFPS (Fig. 3.2). Also, at that moisture content, the abundance of the genes was greater in soils treated with nitrate than in those amended with urea or ammonium (Fig. 3.3). On the other hand, both at 50 and 80% WFPS, the total abundance of the *norB* and *nos*ZI genes in the soil amended with UR + UI was similar to that found in the control soil, but that of *nos*ZII was significantly lower (Figs. 3.2 and 3.3).

The relative abundance of the *amoA* AOA and *amoA* AOB statistically increased during the experimental period in soils treated with urea and ammonium (Table 3.2). Also, the relative abundance of *amoA* from Archaea was 1.3 times higher than that from Bacteria, and the calculated increases at 50% WFPS were significantly higher than those at 80% (Table 3.2). There is also to note that the relative abundance of *amoA* from Archaea significantly increased after amendment of the soil with UR + UI along the experimental period, while that of Bacteria remained similar to the control (Table 3.2).

Finally, the relative abundance of nitrifiers in the control and in the nitrate-treated soils was very scarce (Table 3.2). The relative abundance of denitrifiers at 80% WFPS were 1.2-fold higher than those found at 50% WFPS, and they increased during the 15 days experimental period (Table 3.3). The highest values of relative abundance were found in the soil treated with nitrate followed by those amended with

Table 3.2 Relative abundance of the *amoA* AOB and *amoA* AOA genes in soils treated with urea (UR), ammonium (AS), nitrate (PN) and urea + NBPT (UR + UI) at 50% and 80% WFPS

		amoA AOB/16SB (%)				*amoA* AOA/16SA (%)			
		Time (d)							
Treatment	WFPS (%)	T1	T5	T10	T15	T1	T5	T10	T15
UR	50	0.40cA	0.50dA	1.00 dB	3.16cC	1.26eA	1.58dA	2.00 dB	7.94eC
UR	80	0.18bA	0.22cA	0.40cB	0.47bB	0.56cA	0.67cA	0.81cB	2.59dC
AS	50	0.35cA	0.35dA	0.80 dB	2.51cC	0.79dA	2.00 dB	2.51 dB	6.31eC
AS	80	0.12bA	0.12bA	0.21bB	0.31bB	0.37cA	0.82cB	0.98cB	2.13dC
PN	50	0.01aA	0.01aA	0.01aA	0.01aA	0.01aA	0.01aA	0.01aA	0.01aA
PN	80	0.02aA	0.02aA	0.02aA	0.02aA	0.03aA	0.03aA	0.03aA	0.03aA
UR + UI	50	0.01aA	0.01aA	0.01aA	0.01aA	0.20bA	0.25bA	0.40bB	0.63cB
UR + UI	80	0.02aA	0.02aA	0.02aA	0.02aA	0.14bA	0.17bA	0.26bB	0.37bB
CS	50	0.02aA	0.02aA	0.02aA	0.02aA	0.03aA	0.03aA	0.03aA	0.03aA
CS	80	0.01aA	0.01aA	0.01aA	0.01aA	0.01aA	0.01aA	0.01aA	0.01aA

Unfertilized soil was used as a control (CS). Samples were taken at 1, 5, 10 and 15 days after application of the treatments. For each row and column, values followed by different lowercase or uppercase letters indicate statistical differences among treatments or WFPS, respectively. A Kruskal-Wallis and Conover-Iman tests were performed ($P < 0.05$; $n = 4$)

Table 3.3 Relative abundance of the *norB*, *nosZ*I and *nosZ*II denitrification genes in soils treated with urea (UR), ammonium (AS), nitrate (PN) and urea + NBPT (UR + UI) at 50% and 80% WFPS

		norB/16SB (%)				*nosZ*I/16SB (%)				*nosZ*II/16SB (%)			
		Time (d)											
Treatment	WFPS (%)	T1	T5	T10	T15	T1	T5	T10	T15	T1	T5	T10	T15
UR	50	0.10bA	0.10bA	0.13bA	0.25bB	0.01aA	0.02aA	0.02aA	0.08bB	0.00aA	0.00aA	0.00aA	0.00aA
UR	80	0.98dA	0.63cA	1.57 dB	3.15eC	0.16cA	0.20cA	0.20cA	0.97cB	0.01aA	0.01aA	0.01aA	0.04bB
AS	50	0.08bA	0.13bA	0.13bA	0.18bB	0.01aA	0.02aA	0.02aA	0.03aA	0.00aA	0.00aA	0.01aA	0.01aA
AS	80	0.79dA	1.24 dB	1.62 dB	2.57eC	0.15cA	0.20cA	0.20cA	0.69cB	0.02bA	0.02bA	0.02bA	0.03bA
PN	50	0.20cA	0.25cB	0.30cB	0.32cB	0.06bA	0.08bA	0.12bB	0.13bB	0.00aA	0.00aA	0.01aA	0.04bB
PN	80	2.47eA	3.25eB	4.19eB	9.82fC	0.78dA	0.98 dB	1.24 dB	1.56 dB	0.02bA	0.02bA	0.06cB	0.21cC
UR + UI	50	0.04aA	0.05aA	0.06aA	0.10bB	0.00aA	0.00aA	0.00aA	0.00aA	0.00aA	0.00aA	0.00aA	0.00aA
UR + UI	80	0.33cA	0.43cA	0.59cB	0.92dC	0.01aA	0.01aA	0.01aA	0.01aA	0.00aA	0.00aA	0.00aA	0.00aA
CS	50	0.02aA	0.02aA	0.03aA	0.04aA	0.00aA	0.00aA	0.00aA	0.00aA	0.00aA	0.00aA	0.00aA	0.00aA
CS	80	0.09bA	0.17bB	0.29cC	0.27cC	0.00aA	0.00aA	0.02aA	0.01aA	0.00aA	0.00aA	0.00aA	0.00aA

Unfertilized soil was used as a control (CS). Samples were taken at 1, 5, 10 and 15 days after application of the treatments. For each row and column, values followed by different lowercase or uppercase letters indicate statistical differences among treatments or WFPS, respectively. A Kruskal-Wallis and Conover-Iman tests were performed ($P < 0.05$; $n = 4$)

urea, ammonium and with U + UI (Table 3.3). The relative abundance of *norB* was, on average, 3.5 and 8 times higher than that of *nosZ* at 50 and 80% WFPS, respectively (Table 3.3).

3.4 Discussion

Surface-applied ammonia- and ammonia-based N-fertilizers release nitrogen to the atmosphere via ammonia volatilization. Multiple, and often related, factors such as soil texture, type of N-fertilizer, pH, irrigation regime, etc., make ammonia volatilization variable and difficult to predict under laboratory and field conditions. Here we have used a mesocosm approach to study the rates of ammonia volatilization in soils treated with 3 N-fertilizers (urea, ammonium and nitrate), at 50 and 80% WFPS. The water content was a main driver of ammonia volatilization as NH_3 fluxes were much higher at 50 than at 80% WFPS (Fig. 3.1a, b). It has been suggested that adsorption of NH_4^+ to the soil particles with cation exchange sites is facilitated under high moisture conditions, thus reducing the potential for NH_4^+ release to the atmosphere (Sanz-Cobeña et al. 2011). The moisture content also influenced the time at which the maximal emissions occurred, as the wetter conditions delayed the appearance of the highest emissions (Fig. 3.1a, b). This agrees with the suggestion by Sanz-Cobeña et al. (2011) that high water contents facilitate dissolution of urea in the upper part of the soil.

Different authors have reported values of urea-dependent ammonia volatilization ranging from 5% to 50% of the applied N, probably due to varying soil conditions and fertilization management (Yang et al. 2016 and references therein). Under the experimental conditions in this study, the amount of N lost through NH_3 volatilization during 15 days after urea application accounted for 35.3 and 14.7% at 50 and 80% WFPS, respectively (Table 3.1). After application of the urea together with NBPT, the NH_3 fluxes, and the concomitant cumulative emissions were reduced, so that the final N losses were of 0.9 and 0.7%, at 50 and 80% WFPS, respectively, after the 15-days experimental period (Table 3.1). These results are more likely due to a lower availability of NH_4^+ since the addition of NBPT retarded the hydrolysis of urea by urease, and effect that has been already observed by other authors (Gioacchini et al. 2002; Sanz-Cobeña et al. 2008, 2011; Zaman et al. 2009). Application of ammonium reduced N losses respect to urea fertilizer to about 24 and 11% of the total N applied at 50 and 80%, respectively (Table 3.1). This results may be explained as ammonium does not increase the pH of soil that favors NH_3 volatilization. As expected, when nitrate is applied the NH_3 volatilization was negligible, due to the absence of ammonium in the soil.

While the UR + UI treatment had no effect on the total abundance of non-target soil microbial populations, the addition of any of the N-fertilizers increased the total abundance of the bacterial and archaeal populations (Fig. 3.2). These results do not agree with those that indicate that the application of chemical fertilizers decreased the biomass of members of the Bacteria and Archaea domains (Hallin et al. 2009;

Chan et al. 2013). It is possible that the differences in our study could be assigned to the short duration of the experiment, only 15 days, whereas the above studies were long-term experiments of at least 3 months.

While urea and ammonium produced an increase in the abundance of nitrifiers in the soil, the application of urea together with NBPT maintained their abundances in values similar to those found in the control soil regardless of the water content (Fig. 3.2). NBPT has been shown to form a tridentate ligand with urease so that the activity of the enzyme is reduced (Chen et al. 2008) and, as a consequence, less NH_4^+ would be available for intracellular nitrification.

As expected, urea and ammonium increased the relative abundance of nitrifiers. NBPT, however, had a different effect on the nitrifier communities as it increased the relative abundance of the *amoA* AOA, but not that of *amoA* AOB (Table 3.2). Whether or not this could be associated with a lower susceptibility of Bacteria to the inhibitor cannot be elicited from the present results. In this sense, addition of NBPT affected differently the abundances of the *amoA* AOA and *amoA* AOB genes after a 2-year amendment, depending on the soil type (Shi et al. 2017).

The increases in the total and relative abundance of the *amoA* AOA and *amoA* AOB genes were observed not only at 50 but also at 80% WFPS (Fig. 3.2 and Table 3.2). These results suggest that although nitrification is carried out under aerobic conditions (Sahrawat 2008, and references therein), the nitrifying activity could exist under oxygen-limiting conditions too. In fact, the existence of microniches where nitrification can occur under high moisture contents have been reported (Cardenas et al. 2017). Interestingly, the *amoA* gene from Archaea was more abundant than that from Bacteria in the soils treated with urea or ammonium. This could be associated with the greater affinity of Archaea than Bacteria for NH_4^+ (Leininger et al. 2006, Prosser and Nicol 2012; Daebeler et al. 2015).

It is well established that the denitrification process is favored under O_2-limiting conditions (Thamdrup 2012; Ward and Jensen 2014), a reason why the addition of the N-fertilizers at 80% rather than at 50% increased the total abundance of the *norB*, *nos*ZI and *nos*ZII denitrification genes, albeit the higher increments were found after addition of nitrate (Fig. 3.3 and Table 3.3). Nitrate is readily available for denitrification, whereas urea and ammonia require further conversion to nitrate. Thus, it is possible that the higher abundance of the denitrification genes in nitrate-treated soils could be due to the short 15-days duration of this study. The total and relative abundance of *norB* gene was always higher than that of *nosZ* under all conditions examined (Table 3.3), which agrees with the finding that one-third of the sequenced denitrifier genomes lack the *nosZ* gene (Jones et al. 2008). In this study NBPT did not affect the total abundance of the denitrification genes (Fig. 3.3). This could be explained considering that the treatment with the inhibitor diminished nitrification, thus leading to a shortage of nitrate available for denitrification. In fact, the amount of nitrate in the UR + UI was similar to that found in the control soil. A previous report, however, has shown that NBPT decreases the abundance of the *narG* and *nirK* genes at 60% WFPS after 2-year treatment under field conditions (Shi et al. 2017).

3.5 Conclusions

The soil water content controlled ammonia volatilization and the abundance of nitrification and denitrification genes in soils fertilized with urea, ammonium or nitrate after a 15-days short-term experiment. Losses of the applied N were greater in soils fertilized with urea followed by those treated with ammonium, whereas those due to nitrate were similar to the unfertilized soil. The NH_3 fluxes were reduced under high water content (80% WFPS). The urease inhibitor NBPT decreases ammonia volatilization and is application, whether under low or high moisture conditions could be a good strategy to reduce N-loses due to NH_3 volatilization. The addition of NBPT reduces the abundance of the nitrification genes and does not affect those of the denitrification process.

Acknowledgments This study was supported by the ERDF-cofinanced grants PEAGR2012-1968 from Consejería de Economía, Innovación y Ciencia (Junta de Andalucía, Spain), AGL2015-64582-C3-3-R from MINECO, P2013/ABI-2717 from the Autonomous Community of Madrid, and the MINECO-CSIC Agreement RECUPERA 2020. Authors also thank Sirena Network (AGL2015-68881-REDT) for financial support. ACH is recipient of a grant of MECD (FPU 2014/01633).

References

Abalos, D., Sanz-Cobeña, A., Misselbrook, T., et al. (2012). Effectiveness of urease inhibition on the abatement of ammonia, nitrous oxide and nitric oxide emissions in a non-irrigated Mediterranean barley field. *Chemosphere, 89*, 310–318.

Abalos, D., Jeffery, S., & Sanz-Cobena, A. (2014). Meta-analysis of the effect of urease and nitrification inhibitors on crop productivity and nitrogen use efficiency. *Agriculture, Ecosystems and Environment, 189*, 136–144.

Aneja, V. P., Chauhan, J. P., & Walker, J. (2000). Characterization of atmospheric ammonia emissions from swine waste storage and treatment lagoons. *Journal of Geophysical Research (Atmospheres), 105*, 11,535–11,545.

Bedmar, E. J., Bueno, E., Correa, D., et al. (2013). Ecology of denitrification in soils and plant-associated bacteria. In B. Rodelas & J. González-López (Eds.), *Beneficial plant-microbial interactions: Ecology and applications* (pp. 164–182). Boca Raton: CRC Press.

Braker, G., & Tiedje, J. M. (2003). Nitric oxide reductase (*norB*) genes from pure cultures and environmental samples. *Applied and Environmental Microbiology, 69*, 3476–3483.

Butterbach-Bahl, K., Baggs, E. M., Dannenmann, M., et al. (2013). Nitrous oxide emissions from soils: How well do we understand the processes and their controls? *Philosophical Transactions of the Royal Society of London. Series B, Biological Sciences, 368*, 91–97.

Cardenas, L., Bol, R., Lewicka-Szczebak, D., et al. (2017). Effect of soil saturation on denitrification in a grassland soil. *Biogeosciences, 14*, 4691–4710.

Castellano-Hinojosa, A., González-López, J., & Bedmar, E. J. (2018). Distinct effect of nitrogen fertilisation and soil depth on nitrous oxide emissions and nitrifiers and denitrifiers abundance. *Biology and Fertility of Soils, 54*, 829–840.

Chan, Y. K., Mccormick, W. A., & Ma, B. (2013). Effects of inorganic fertilizer and manure on soil archaeal abundance at two experimental farms during three consecutive rotation-cropping seasons. *Applied Soil Ecology, 68*, 26–35.

Chen, D. L., Suter, H., Islam, A., et al. (2008). Prospects of improving efficiency of fertilizer nitrogen in Australian agriculture a review of enhanced efficiency fertilizers. *Australian Journal of Soil Research, 46*, 289–301.

Correa-Galeote, D., Tortosa, G., & Bedmar, E. J. (2014). Quantification of functional microbial nitrogen cycle genes in environmental samples. In D. E. Marco (Ed.), *Metagenomics of the microbial nitrogen cycle: Theory, methods and applications* (pp. 65–85). Norwich: Caister Academic Press.

Daebeler, A., Bodelier, P. L. E., Hefting, M. M., et al. (2015). Ammonia-limited conditions cause of Thaumarchaeal dominance in volcanic grassland soil. *FEMS Microbiology Ecology, 91*, 1–7.

Danielson, R. E., & Sutherland, P. L. (1986). Porosity. In *Methods of soil analysis part 1. Physical and mineralogical methods* (Agronomy Monograph no. 9) (pp. 443–461). Madison: Soil Science Society of America.

Erisman, J. W., Galloway, J. N., Dice, N. B., et al. (2015). *Nitrogen: Too much of a vital resource* (Science Brief). Zeist: WWF Netherlands.

FAO. (2017). *Soil organic carbon: The hidden potential*. Rome: Food and Agriculture Organization of the United Nations.

Gill, J. S., Khind, B. S., & Yadvinder-Singh, C. S. (1999). Efficiency of N-(n-butyl) thiophosphoric triamide in retarding hydrolysis of urea and ammonia volatilization losses in a flooded sandy loam soil amended with organic materials. *Nutr Cycl Agroecosy, 53*, 203–207.

Gioacchini, P., Nastri, A., Marzadori, C., et al. (2002). Influence of urease and nitrification inhibitors on N losses from soils fertilized with urea. *Biology and Fertility of Soils, 36*, 129–135.

Giovannini, C., Garcia-Mina, J. M., Ciavatta, C., et al. (2009). Ureic nitrogen transformation in multi-layer soil columns treated with urease and nitrification inhibitors. *Journal of Agricultural and Food Chemistry, 57*, 4883–4887.

González-Martínez, A., Rodríguez-Sánchez, A., García-Ruiz, M. J., et al. (2016). Performance and bacterial community dynamics of a CANON bioreactor acclimated from high to low operational temperatures. *Chemical Engineering Journal, 287*, 557–567.

Hallin, S., Jones, C. M., Schloter, M., et al. (2009). Relationship between N-cycling communities and ecosystem functioning in a 50-year-old fertilization experiment. *The ISME Journal, 3*, 597–605.

Heffer, P., & Prud'homme, M. (2011). *Fertilizer outlook 2011–2015*. Paris: International Fertilizer Industry Association e IFA. http://www.fertilizer.

Henry, S., Bru, D., Stres, B., et al. (2006). Quantitative detection of the *nosZ* gene, encoding nitrous oxide reductase, and comparison of the abundances of 16S rRNA, *narG*, *nirK*, and *nosZ* genes in soils. *Applied and Environmental Microbiology, 72*, 5181–5189.

Jones, C. M., Stres, B., Rosenquist, M., et al. (2008). Phylogenetic analysis of nitrite, nitric oxide, and nitrous oxide respiratory enzymes reveal a complex evolutionary history for denitrification. *Molecular Biology and Evolution, 25*, 1955–1966.

Jones, C. M., Graf, D. R. H., Bru, D., et al. (2013). The unaccounted yet abundant nitrous oxide-reducing microbial community: A potential nitrous oxide sink. *The ISME Journal, 7*, 417–426.

Kaplan, W. A., Wofsy, S. C., Keller, M., et al. (1988). Emission of NO and deposition of O_3 in a tropical forest system. *Journal of Geophysical Research, 93*, 1389–1395.

Kieft, T. L., Soroker, E., & Firestone, M. K. (1987). Microbial biomass response to a rapid increase in water potential when dry soil is wetted. *Soil Biology and Biochemistry, 19*, 119–126.

Leininger, S., Urich, T., Schloter, M. S., et al. (2006). Archaea predominate among ammonia oxidizing prokaryotes in soils. *Nature, 442*, 806–809.

López-Gutiérrez, J. C., Henry, S., Hallet, S., et al. (2004). Quantification of a novel group of nitrate-reducing bacteria in the environment by real-time PCR. *Journal of Microbiological Methods, 57*, 399–407.

Modolo, L. V., de Souza, A. X., Hora, L. P., et al. (2015). An overview on the potential of natural products as ureases inhibitors: A review. *Journal of Advanced Research, 6*, 35–44.

Nannipieri, P., Ceccanti, B., Cervelli, S., et al. (1980). Extraction of phosphatases, urease, proteases, organic carbon and nitrogen from soil. *Soil Science Society of America Journal, 44*, 1011–1016.

Ochsenreiter, T., Selezi, D., Quaiser, A., et al. (2003). Diversity and abundance of Crenarchaeota in terrestrial habitats studied by 16S RNA surveys and real time PCR. *Environmental Microbiology, 5*, 787–797.

Prosser, J. I., & Nicol, G. W. (2012). Archaeal and bacterial ammonia-oxidisers in soil: the quest for niche specialisation and differentiation. *Trends in Microbiology, 20*, 522–531.

Rodríguez-Soares, J. R., Cantarella, H., & Menegale, L. C. (2012). Ammonia volatilization losses from surface-applied urea with urease and nitrification inhibitors. *Soil Biology and Biochemistry, 52*, 82–89.

Rotthauwe, J. H., Witzel, K. P., & Liesack, W. (1997). The ammonia monooxygenase structural gene *amoA* as a functional marker: Molecular fine-scale analysis of natural ammonia-oxidizing populations. *Applied and Environmental Microbiology, 63*, 4704–4712.

Sahrawat, K. L. (2008). Factors affecting nitrification in soils. *Communications in Soil Science and Plant Analysis, 39*, 1436–1446.

Sanz-Cobeña, A., Misselbrook, T. H., Arce, A., et al. (2008). An inhibitor of urease activity effectively reduces ammonia emissions from soil treated with urea under Mediterranean conditions. *Agriculture, Ecosystems and Environment, 126*, 243–249.

Sanz-Cobeña, A., Misselbrook, T., Camp, V., et al. (2011). Effect of water addition and the urease inhibitor NBPT on the abatement of ammonia emission from surface applied urea. *Atmospheric Environment, 45*, 1517–1524.

Shi, X., Hu, H. W., Kelly, K., et al. (2017). Response of ammonia oxidizers and denitrifiers to repeated applications of a nitrification inhibitor and a urease inhibitor in two pasture soils. *Journal of Soils and Sediments, 17*, 974–984.

Sutton, M. A., Oenema, O., Erisman, J. W., et al. (2011). Too much of a good thing. *Nature, 472*, 159–161.

Thamdrup, B. (2012). New pathways and processes in the global nitrogen cycle. *Annual Review of Ecology, Evolution, and Systematics, 43*, 407–428.

Torres, M. J., Simon, J., Rowley, G., et al. (2016). Nitrous oxide metabolism in nitrate-reducing bacteria: Physiology and regulatory mechanisms. *Advances in Microbial Physiology, 68*, 353–432.

Tourna, M., Freitag, T. E., Nicol, G. W., et al. (2008). Growth, activity and temperature responses of ammonia-oxidizing archaea and bacteria in soil microcosms. *Environmental Microbiology, 10*, 1357–1364.

Walker, J. T., Geron, C. D., Vose, J. M., et al. (2002). Nitrogen trace gas emissions from a riparian ecosystem in southern Appalachia. *Chemosphere, 49*, 1389–1398.

Ward, B. B., & Jensen, M. M. (2014). The microbial nitrogen cycle. *Frontiers in Microbiology, 5*, 553.

Yang, M., Fang, Y., Sun, D., et al. (2016). Efficiency of two nitrification inhibitors (dicyandiamide and 3,4-dimethypyrazole phosphate) on soil nitrogen transformations and plant productivity: A meta-analysis. *Scientific Reports, 6*, 22075.

Zaman, M., & Blennerhassett, J. D. (2010). Effect of the different rates of urease and nitrification inhibitors on gaseous emissions of ammonia and nitrous oxide, nitrate leaching and pasture production from urine patches in an intensive grazed pasture system. *Agriculture, Ecosystems and Environment, 136*, 236–246.

Zaman, M., Saggar, S., Blennerhassett, J. D., et al. (2009). Effect of urease and nitrification inhibitors on N transformation, gaseous emissions of ammonia and nitrous oxide, pasture yield and N uptake in grazed pasture system. *Soil Biology and Biochemistry, 41*, 1270–1280.

Chapter 4
Azospirillum as Biofertilizer for Sustainable Agriculture: *Azospirillum brasilense* AZ39 as a Model of PGPR and Field Traceability

Anahí Coniglio, Verónica Mora, Mariana Puente, and Fabricio Cassán

Abstract *Azospirillum* is one of the best studied genus of plant growth promoting rhizobacteria at present. These bacteria are able to colonize hundreds of plant species and significantly improve their growth, development and productivity under field conditions. Besides nitrogen fixation, the most studied mechanism proposed for *Azospirillum* to explain plant growth promotion of inoculated plants has been related to its ability to produce several phytohormones, mainly auxins and particularly indole-3-acetic acid. Although different capacities have been described to explain the plant growth regulation by *Azospirillum* one single mechanism is not quite extensive to explain the full effect observed on inoculated plants. The bacterial mode of action is currently better explained as the result of additive and selective effects. One of the most important achievements obtained thus far is the utilization of azospirilla as commercial inoculants in approximately 7.0 million doses and 5.0 million ha, mainly cultivated with cereal crops and legumes in South America. Different inoculation practices (farmer applied or industrial seed treatments, in-furrow, foliar or soil sprayed applications) have been developed and improved in the last two decades for a wide range of crops, in field conditions. Particularly, the combined inoculation of legumes with rhizobia and azospirilla, could over improve the performance of the plants compared with a single inoculation, due to the complementary biological processes of both microbes. The development and validation of specific novel methodologies for identification of *A. brasilense*, and particularly the strain Az39 in both bio-products and inoculated samples (i.e. soil, rhizosphere, seeds or plant tissues) offer a precise tool to evaluate the functionality and traceability

A. Coniglio · V. Mora · F. Cassán (✉)
Laboratorio de Fisiología Vegetal y de la Interacción Planta-Microorganismo (LFVIPM), Departamento de Ciencias Naturales, FCEFQyN, Universidad Nacional de Río Cuarto, Córdoba, Argentina
e-mail: fcassan@exa.unrc.edu.ar

M. Puente
Instituto Nacional de Tecnología Agropecuaria (INTA), Instituto de Microbiología y Zoología Agrícola (IMYZA), Buenos Aires, Argentina

D. Zúñiga-Dávila et al. (eds.), *Microbial Probiotics for Agricultural Systems*, Sustainability in Plant and Crop Protection,
https://doi.org/10.1007/978-3-030-17597-9_4

of these microorganisms in the environment. In this chapter, we explore some classical mechanisms of plant growth promotion in *A. brasilense* Az39, one of the most widely used PGPR strains for inoculant production in South America. Additionally, we discuss some novel molecular tools designated to identify this strain in both bio-products and field conditions.

4.1 Introduction

Azospirillum is a genus of α-Proteobacteria belonging to the family *Rhodospirillaceae*. It includes a species previously named *Spirillum lipoferum* (Beijerinck 1925). The growth of a spirillum-like bacterium was first observed by Beijerinck in 1922, when a nitrogen-deficient malate-based medium was inoculated with garden soil. This new bacterium increased the nitrogen content of the medium and was named *Azotobacter spirillum*. Three years later, the genus was renamed as *Spirillum* and little attention was given to this organism until 1974, when root samples of several grasses (*Digitaria*, *Panicum maximum*, *Cynodon dactylon*), grain crops (wheat, maize, sorghum, millet, rice) and soil samples were used by Döbereiner and collaborators to isolate this organism. The genus *Azospirillum* was first proposed by Tarrand et al. (1978) and two species were initially described, *A. lipoferum* and *A. brasilense*. Since then, 20 species belonging to this bacterial genus were described, namely *A. halopraeferens* (Reinhold et al. 1987), *A. largimobile* (Ben Dekhil et al. 1997), *A. doebereinerae* (Eckert et al. 2001), *A. oryzae* (Xie and Yokota 2005), *A. melinis* (Peng et al. 2006), *A. canadense* (Mehnaz et al. 2007a), *A. zeae* (Mehnaz et al. 2007b), *A. rugosum* (Young et al. 2008), *A. massiliensis* (Pagnier et al. 2008), *A. picis* (Lin et al. 2009), *A. palatum* (Zhou et al. 2009), *A. thiophilum* (Lavrinenko et al. 2010), *A. formosense* (Lin et al. 2012), *A. humicireducens* (Zhou et al. 2013), *A. fermentarium* (Lin et al. 2013), *A. himalayense* (Tyagi and Singh 2014), *A. soli* (Lin et al. 2015) and *A. agricola* (Young et al. 2016). Additionally, *A. amazonense* (Falk et al. 1985) and *A. irakense* (Khammas et al. 1989) were reclassified into two different genera *Nitrospirillum* and *Niveispirillum*, respectively (Lin et al. 2014). The free-living nitrogen fixer (diazothrophic) *Azospirillum brasilense* is able to colonize the soil, rhizosphere, rhizoplane endosphere and internal tissues of numerous plants, enhancing their growth, development and yield under controlled or agronomic conditions (Kloepper et al. 1989; Okon and Labandera-Gonzalez 1994; Bashan and de-Bashan 2005; Compant et al. 2010). In thousands of reports it has been shown that PGPR and particularly *Azospirillum* can enhance plant growth by direct or indirect mechanisms. Direct mechanisms include the increase of nutrients bio-availability for plants (i.e. nitrogen, phosphorous, iron, potassium) or the production of enzymes and plant growth regulators (phytohormones). Indirect mechanisms include the induction of both, plant tolerance against abiotic stress or plant resistance against pathogens (Kloepper and Schroth 1981; Glick 2012). The first mechanism proposed to explain the bacterial plant growth promotion ability of *Azospirillum* has been associated, almost exclusively, with the nitrogen status of inoculated plants (Okon et al. 1983), through the biological nitrogen fixation (BNF) or nitrate reductase activity. However, these mechanisms showed less agronomic significance than initially expected (Kennedy et al. 1997) and new hypothesis were proposed.

4.1.1 The Versatility of the Genus Azospirillum

Besides BFN, *Azospirillum* has been widely studied due its capacity to produce several phytohormones and plant growth regulators, such as indole-3-acetic acid (IAA) (Bashan et al. 2004; Spaepen et al. 2007), gibberellins (GAs) (Bottini et al. 2004; Cassán et al. 2009), abscisic acid (ABA) (Tien et al. 1979; Kolb and Martin 1985), polyamines (Thuler et al. 2003; Perrig et al. 2007; Cassán et al. 2009), cytokinins (Tien et al. 1979; Strzelczyk et al. 1994), ethylene (Perrig et al. 2007), and nitric oxide (Molina-Favero et al. 2007). Many authors additionally reported that *Azospirillum* is able to enhance plant growth not only by self-production of hormones, but also by inducing the synthesis of these compounds in plant tissues (Chamam et al. 2013; Duca et al. 2014; Cohen et al. 2015). Although many mechanisms have been described to explain the plant growth promotion by *Azospirillum*, one single mechanism is mostly not uniquely responsible for the full effect. Therefore, the bacteria mode of action could be better explained by the use of the "additive hypothesis" which allows explaining the plant growth promoting effect due to inoculation. This hypothesis was suggested more than 20 years ago (Bashan and Levanony 1990) and considers multiple mechanisms are involved rather than only one, participating in the successful association of *Azospirillum* with plants.

4.1.2 From Laboratory to the Field

The positive effects of *Azospirillum* on plants have attracted attention of researcher due to the need to reduce agricultural chemicals in the context of sustainable agriculture. Inoculants based on *Azospirillum* have already been commercialized in several countries including Argentina, Brazil, Mexico, Italy, India, France, among others (Díaz-Zorita and Fernández-Canigia 2009). Biological products containing *Azospirillum* are mainly applied on maize and wheat crops, followed by sorghum, sunflower, soybean, grasses and winter cereals for grazing, rice, barley, cotton, oats, sugar cane, tobacco and lettuce (Cassán and Díaz-Zorita 2016). This approach represents an important component for sustainable agriculture. On the one hand, inoculants generate a significant economic impact by increasing yields of agricultural crops. On the other hand, they represent an ecologically acceptable contribution, due to the use of soil microorganisms.

Field crop inoculation with *Azospirillum* is carried out by either pre-treating the seeds (seed inoculation) with liquid or solid carriers containing microorganisms or applying them directly in the sowing furrow. Inoculation in-furrow with *A. brasilense*, via soil spray at sowing and via foliar spray after seedlings emergence, were described (Fukami et al. 2016; Puente et al. 2017). The foliar application of *Azospirillum* and their metabolites improved not only plant growth, but also up-regulated plant genes related to defense mechanisms (Puente et al. 2017, 2018).

At present, available bibliography about changes on crop yields due to the inoculation with *A. brasilense* is massive. The presence of these bacteria has been related with changes in plant growth during early stages of development, and mainly under environmental stress conditions (i.e. nutrient deficiencies, water stress, etc.). The mean grain response due to inoculation with *Azospirillum* was 10.0%, with greater yields in winter cereals (14.0%) rather than in summer cereals (9.5%) or legumes (6.6%) (Cassán and Díaz-Zorita 2016). Moreover, a meta-analysis on several published articles concerning the effects of *Azospirillum* inoculation on grain yield of wheat crops was also described (Veresoglou and Menexes 2010). A large number of studies on field conditions in the Pampas region of Argentina, showed that the effects of cereals and legumes inoculation or co-inoculation with *A. brasilense* Az39 are considered as a consolidated practice (Díaz-Zorita and Fernández-Canigia 2009). Currently, Az39 strain is the active principle in 75% of inoculants in Argentina and is also used in Brazil and Uruguay (Cassán and Díaz-Zorita 2016).

4.1.3 Biological Products with Azospirillum *in South America*

In the South cone of South America there are around 106 biological products containing *Azospirillum* available for commercialization. The production is performed by more than 70 companies and mainly formulated in liquid carriers. Most of the bioformulations (90) are produced in Argentina, 14 in Brazil and 2 in Uruguay. All products available commercially are produced in Argentinaor imported from Argentina in Brazil and Uruguay. The Az39 strain of *A. brasilense* is the active principle in 75% (79 products) of these inoculants. In 13 products, it is combined with other strains of *A. brasilense* (1 product, containing CFN535), *Psudomonas fluorescens* (1) or *Bradyrhizobium* sp. (11 products). The combination of *A. brasilense* strains Abv5 and Abv6 is used for the formulation of 18 products. Az78 and Az70 are employed in 3 products. The other azospirilla inoculants are formulated with single strains (Abv5, AzM3, AzT5, 1003, Tuc 27/85, Tuc 10/1 and 11,005). There are no other species of *Azospirillum* used for the inoculants formulation in the region.

Liquid carriers are the most used for the bioformulation products (94%) while 6% of them are formulated on solid carriers like peat or bentonite. The shelf life of registered products is below 6 months 1.0^7 cfu ml^{-1} in Argentina and 1.0^8 in Brazil and Uruguay. Although 16 crops have been recommended for use of these biological products, the registration is mainly for wheat (67 products) and corn (65). Sunflower (15), soybean (12), sorghum (*Sorghum bicolor*) (9), rice (5), grasses and winter cereals for grazing (4), barley (3), cotton (*Gossypium hirsutum*) (3), oats (*Avena sativa*) (2), sugar cane (*Saccharum officinarum*) (1), tobacco (*Nicotiana tabacum*) (1) and lettuce (*Lactuca sativa*) (1) are the other crops for which the application of commercial products containing *A. brasilense* is recommended. In Argentina, during 2016, most of the products containing *A. brasilense* were used in winter grasses, soybean, corn and wheat covering almost 1 million ha (0.7 with soybean). In Brazil, most of the commercialized products were allocated in the corn and soybean grain production market. Based on 2016 data, approximately 6.0 million doses of azospirilla inoculants were commercialized covering almost 4.0 million ha. Almost 2000 ha cultivated with corn have been treated with azospirilla inoculants during the 2016 growing season in Uruguay.

4.1.4 The Model Strain A. brasilense *Az39*

In Argentina, comparative studies were performed in the 80's at the Instituto de Microbiología y Zoología Agrícola (IMyZA), belonging to the Instituto Nacional de Tecnología Agropecuaria (INTA), located in Castelar, Buenos Aires. These studies showed that *A. brasilense* Az39 was the most effective strain to promote plant growth and increase crop yields of maize and wheat, under experimental agronomic conditions. *A. brasilense* Az39 was isolated in 1982 from surface-sterilized wheat seedlings in Marcos Juarez, Argentina and selected for the inoculants formulation, due to its effectiveness to increase the maize and wheat grain production from 13% to 33% (Rodriguez-Cáceres et al. 1994; Okon and Labandera-Gonzalez 1994). Thus, the Servicio Nacional de Sanidad Agropecuaria (SENASA), together with INTA and several national bio-products companies decided to recommend *A. brasilense* Az39 for the development of bio-products for seeds biological treatment of corn, wheat and other non-legumes species (Díaz-Zorita and Fernández-Canigia 2009).

While the potential mechanisms responsible for the growth promotion by Az39 have been partially unraveled *in vitro* (Perrig et al. 2007) and *in vivo* (Cassán et al. 2009; Rodriguez-Cáceres et al. 2008), the Laboratorio de Fisiología Vegetal y de la Interacción Planta-Microorganismos (LFVIPM) at the Universidad Nacional de Río Cuarto decided to sequence the genome of that strain in order to access to the genetic information of this bacterium (Rivera et al. 2014) and to improve the current knowledge on the model of *Azospirillum*-plant interaction. This task was afforded with the Instituto de Agrobiotecnología Rosario (INDEAR), Argentina; The Genome Analysis Centre (TGAC), Norwich (UK); Katholieke Universiteit Leuven (Belgium); Max Planck Institute for Plant Breeding Research, Cologne (Germany); Ecologie Microbienne, Université Lyon, Villeurbanne (France) and Grupo de Ecología Genética de la Rizósfera, Estación Experimental del Zaidín (CSIC), Granada (Spain).

4.2 The Genomic Era and *Azospirillum*

To date, six whole *Azospirillum* genomes have been published: *A. thiophilum* BV-S (PRJNA292868) (Fomenkov et al. 2016), *A. brasilense* Sp7 (PRJNA293508) (Tarrand et al. 1978), *A. brasilense* Sp245 (PRJEA70627) (Wisniewski-Dyé et al. 2011), *A. brasilense* Az39 (PRJNA238398) (Rivera et al. 2014), *A. lipoferum* 4B (PRJEA50367) (Wisniewski-Dyé et al. 2011) and *Azospirillum* B510 (PRJDA32551) (Kaneko et al. 2010). Although there are 22 genome sequencing and assembly projects for *Azospirillum* spp., a large amount of information remains still unpublished (e.g. *A. brasilense* CBG497). In 2018, five projects have been registered, including REC3 (PRJNA429356), ABV6 (PRJNA429631) and ABV5 (PRJNA429443) strains of *A. brasilense* and two strains of *Azospirillum* sp.

For its part, *A. amazonense* Y2 strain (Magalhães et al. 1983) (PRJNA224116) was reclassified into a new genus *Nitrospirillum* (Lin et al. 2014). *A. brasilense* Sp245 is considered the type strain for this species. It was isolated from surface-sterilized wheat (*Triticum aestivum* L.) roots in Paraná state from the South region of Brazil (Baldani et al. 1986). It was one of the most promising strains for wheat inoculation in Brazil during the 80's decade of past century. *Azospirillum lipoferum* 4B was isolated from the rhizosphere of rice (*Oryza sativa* L.) (Bally et al. 1983). It was successfully used as inoculant to increase rice yields under field conditions (Charyulu et al. 1985). *Azospirillum* B510 is an endophytic bacterium isolated from surface-sterilized stems of rice plants in Kashimadai, Japan (Elbeltagy et al. 2001). *Azospirillum brasilense* strain CBG497 was isolated from the rhizosphere of maize (*Zea mays* L.) plants grown on soils with pH 8.0, from the North-East region of Mexico (Nelson and Knowles 1978). *Azospirillum brasilense* Sp7 (ATCC 29145) was isolated by Joana Döbereiner from pangola grass (*Digitaria decumbens*) plants from Rio de Janeiro in Brazil and proposed by Tarrand et al. (1978) as the type strain for the species. Finally, *A. brasilense* FP2 is a spontaneous mutant of Sp7 (Nalr and Smr) obtained by Pedrosa and Yates (1984).

4.2.1 Genomic Analysis of A. brasilense *Az39*

The genome sequence of this strain was obtained in 2014 using a combined whole genome shotgun and 8-kb paired-end strategy with a 454 GS FLX Titanium pyrosequencer at INDEAR (Argentina), resulting in a 21-fold genome coverage. Sequencing reads were *de novo* assembled (Newbler v 2.8), resulting in 6 scaffolds (>160 kbp each; N50, 1,908,534 bp). The closure of the gap intra- and interscaffolds was achieved at the Estación Experimental del Zaidín (Spain) by detailed observation of relevant sequencing reads using the Geneious R7 software platform (Kearse et al. 2012). Optical mapping analysis was performed with an OpGen Argus optical mapper at TGAC (United Kingdom) to validate the final assembly. In agreement with the bioinformatic data, a pulsed-field gel electrophoresis (PFGE) analysis of

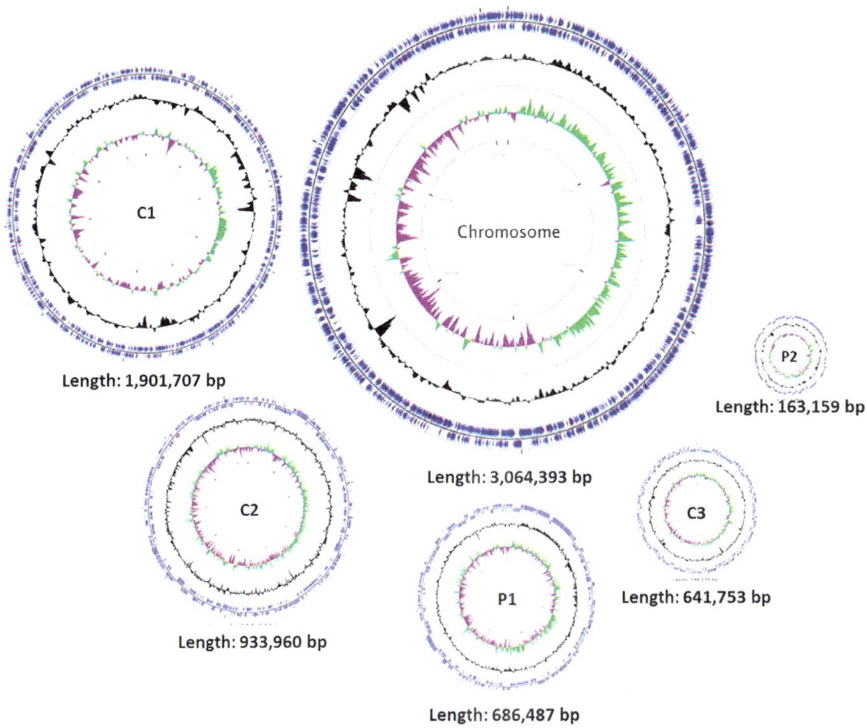

Fig. 4.1 Gene map of six genome replicons of *A. brasilense* Az39 generated by CGView Comparison Tool showing dimensions of one chromosome, three chromides, and two plasmids for a total genome size of 7,391,279 bp (7.39 Mbp). The two outer circles show coding sequences (blue arrows) present in the sense (external) and antisense strands. Inner circles indicate the GC content (in black) and the GC deviation (green/purple) in each of the replicons. (Image credit: Rivera, D. et al. 2018)

total DNA performed at the Université Lyon (France) revealed the presence of six replicons in *A. brasilense* Az39, one chromosome (3,064,393 bp), three chromids (1901.707; 933,960; 641.573 pb) and two plasmids (686.487 bp; 163.159 bp).

The genome consists of 7.39 Mbp and 6311 protein-coding sequences: 2763 on the chromosome, 1605 on AbAZ39_p1, 744 on AbAZ39_p2, 534 on AbAZ39_p3, 557 on AbAZ39_p4, and 108 on AbAZ39_p5 (Rivera et al. 2014) (Fig. 4.1). From innumerable biochemical tests and *in silico* analyses of genes found in A39 using RAST (Aziz et al. 2008) and KEGG (Kanehisa et al. 2004), their participation in multiple mechanisms of plant growth-promotion was reconfirmed. The classic mechanisms in which Az39 participates are related to the capacity to fix nitrogen, biosynthesis of phytohormones such as the most representative auxin, the indole-3-acetic acid (IAA); cytokinins, gibberellins and polyamines. Identical mechanisms and similar genes encoding for homolog proteins were previously identified in *A. brasilense* Sp245 genome (Wisniewski-Dyé et al. 2011) and later in *A. brasilense* Sp7 (Tarrand et al. 1978 and completely sequenced in 2015), *A. brasilense* FP2 and

CBG497 genomes (unpublished). By contrast, in the case of *A. lipoferum* 4B (Wisniewski-Dyé et al. 2011), *Azospirillum* B510 (Kaneko et al. 2010) and *N. amazonense* Y2 (Lin et al. 2014), their genome sequences reveal differences at level of IAA biosynthesis, and other mechanisms. As part of the comparison strategy, post-sequencing analyses defining the "core genome" of the genus *Azospirillum* and *A. brasilense* species were performed. A total of 4216 genes were identified as mutual in all genome sequences of several strains belonging to this genus obtained from different databases. Later, a second analysis was developed between strain Az39, FP2, Sp7, CBG497 and Sp245 belonging to *A. brasilense*. The result of that comparison was represented in a Venn diagram (Rivera D, personal communication).

As summarized Fig. 4.2, the number of unique sequences found in Sp245 were 2099, while in FP2 were 1778, 1140 for Az39, 213 for Sp7 and 182 for CBG497. The comparative analysis shows that Az39 and Sp245 share 5350 genes, Az39 and Sp7 4976 genes, Az39 and CBG497 4954 genes, and Az39 with FP2 share 4794 genes. A fivefold increase in the number of unique genes for *A. brasilense* Az39 was

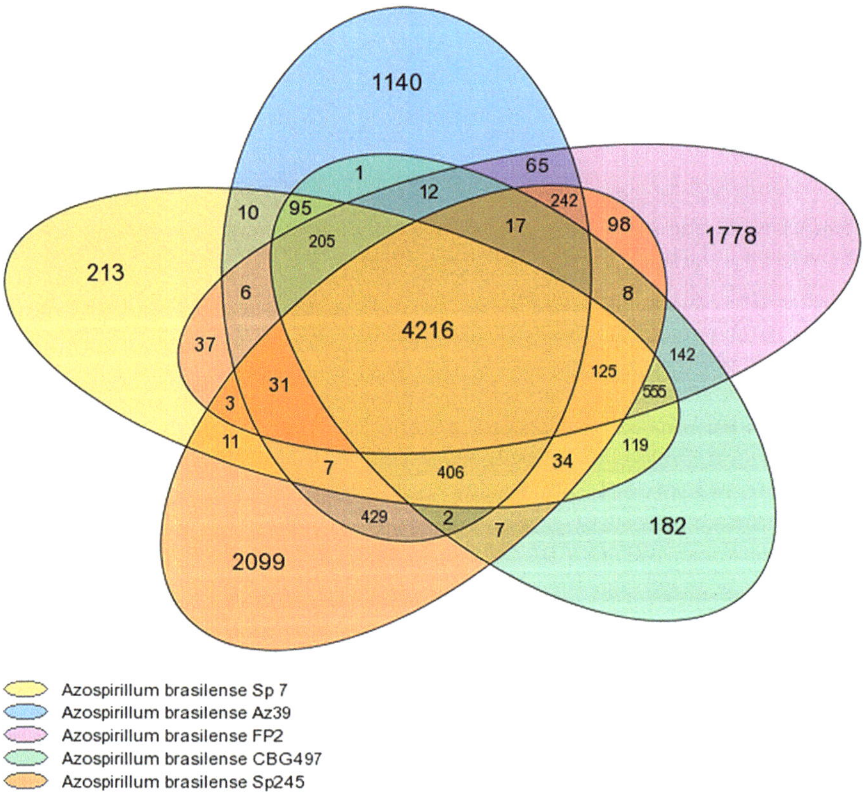

Fig. 4.2 Venn diagram representing the core genome of *A. brasilense* strains. The numbers at the edges of circles represent the unique genes belonging to each strain analyzed. (Image credit: Rivera, D. et al. 2018)

observed as an example of the comparison established within this bacterial genus. Considering that *A. brasilense* Az39 is one of the strains mostly successfully used for agriculture during the last 50 years in diverse environments of Argentina and South America, and the full gapless sequence of its genome was recently obtained. It became hence a genuine tool for molecular studies and an excellent model for in-depth comparative genome analyses and/or *Azospirillum*-plant interactions.

4.3 Functional Analysis of Plant Growth Promotion by *Azospirillum*

Azospirillum has been associated to several mechanisms promoting plant growth. A range of studies detailing the beneficial effects of inoculation with these beneficial rhizobacteria have been undertaken. The improvement of plant growth by *A. brasilense* has been mostly attributed to the bacterial capacity to fix nitrogen and produce phytohormones and not to: the micronutrients bio-disposition., the expression of enzymes; the synthesis of molecules related to plant stress mitigation, and the less understood competition against phytopathogens. Nevertheless, phytohormones production has attracted the attention of researchers around the world. Tien et al. (1979) were the first to suggest that *Azospirillum* could enhance plant growth by auxins and cytokinins. Subsequent studies showed the capacity of this genus to produce a wide range of phytohormones (Table 4.1) and related plant growth regulators apart of auxins and cytokinins, such as gibberellins, abscisic acid, ethylene, NO and polyamines, among others (Cassán et al. 2014).

Table 4.1 Plant growth regulators produced *in vitro* by *Azospirillum* sp. based on their class and its hierarchy

Class	Hierarchy	Molecules[a]	References
Auxins	1st	IAA, PAA, IBA	Prinsen et al. (1993), Martínez-Morales et al. (2003), Bashan et al. (2004), Somers et al. (2005), and Spaepen et al. (2007)
Gibberellins	4th	GA3, GA1	Bottini et al. (1989, 2004), Piccoli et al. (1996), and Cassán et al. (2009)
Cytokinins	3rd	iP, iPr, Z, t-Zr	Horemans et al. (1986) and Esquivel-Cote et al. (2010)
Ethylene	5th	Et	Strzelczyk et al. (1994) and Perrig et al. (2007)
ABA	6th	ABA	Tien et al. (1979) and Kolb and Martin (1985)
Nitric oxide	2nd	NO	Creus et al. (2005) and Molina-Favero et al. (2007)
Polyamines	7th	Cad, Spm, Spd, Put	Thuler et al. (2003), Perrig et al. (2007), and Cassán et al. (2009)

Adapted from Cassán et al. (2014)

[a]*IAA* indole-3-acetic acid, *PAA* phenylacetic acid, *IBA* indole-3-butyric acid, *NO* nitric oxide, *iP* isopentenyl adenine, *iPr* isopentenyl adenine riboside, *Z* zeatin, *t-Zr* trans zeatin riboside, *GA3,1* gibberellins, *Et* ethylene, *ABA* abscisic acid, *Cad* cadaverine, *Spm* spermine, *Spd* spermidine, *Put* putrescine

4.3.1 *Auxins Production by* Azospirillum *A Classical Model of Plant Growth Promotion*

Auxins, such as the indole-3-acetic acid (IAA), represent a group of compounds characterized by their ability to induce plant cell elongation in the sub-apical region of the stem, gravitropism and phototropism; vascular tissue differentiation; apical dominance; lateral and adventitious root initiation and stimulation of cell division, (Teale et al. 2006). Tien et al. (1979) proved that the exogenous application of IAA and kinetin in pearl millet produced changes in root morphology similar to those found in seedlings inoculated with *A. brasilense*.

Currently, members of *Azospirillum* have provided an excellent experimental model for investigating the physiological role of auxins in *Azospirillum*-plant interactions. Several naturally occurring auxin-like molecules have been described as products of bacterial metabolism. In addition to IAA, other molecules such as indole-butyric acid (IBA) (Martínez-Morales et al. 2003), phenyl acetic acid (PAA) (Somers et al. 2005), indole-3-lactic acid (ILA), indole-3-ethanol and indole-3-methanol (Crozier et al. 1988), indole-3-acetamide (IAM) (Hartmann et al. 1983), indole-3-acetaldehyde (Costacurta et al. 1994), tryptamine (TAM) and anthranilate (Hartmann et al. 1983), have been identified in *Azospirillum* culture media.

At least four different IAA biosynthesis pathways have been proposed in *Azospirillum*: IPyA (indole 3-pyruvic acid), IAM (indole acetamide) and TAM (tryptamine), considered tryptophan-dependent pathways, and the putative tryptophan-independent pathway (Prinsen et al. 1993). Despite this diversity, IPyA is considered the most important pathway for IAA biosynthesis in this genus since it is responsible for the plant growth promotion effects. The key enzyme of this pathway is indole-3-pyruvate decarboxylase, encoded by the *ipdC* gene (Vande Broek et al. 1999).

From a bioinformatic point of view, the genome sequence of *A. brasilense* Az39 revealed the existence of all the genes involved in the IPyA pathway: *hisC1, ipdC* and the aldehyde dehydrogenase gene (Table 4.2). For strains Sp245 and CBG497, only the *hisC1* and *ipdC* genes were identified with no evidence for the aldehyde dehydrogenase existence in these genomes. Considering that *A. brasilense* Sp245 and Az39 genome sequences are very similar, it is not surprising that all genes encoding for the IPyA pathway are highly similar in both strains. No evidence has been found for the existence of *ipdC* or aldehyde dehydrogenase in the genome sequence of *A. lipoferum* 4B (Table 4.2). Only a putative aromatic amino transferase sequence with homology to AAT1 from *A. brasilense* Sp7 was identified (Wisniewski-Dyé et al. 2012). *Azospirillum* sp. B510 genome sequence analysis revealed a putative aromatic amino transferase with homology to AAT1 from *A. brasilense* Sp7 (Wisniewski-Dyé et al. 2012). Two candidate genes were proposed to be involved in the IAM pathway (Kaneko et al. 2010), although their role in IAA biosynthesis was questioned, due to low similarity (especially for the putative *iaaM* gene) with known *iaaM* and *iaaH* genes (Table 4.2).

Table 4.2 Putative genes involved in IAA biosynthesis in *Azospirillum* sp.

IAA biosynthesis pathway	IPyA			IAM		IAN	
Enzyme name	Indole-3-pyruvate decarboxylase	Aromatic amino transferase	Aldehyde dehydrogenase	Tryptophan monooxygenase	Indole-3-acetamide hydrolase	Nitrilase	
Gene name	*ipdC*[a]	*hisC1*[b]		*iaaM*	*iaaH*	*NIT1*[c]	*NIT2*[c]
A. brasilense Sp245	+	+	+	–	–	+	+
A. brasilense Az39	+	+	+	–	–	+	+
A. brasilense CBG497	+	+	–	–	–	–	–
A. lipoferum 4B	–	+	–	–	–	–	–
Azospirillum sp. B510	–	+	–	+ (*)	+ (*)	–	–
A. amazonense Y2	–	–	–	–	–	+	+

From Cassán et al. (2014)

The genome sequence of *A. brasilense* Sp245, Az39, and CBG497, *A. lipoferum* 4B, *Azospirillum* sp. B510, and *A. amazonense* Y2 was searched by BLAST for putative genes involved in IAA biosynthesis

References: (*) homology Sp7; [a]*ipdC* from *A. brasilense* Sp245; [b]*hisC1* from *A. brasilense* Sp7; [c]*NIT1* and *NIT2* from *Arabidopsis thaliana*

The whole IAA metabolism has been recently evaluated by Rivera et al. (2018) in several strains of *A. brasilense*. In their study they confirmed that *A. brasilense* strains Sp245, Az39 and Cd produces IAA only in presence of L-trp, are not able to degrade auxins (catabolism), conjugate IAA with sugars and/or L-amino acids (conjugation), or hydrolize such conjugates to release free IAA (hydrolysis). Additionally, they find that some amino acids such as L-met, L-val, L-cys and L-ser inhibite bacterial growth and decreased IAA biosynthesis; while the expression of *ipd*C and IAA biosynthesis, but not bacterial growth are affected by L-leu, L-phe, L-ala, L-ile, L-pro. Finally, L-arg, L-glu, L-his, L-lyis, L-asp and L-thr do not affect either bacterial growth or IAA biosynthesis or *ipd*C gene expression, with some impact in the rhizosphere during plant-microbe interactions. Finally, IAA biosynthesis was evaluated in *A. brasilense* under abiotic and biotic stress conditions by Molina et al. (2018).

4.3.2 The IAA Model of Plant Growth Promotion

As it was mentioned before, the major effect on plant growth by *Azospirillum* occurs at the roots level by promoting morphological changes on the organ architecture. Inoculation promotes development of lateral and adventitious roots, root elongation, and branching of root hairs (Dobbelaere et al. 1999; Creus et al. 2005; Molina-Favero et al. 2008; Okon and Kapulnik 1986; Jain and Patriquin 1985). In this sense, *A. brasilense* Sp245 and *A. brasilense* Az39 induced drastic changes at root hairs growth of *Arabidopsis thaliana* under *in vitro* conditions (Fig. 4.3) (Lopez G. personal communication). As a consequence of the architectural changes, plants enhance their mineral- and water uptake, increasing their whole growth. These effects have been attributed to the bacterial production of phytohormones, but mainly IAA (see Bashan and de-Bashan 2010).

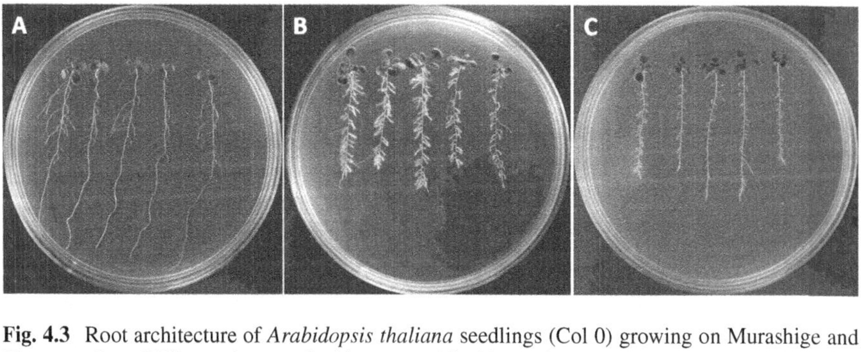

Fig. 4.3 Root architecture of *Arabidopsis thaliana* seedlings (Col 0) growing on Murashige and Skoog medium (MS) supplemented with sucrose 1%. (**a**) non-inoculated seedlings, (**b**) inoculated with *Azospirillum brasilense* Sp245, (**c**) inoculated with *Azospirillum brasilense* Az39. Seedlings were inoculated (1×10^8 CFU · ml^{-1}) 7 d after germination and images were taken at 3 d post inoculation. (Photography credit: Lopez, G.)

The IAA-dependent positive effects of inoculation with *Azospirillum* have been demonstrated in different plant species, both legume and non-legume. Remans et al. (2007, 2008a, b) evidenced that *Azospirillum* strains have potential to enhance nodulation and plant growth of several legumes, including common bean, when co-inoculated with *Rhizobium* by the production of IAA. Co-inoculation of *Sinorhizobium meliloti* (inefficient IAA producer) with *A. brasilense* (efficient IAA producer) on alfalfa seeds significantly increased the number of root nodules in the primary root. The increase was correlated with the inoculum size. This response could be mimicked by the addition of exogenous IAA (Schmidt et al. 1988). Direct evidence of the role of IAA-promoting effects in co-inoculation studies of *A. brasilense* and *R. etlii* on common bean was also provided by Remans et al. (2008a, b) with the use of the *ipdC* knockout mutant of *A. brasilense*. IAA production by *A. brasilense* Sp7 was the key component for enhancement of secretion of *nod*-gene-inducing flavonoids by legume roots (Star et al. 2012).

Recently, two reports showed the importance of the IAA produced by *Azospirillum* in the *Bradyrhizobium*-soybean symbiosis. Puente et al. (2017) (Fig. 4.4) proposed

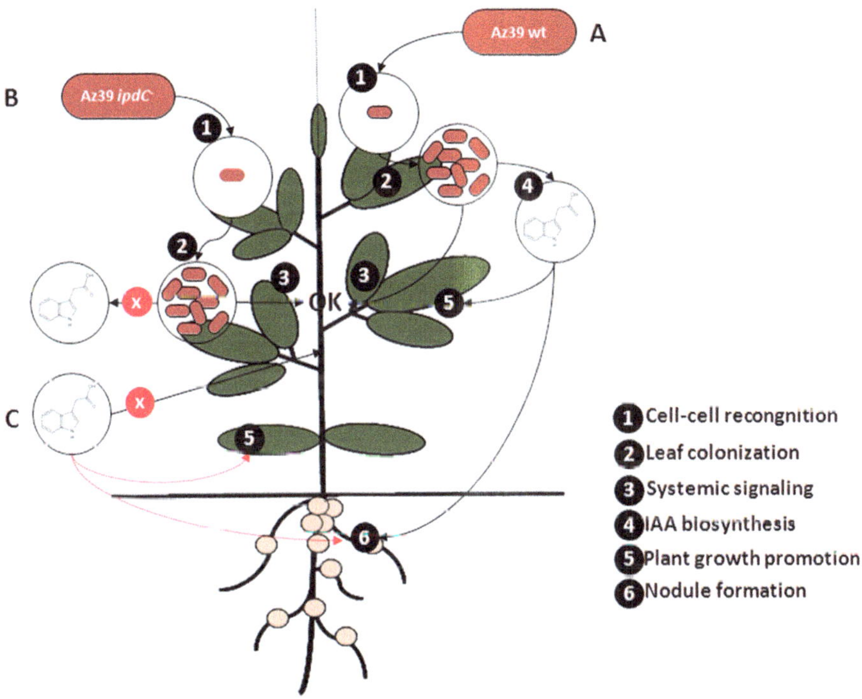

Fig. 4.4 Biological model explaining the effects of the foliar inoculation with *A. brasilense* on soybean plants. The figure shows the plant response to the inoculation with an IAA-producing *A. brasilense* Az39 strain (A); the inoculation with the IAA deficient mutant *ipd*C (B) and the exogenous application of pure IAA (C). According to this model, the presence of bacteria in host tissues enhances the activity of the bacterial IAA on soybean tissues and its symbiotic interaction with *B. japonicum*. The physical recognition of the bacteria by the plant induces growth promotion, even in absence of IAA, but the most effective response occurs in presence of both, IAA and *A. brasilense*. (Figure published in Puente et al. 2017)

a biological model of response to foliar co-inoculation of soybean with *A. brasilense* Az39. This model shows that both the presence of microorganism as part of the colonization process and the production of IAA *in situ* are co-responsible, via plant signaling molecules, for the positive effects on growth and establishment of symbiosis.

In a second report Puente et al. (2018) reported that foliar inoculation with *A. brasilense* Az39 on soybean growth and nodulation, showed higher nitrogen and protein contents in grains harvested from foliar-inoculated plants than in those from plants co-inoculated at sowing. The results from 21 field trials with alfalfa pastures performed in the pampas region of Argentina showed that the application of biological seed treatments combining *S. meliloti* and *A. brasilense* enhanced both the establishment of plants and the forage production (Díaz-Zorita et al. 2012).

Studies related to non-legume crop responses to *Azospirillum* inoculation have been carried out mainly for cereal crops, particularly wheat, sorghum and maize. In wheat plants, inoculation with *A. brasilense* simulated the effect of exogenous IAA treatments on the growth pattern of stems and roots (Kucey 1988). Furthermore, the exogenous addition of IAA and nitrate to wheat plants could (partly to fully) be replaced by the inoculation with *A. brasilense* (Zimmer et al. 1988). The levels of free IAA and IBA in seedlings of maize inoculated with *A. brasilense* Cd were higher than in not inoculated roots (Falik et al. 1989).

Further evidence has been provided by the use of strains altered in IAA biosynthesis: inoculation with a wild-type IAA-producing strain of *A. brasilense* increased the number and length of lateral roots of wheat. In contrast, inoculation with a bacterial mutant with lower IAA production did not modify the root morphology (Barbieri et al. 1988). The role of IAA in the phytostimulatory effect upon *Azospirillum* inoculation was further demonstrated by comparing the plant-growth-promoting effects of a wild-type, auxin-impaired (*ipd*C knockout) mutant, and strains altered in auxin production. Inoculation with the wild-type strain resulted in a decrease in root length but an increase in root hair length and density, whereas inoculation with the *ipd*C mutant did not cause these morphological changes, providing direct evidence for the role of IAA production. In addition, altered IAA production caused by exchanging the native promoter of the *ipd*C gene with a constitutive or plant-inducible one, further pronounced the effect of inoculation (Dobbelaere et al. 1999; Spaepen et al. 2008).

4.4 New Strategies for *Azospirillum* Traceability

4.4.1 Molecular Markers

Currently, molecular characterization using genus-specific molecular markers helps in the rapid detection and identification of *Azospirillum* species starting from both, pure cultures or soil samples. In this sense, isolation and molecular identification of

Azospirillum strains from inoculants, seeds, plants, rizhosphere or soil is widely necessary, being correctly identified the microorganism. The advances of DNA sequencing and fingerprinting methods have permitted the identification of *Azospirillum* bacteria from different sources.

The amplification of 16S rDNA from extracted DNA using universal primers is a very useful tool for routine molecular characterization of *Azospirillum* spp. Lin et al. (2011) developed an *Azospirillum*-specific primer pair, Azo494-F/Azo756-R, which is used for rapid discrimination of the genus *Azospirillum* from other genera (Table 4.3). This genus-specific PCR-based technique allows for a rapid detection and identification of members of *Azospirillum* from pure cultures and soil samples, or detects their presence in plants after inoculation (Table 4.3). Likewise, there are also genotypic identification techniques using labeled probes based on sequences of 23S rRNA genes (Kirchhof and Hartmann 1992). Fluorophore-labelled oligonucleotide probes that hybridize to the 16 s rDNA of *Azospirillum* strains (Stoffels et al. 2001) were also developed. However, due to the high heterogeneity of rRNA operons (*rrn*) with multiple inter-genic variability within *Azospirillum* genomes (Maroniche et al. 2017), the presence of 16S rRNA genes in multiple replicons in *A. brasilense* (Caballero-Mellado et al. 1999), the occurrence of homologous recombination (Teyssier et al. 2003), and horizontal gene transfer (HGT) events (Tian et al. 2015), these techniques have considerably limited specificity (Table 4.3).

Alternative sequences for genus-specific identification of *Azospirillum* have been searched. Baudoin et al. (2010) employed the ribosomal intergenic spacer region (ISR) located between the 16S rRNA and 23S rRNA genes to design PCR primers (fAZO/rAZO) targeting *Azospirillum* spp. in soil samples. Additionally, PCR primers based on the sequences of *Azospirillum* 16S rRNA, indolepyruvate-3-decarboxylase

Table 4.3 Molecular methodologies applied for *Azospirillum* spp. identification

Sequences target	Level identification	Methodology	Author
16S rDNA	Genus	PCR	Lin et al. (2011)
ISR	Genus	Nested PCR	Baudoin et al. (2010)
23S rDNA	Genus	Hybridization with fluorescently labeled	Kirchhof and Hartmann (1992)
16S rRNA, *ipd*C, *nif*A and *nif*H	Genus	PCR and sequencing	Shime-Hattori et al. (2011)
16S rDNA	Genus	Hybridization with fluorescently labeled	Stoffels et al. (2001)
*rpo*D	Genus/specie	PCR and sequencing	Maroniche et al. (2016)
*nif*H, *nif*D and *nif*K	Specie	PCR	Sahoo et al. (2014)
ipdC	*Azospirillum brasilense* Specie	PCR and sequencing	Jijón-Moreno et al. (2015)
hisC	*A. brasilense* and *A. lipoferum*	PCR and sequencing	Jijón-Moreno et al. (2015)
SCAR markers	Strain	PCR/qPCR	Priya et al. (2016) and Couillerot et al. (2010a, b)

(*ipdC*) and nitrogen fixation (*nif*A, *nif*H) were designed by Shime-Hattori et al. (2011). Other primers from sequences of the *nif* gene cluster (*nif*H, *nif*D and *nif*K) of *A. lipoferum* were also designed (Sahoo et al. 2014). Auxin-related genes *ipdC*, *hisC1* and *hisC2* were identified in several *Azospirillum* strains by Jijón-Moreno et al. (2015). In this previous study, *A. brasilense* and *A. lipoferum* were distinguished using a primers pair for *hisC* gene.

Nevertheless, although all these procedures allow the detection and identification of *Azospirillum* sp. isolates within populations of rhizosphere bacteria, plants or complex inoculants samples, the used sequences have some limitations due to the high genome plasticity of these bacteria (i.e. there are multiple and heterogeneous copies in the genome of *Azospirillum*, other sequences are short to design phylogenetic trees, or they have similarities with other species or genera). In this regard, it is necessary to complement them with other confident and sensitivity identification strategies. Recently, Maroniche et al. (2017) proposed alternative genetic markers for a molecular identification of *Azospirillum* employing the sequencing of *rpo*D gene, which is a single-copy gene encoded in the main chromosome, showing a correlation to genome similarity. These authors developed a PCR assay to obtain partial amplification of *rpoD* gene using one forward primer named Azo.rpoDF (5′-GAGATGGGCATCAACATCGT-3′) and two reverse primers: Azo.rpoD R (5′-CCTTCATCAGGCCGATGTTGCC-3′) and Azo.rpoD R2 (5′-ACCTTCT CCAGCGGCATCATC-3′). The sequence of these amplicons allows a species-level identification of *Azospirillum* strains.

4.4.2 qPCR Protocols

Quantitative PCR (qPCR) is a valuable technique to quantitatively monitor populations of unlabeled bacteria in greenhouse experiments. However, the application of strain-specific primers in field experiments is difficult due to the presence of indigenous bacteria that may interfere with the amplification and quantification. For strain-specific molecular monitoring, Sequence-Characterized Amplified Region (SCAR) markers obtained from fingerprinting techniques were applied to design primers for the qPCR quantification of *A. brasilense* and *A. lipoferum* (Couillerot et al. 2010a, b). SCAR markers are employed as a quality standard of commercial products based on *Azospirillum* (Priya et al. 2016) and they are applicable to uncharacterized genomes (Felici et al. 2008) allowing for a rapid identification from a complex sample. An advantage is that SCAR markers can be obtained from all fingerprinting methodologies such as ERIC (Enterobacterial Repetitive Intergenic Consensus)-, BOX (BOX Filament)-, RAPD (Random Amplified Polymorphic DNA)-, or ARDRA (Amplified Ribosomal DNA Restriction Analysis)-PCR (Priya et al. 2016; Holmberg et al. 2009). These fingerprinting techniques are highly specific and sensitive to discriminate *A. lipoferum* and *A.*

brasilense from native isolates (Couillerot et al. 2010a, b, Priya et al. 2016). A disadvantage of SCAR markers is that they do not distinguish between DNA from viable or dead cells (Pujol et al. 2006). In these cases, the amplification of DNA from mRNA, present only in viable cells through a real time-PCR (RT-PCR), is a good alternative to avoid false positives.

4.4.3 *Molecular Markers for* A. brasilense *Az39*

Bioinformatics becomes the preferential tool when whole-genome sequence is available in a database, which allows finding strain-specific DNA sequences for the comparison of genomes of two closely related strains (Stets et al. 2015). This is the case of *A. brasilense* Az39, whose complete genome sequence was published by Rivera et al. (2014). In this regard, 12 non-homologous genomic fragments have been shown as unique of Az39 in relation to the sequences available at National Center for Biotechnology Information (NCBI) (Coniglio et al., previously unpublished results). For that, a bioinformatic search of DNA fragments obtained by *in silico* analysis of Amplified Fragment Length Polymorphism (AFLP-PCR) using the software available at http://insilico.ehu.eus (Bikandi et al. 2004) was performed. DNA sequences, previously subjected to alignments using BLASTn (Altschul et al. 1990), showed to be present only in *A. brasilense* Az39. For the detection of these sequences, primers that amplified under conventional PCR, specifically in the interest strain, were designed. One of the new primers pairs denominated TP5 has been most used based in its specificity. The forward TP5-F (5′-GGCCGTTGTTCTCGGTCTCTAT-3′) and reverse TP5-R (5′-GCGATCTCCATTTATTTCGCCCT-3′) amplify one fragment of 307 pb of the Az39 chromosome (Fig. 4.5). The detection limit of the amplification was 10^5 CFU·ml^{-1} with genomic DNA extracted by an ion exchange resin (Chelex100® Bio-Rad Laboratories 2000) from pure cultures or bacterial mix (Coniglio et al. unpubl.), unlike traditional or immunological methods which are not able to detect low levels of microorganisms (<10^4–10^5 CFU · ml^{-1}). The use of these molecular tools does not only reveal the identification of bacterial isolates, but is also of great interest for the quality control of commercial inoculants used in the field. All of the above reveals that the use of molecular tools based on the DNA sequences allows the effective and precise identification of certain microorganisms such as *A. brasilense*.

Various authors have advanced developing methodologies for the identification and quantification at strains-level within the same species, as occurred through a simple and accurate methodology developed for the identification from pure cultures of *A. brasilense* strain Az39 (Coniglio et al., unpubl.).

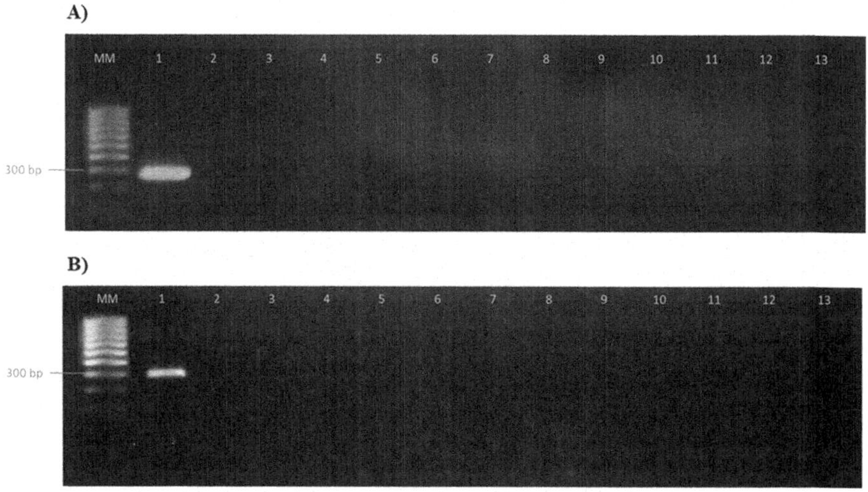

Fig. 4.5 Agarose gel electrophoresis (1%) of the PCR products showing the specificity of the TP5 primers for *A. brasilense* Az39. (**a**) Lanes from left to right: (MM) 100 bp Molecular Marker; (1) *A. brasilense* Az39; (2) *A. brasilense* Sp245; (3) *A. brasilense* FP2; (4) *A. brasilense* Abv5; (5) *A. brasilense* Abv6 (6); *A. brasilense* Az 19; (7) *A. brasilense* Az 36; (8) *A. brasilense* Az 45; (9) *A. brasilense* Az 48; (10) *A. brasilense* Az 63; (11) *A. brasilense* Az 65; (12) *Pseudomonas fluorescens* CHA0; (13) *Bradyrhizobium japonicum* E109; (**b**) Lanes from left to right: (MM) 100 bp Molecular Marker; (1) *A. brasilense* Az39; (2) *A. formosense* Nfb7; (3) *A. fermentarium* LY743; (4) *A. halopraeferens* Au4; (5) *A. oryzae* COC8; (6) *A. palatum* WW10; (7) *A. melinis* TMYCO552; (8) *A. lipoferum* Sp59b; (9) *A. canadense* DS2; (10) *A. rugosum* IMMIB AFH-6; (11) *A. agrícola* CC-HIH038; (12) *A. soli* CC-Ly788; (13) *A. zeae* N7

4.5 Conclusions

The use of inoculants containing *Azospirillum* sp. is an increasing practice in the agriculture of many countries. The improvement of plant growth and crop yields by *A. brasilense* was initially attributed to the bacterial capacity to fix atmospheric nitrogen and produce several phytohormones. The production of phytohormones, mainly auxins and particularly IAA, seems in fact to be the most significant mechanism responsible for the benefic effects observed after inoculation. The mean grain response due to inoculation with *Azospirillum* in South America had an average between 7.0% and 10.0% in the last 20 years, its variation depending on plant species and inoculation conditions. *Azospirillum brasilense* Az39 is one of the most representative strains for the formulation of bioproducts in Argentina and South America and slowly became an agronomic and research study model. The full and gapless genome sequence of Az39 was obtained in LFVIPM, at UNRC in 2014 and was analyzed for identifying plant growth-promotion and lifestyle mechanisms. Bioproducts represent essential components of the sustainable crop production, since they constitute an economically and ecologically acceptable way to reduce costs and improve quantity and quality of resources, through the use of soil microorganisms properly selected. Improving bioproducts quality and functionality we are improving the crop production and the environment sustainability. The utilization of molecular techniques for identification of active principles allows the correct identification of the microorganisms present in such products, increasing our confidence in their use. In this sense, some molecular methods have been illustrated and discussed in this chapter allowing phylogenetic relationships among bacteria, increasing detection precision or allowing differentiation at species- or even strains-levels.

Acknowledgements We thank Universidad Nacional de Río Cuarto (UNRC), Consejo Nacional de Investigaciones Científicas y Tecnológicas (CONICET), and Fondo Nacional para la Investigación Científico tecnológica (FONCyT).

References

Altschul, S. F., Gish, W., Miller, W., Myers, E. W., & Lipman, D. J. (1990). Basic local alignment search tool. *Journal of Molecular Biology, 215*(3), 403–410.

Aziz, R. K., Bartels, D., Best, A., DeJongh, M., Disz, T., Edwards, R. A., Formsma, K., Gerdes, S., Glass, E. M., Kubal, M., Meyer, F., Olsen, G. J., Olson, R., Osterman, A. L., Overbeek, R. A., McNeil, L. K., Paarmann, D., Paczian, T., Parrello, B., Pusch, G. D., Reich, C., Stevens, R., Vassieva, O., Vonstein, V., Wilke, A., & Zagnitko, O. (2008). The RAST server: Rapid annotations using subsystems technology. *BMC Genomics, 9*, 75.

Baldani, V., Alvarez, M., Baldani, J., & Döbereiner, J. (1986). Establishment of inoculated *Azospirillum* spp. in the rhizosphere and in roots of field grown wheat and sorghum. *Plant and Soil, 90*, 35–46.

Bally, R., Thomas-Bauzon, D., Heulin, T., Balandreau, J., Richard, C., & De Ley, J. (1983). Determination of the most frequent N_2-fixing bacteria in a rice rhizosphere. *Canadian Journal of Microbiology, 29*, 881–887.

Barbieri, P., Bernardi, A., Galli, E., & Zanetti, G. (1988). Effects of inoculation with different strains of *Azospirillum brasilense* on wheat roots development. In W. Klingmüller (Ed.), *Azospirillum IV: Genetics, physiology, ecology* (pp. 181–188). Berlin: Springer.

Bashan, Y., & de-Bashan, L. E. (2005). Bacteria/plant growth-promoting. In D. Hillel (Ed.), *Encyclopedia of soils in the environment* (Vol. 1, pp. 103–115). Oxford: Elsevier.

Bashan, Y., & de-Bashan, L. E. (2010). How the plant growth-promoting bacterium *Azospirillum* promotes plant growth. A critical assessment. *Advances in Agronomy, 108*, 77–136.

Bashan, Y., & Levanony, H. (1990). Current status of *Azospirillum* inoculation technology: *Azospirillum* as a challenge for agriculture. *Canadian Journal of Microbiology, 36*, 591–608.

Bashan, Y., Holguin, G., & de-Bashan, L. E. (2004). Azospirillum-plant relationships: physiological, molecular, agricultural, and environmental advances (1997–2003). *Canadian Journal of Microbiology, 50*, 521–577.

Baudoin, E., Couillerot, O., Spaepen, S., Moënne-Loccoz, Y., & Nazaret, S. (2010). Applicability of the 16S-23S rDNA internal spacer for PCR detection of the phytostimulatory PGPR inoculant *Azospirillum lipoferum* CRT1 in field soil. *Journal of Applied Microbiology, 108*, 25–38.

Beijerinck, M. W. (1922). *Azotobacter chroococcum* als indikator van de vruchtbarrheid van den ground. *K Ned Akab Wet Versl Gewone Vergad Afd Natuurkd, 30*, 431–438.

Beijerinck, M. W. (1925). Über ein *Spirillum*, welches freien Stick-stoff binden kann. *Zentralblatt für Bakteriologie, 63*, 353–359.

Ben Dekhil, S., Cahill, M., Stackebrandt, E., & Sly, L. I. (1997). Transfer of Conglomeromonas largomobilis subsp. largomobilis to the Genus *Azospirillum* as *Azospirillum largomobile* comb.

nov., and elevation of *Conglomeromonas largomobilis* subsp. parooensis to the New Type Species of *Conglomeromonas*, *Conglomeromonas parooensis* sp. nov. *Systematic and Applied Microbiology, 20*, 72–77.
Bikandi, J., San Millán, R., Rementeria, A., & Garaizar, J. (2004). In silico analysis of complete bacterial genomes: PCR, AFLP-PCR and endonuclease restriction. *Bioinformatics, 20*(5), 798–799.
Bottini, R., Cassán, F., & Piccoli, P. (2004). Gibberellin production by bacteria and its involvement in plant growth promotion and yield increase. *Applied Microbiology and Biotechnology, 65*, 497–503.
Bottini, R., Fulchieri, M., Pearce, D., & Pharis, R. P. (1989). Identification of gibberellins A1, A3, and isoA3 in cultures of Azospirillum lipoferum. *Plant Physiology, 90*, 45–47.
Caballero-Mellado, J., López-Reyes, L., & Bustillos-Cristales, R. (1999). Presence of 16S rRNA genes in multiple replicons in *Azospirillum brasilense*. *FEMS Microbiology Letters, 178*(2), 283–288.
Cassán, F., & Díaz-Zorita, M. (2016). *Azospirillum* sp. in current agriculture: From the laboratory to the field. *Soil Biology and Biochemistry, 103*, 117–130.
Cassán, F., Maiale, S., Masciarelli, O., Vidal, A., Luna, V., & Ruiz, O. (2009). Cadaverine production by *Azospirillum brasilense* and its possible role in plant growth promotion and osmotic stress mitigation. *European Journal of Soil Biology, 45*, 12–19.
Cassán, F., Vanderleyden, J., & Spaepen, S. (2014). Physiological and agronomical aspects of phytohormone production by model plant growth promoting rhizobacteria (PGPR) belonging to the genus *Azospirillum*. *Journal of Plant Growth Regulation, 33*, 440–459.
Chamam, A., Sanguin, H., Bellvert, F., Meiffren, G., Comte, G., Wisniewski-Dyé, F., Bertrand, C., & Prigent-Combaret, C. (2013). Plant secondary metabolite profiling evidences strain dependent effect in the *Azospirillum-Oryza sativa* association. *Phytochemistry, 87*, 65–77.
Charyulu, P., Fourcassie, F., Barbouche, A., Rondro Harisoa, L., Omar, A., Weinhard, P., Marie, R., & Balandreau, J. (1985). Field inoculation of rice using in vitro selected bacterial and plant genotypes. In W. Klingmüller (Ed.), *Azospirillum III: Genetics, physiology, ecology* (pp. 163–179). Berlin: Springer.
Cohen, A., Bottini, R., Pontin, M., Berli, F., Moreno, D., Boccanlandro, H., Travaglia, C., & Piccoli, P. (2015). *Azospirillum brasilense* ameliorates the response of Arabidopsis thaliana to drought mainly via enhancement of ABA levels. *Physiologia Plantarum, 153*, 79–90.
Compant, S., Clément, C., & Sessitsch, A. (2010). Plant growth-promoting bacteria in the rhizo- and endosphere of plants: Their role, colonization, mechanisms involved and prospects for utilization. *Soil Biology and Biochemistry, 42*, 669–678.
Costacurta, A., Keijers, V., & Vanderleyden, J. (1994). Molecular cloning and sequence analysis of an *Azospirillum brasilense* indole-3-pyruvate decarboxylase gene. *Molecular Genetics and Genomics, 243*, 463–472.
Couillerot, O., Bouffaud, M. L., Baudoin, E., Muller, D., Caballero-Mellado, J., & Moënne-Loccoz, Y. (2010a). Development of a real-time PCR method to quantify the PGPR strain *Azospirillum lipoferum* CRT1 on maize seedlings. *Soil Biology and Biochemistry, 42*, 2298–2305.
Couillerot, O., Poirier, M. A., Prigent-Combaret, C., Mavingui, P., Caballero-Mellado, J., & Moënne-Loccoz, Y. (2010b). Assessment of SCAR markers to design real-time PCR primers for rhizosphere quantification of *Azospirillum brasilense* phytostimulatory inoculants of maize. *Journal of Applied Microbiology, 109*, 528–538.
Creus, C., Graziano, M., Casanovas, E., Pereyra, A., Simontacchi, M., Puntarulo, S., Barassi, C., & Lamattina, L. (2005). Nitric oxide is involved in the *Azospirillum brasilense*-induced lateral root formation in tomato. *Planta, 221*, 297–303.
Crozier, A., Arruda, P., Jasmin, J. M., Monteiro, A. M., & Sandberg, G. (1988). Analysis of indole-3-acetic acid and related indoles in culture medium from *Azospirillum lipoferum* and *Azospirillum brasilense*. *Applied and Environmental Microbiology, 54*, 2833–2283.
Díaz-Zorita, M., & Fernández-Canigia, M. V. (2009). Field performance of a liquid formulation of *Azospirillum brasilense* on dryland wheat production. *European Journal of Soil Biology, 45*, 3–11.

Díaz-Zorita, M., Balína, R. M., & Fernández-Canigia, M. V. (2012). *Azospirillum brasilense* enhances alfalfa productivity: Field evaluation. In J. C. Perez, J. Soler Arango, & L. F. Posada Uribe (Eds.), *PGPR. 9th International and 1st Latinamerican PGPR Workshop. "Returning to Our Roots". Quirama, Medellín (Colombia), June 3–8.*

Dobbelaere, S., Croonenborghs, A., Thys, A., Vande Broek, A., & Vanderleyden, J. (1999). Phytostimulatory effect of *Azospirillum brasilense* wild type and mutant strains altered in IAA production on wheat. *Plant and Soil, 212*, 155–164.

Duca, D., Lorv, J., Patten, C. L., Rose, D., & Glick, B. R. (2014). Indole-3-acetic acid in plant-microbe interactions. *Antonie Van Leeuwenhoek, 106*, 85–125.

Eckert, B., Weber, O. B., Kirchhof, G., Halbritter, A., Stoffels, M., & Hartmann, A. (2001). *Azospirillum doebereinerae* sp. nov., a nitrogen-fixing bacterium associated with the C4-grass *Miscanthus*. *International Journal of Systematic and Evolutionary Microbiology, 51*(1), 17–26.

Elbeltagy, A., Nishioka, K., Sato, T., Suzuki, H., Ye, B., Hamada, T., Isawa, T., Mitsui, H., & Minamisawa, K. (2001). Endophytic colonzation and in planta nitrogen fixation by a *Herbaspirillum* sp. isolated from wild rice species. *Applied and Environmental Microbiology, 67*, 5285–5293.

Esquivel-Cote, R., Ramı'rez-Gama, R., Tsuzuki-Reyes, G., Orozco-Segovia, A., & Huante, P. (2010). Azospirillum lipoferum strain AZm5 containing 1-aminocyclopropane-1-carboxylic acid deaminase improves early growth of tomato seedlings under nitrogen deficiency. *Plant and Soil, 337*, 65–75.

Falik, E., Okon, Y., & Epstein, E. (1989). Identification and quantification of IAA and IBA in *Azospirillum brasilense*-inoculated maize roots. *Soil Biology and Biochemistry, 21*, 147–153.

Falk, E. C., Döbereiner, J., Johnson, J. L., Krieg, N. R., et al. (1985). Deoxyribonucleic acid homology of *Azospirillum amazonense* Magalhães, 1984 and emendation of the description of the genus *Azospirillum*. *International Journal of Systematic Bacteriology, 35*, 117–118.

Felici, C., Vettori, L., Giraldi, E., Forino, L. M. C., Toffanin, A., Tagliasacchi, A. M., & Nuti, M. (2008). Single and co-inoculation of *Bacillus subtilis* and *Azospirillum brasilense* on *Lycopersicon esculentum*: Effects on plant growth and rhizosphere microbial community. *Applied Soil Ecology, 40*, 260–270.

Fomenkov, A., Vincze, T., Grabovich, M., Anton, B. P., Dubinina, G., Orlova, M., Belousova, E., & Roberts, R. J. (2016, January 7). Complete genome sequence of a strain of *Azospirillum thiophilum* isolated from a sulfide spring. *Genome Announcements, 4*(1). pii: e01521-15.

Fukami, J., Nogueira, M., Silva Araujo, R., & Hungria, M. (2016). Accessing inoculation methods of maize and wheat with *Azospirillum brasilense*. *AMB Express, 6*, 3.

Glick, B. (2012). Plant growth-promoting bacteria: Mechanisms and applications. *Scientifica, 2012*, Article ID963401.

Hartmann, A., Singh, M., & Klingmüller, W. (1983). Isolation and characterization of *Azospirillum* mutants excreting high amounts of indoleacetic acid. *Canadian Journal of Microbiology, 29*, 916–923.

Holmberg, A. I. J., Melin, P., Levenfors, J. P., & Sundh, I. (2009). Development and evaluation of SCAR markers for a *Pseudomonas brassicacearum* strain used in biological control of snow mould. *Biological Control, 48*, 181–187.

Horemans, S., Koninck, K., Neuray, J., Hermans, R., & Vlassak, K. (1986). Production of plant growth substances by Azospirillum sp. and other rhizophere bacteria. *Symbiosis, 2*, 341–346.

Jain, D. K., & Patriquin, D. G. (1985). Characterization of a substance produced by *Azospirillum* which causes branching of wheat root hairs. *Canadian Journal of Microbiology, 31*, 206–2010.

Jijón-Moreno, S., Marcos-Jiménez, C., Pedraza, R. O., Ramírez-Mata, A., García de Salamone, I. G., Fernández-Scavino, A., Vásquez-Hernández, C. A., Soto-Urzúa, L., & Baca, B. E. (2015). The ipdC, hisC1 and hisC2 genes involved in indole-3-acetic production used as alternative phylogenetic markers in *Azospirillum brasilense*. *Antonie Van Leeuwenhoek, 107*(6), 1501–1517.

Kanehisa, M., Goto, S., Kawashima, S., Okuno, Y., & Hattori, M. (2004). The KEGG resource for deciphering the genome. *Nucleic Acids Research, 32*(1), D277–D280.

Kaneko, T., Minamisawa, K., Isawa, T., Nakatsukasa, H., & Mitsui, H. (2010). Complete genomic structure of the cultivated rice endophyte *Azospirillum* sp. B510. *DNA Research, 17*, 37–50.

Kearse, M., Moir, R., Wilson, A., Stones-Havas, S., Cheung, M., Sturrock, S., Buxton, S., Cooper, A., Markowitz, S., Duran, C., Thierer, T., Ashton, B., Meintjes, P., & Drummond, A. (2012). Geneious basic: An integrated and extendable desktop software platform for the organization and analysis of sequence data. *Bioinformatics, 28*(12), 1647–1649.

Kennedy, I., Pereg-Gerk, L., Wood, C., Deaker, R., Gilchrist, K., & Katupitiya, S. (1997). Biological nitrogen fixation in non-leguminous field crops: Facilitating the evolution of an effective association between *Azospirillum* and wheat. *Plant and Soil, 194*(1/2), 65–79.

Khammas, K. M., Ageron, E., Grimont, P. A. D., & Kaiser, P. (1989). *Azospirillum irakense* sp. nov., a nitrogen-fixing bacterium associated with rice roots and. *Research in Microbiology, 140*(8), 679–693.

Kirchhof, G., & Hartmann, A. (1992). Development of gene probe for *Azospirillum* based on 23S-rRNA sequences. *Symbiosis, 13*, 27–35.

Kloepper, J., & Schroth, M. (1981). Plant growth-promoting rhizobacteria and plant growth under gnotobiotic conditions. *Phytopatology, 71*, 642–644.

Kloepper, J., Lifshitz, R., & Zablotowicz, R. (1989). Free-living bacteria inocula for enhancing crop productivity. *Trends in Biotechnology, 7*(2), 39–44.

Kolb, W., & Martin, P. (1985). Response of plant roots to inoculation with *Azospirillum brasilense* and to application of indoleacetic acid. In W. Klingmüller (Ed.), *Azospirillum III: Genetics, physiology, ecology* (pp. 215–221). Berlin: Springer.

Kucey, R. M. N. (1988). Plant growth-altering effects of *Azospirillum brasilense* and *Bacillus* C-11-25 on two wheat cultivars. *The Journal of Applied Bacteriology, 64*, 187–196.

Lavrinenko, K., Chernousova, E., Gridneva, E., Dubinina, G., Akimov, V., Kuever, J., Lysenko, A., & Grabovich, M. (2010). Azospirillum thiophilum sp. nov., a diazotrophic bacterium isolated from a sulfide spring. *International Journal of Systematic and Evolutionary Microbiology, 60*, 2832–2837.

Lin, S. Y., Young, C. C., Hupfer, H., Siering, C., Arun, A. B., Chen, W. M., Lai, W. A., Shen, F. T., Rekha, P. D., & Yassin, A. F. (2009). *Azospirillum picis* sp. nov., isolated from discarded tar. *International Journal of Systematic and Evolutionary Microbiology, 59*, 761–765.

Lin, S. Y., Shen, F. T., & Young, C. C. (2011). Rapid detection and identification of the free-living nitrogen fixing genus *Azospirillum* by 16S rRNA-gene-targeted genus-specific primers. *Antonie Van Leeuwenhoek, 99*(4), 837–844.

Lin, S. Y., Shen, F. T., Young, L., Sen Zhu, Z. L., Chen, W. M., & Young, C. C. (2012). *Azospirillum formosense* sp. nov., a diazotroph from agricultural soil. *International Journal of Systematic and Evolutionary Microbiology, 62*, 1185–1190.

Lin, S. Y., Liu, Y. C., Hameed, A., Hsu, Y. H., Lai, W. A., Shen, F. T., & Young, C. C. (2013). *Azospirillum fermentarium* sp. nov., a nitrogen-fixing species isolated from a fermenter. *International Journal of Systematic and Evolutionary Microbiology, 63*, 3762–3768.

Lin, S. Y., Hameed, A., Shen, F. T., Liu, Y. C., Hsu, Y. H., Shahina, M., Lai, W. A., & Young, C. C. (2014). Description of *Niveispirillum fermenti* gen. nov., sp. nov., isolated from a fermentor in Taiwan, transfer of *Azospirillum irakense* (1989) as *Niveispirillum irakense* comb. nov., and reclassification of *Azospirillum amazonense* (1983) as *Nitrospirillum amazonense* gen. nov. *Antonie Van Leeuwenhoek, 105*, 1149–1162.

Lin, S. Y., Hameed, A., Liu, Y. C., Hsu, Y. H., Lai, W. A., Shen, F. T., & Young, C. C. (2015). *Azospirillum soli* sp. nov., a nitrogen-fixing species isolated from agricultural soil. *International Journal of Systematic and Evolutionary Microbiology, 65*(12), 4601–4607.

Magalhães, F., Baldani, J., Souto, S., et al. (1983). A new acid-tolerant Azospirillum species. *Anais da Academia Brasileira de Ciências, 55*, 417–430.

Maroniche, G. A., García, J. E., Salcedo, F., & Creus, C. M. (2016). Molecular identification of Azospirillum spp.: Limitations of 16S rRNA and qualities of rpoD as genetic markers. *Microbiological Research, 195*, 1–10.

Maroniche, G. A., García, J. E., Salcedo, F., & Creus, C. M. (2017). Molecular identification of Azospirillum spp.: Limitations of 16S rRNA and qualities of rpoD as genetic markers. *Microbiological Research, 195*, 1–10.

Martínez-Morales, L., Soto-Urzua, L., Baca, B., & Sanchez-Ahedo, J. (2003). Indole-3- butyric acid (IBA) production in culture medium by wild strain *Azospirillum brasilense*. *FEMS Microbiology Letters, 228*, 167–173.

Mehnaz, S., Weselowski, B., & Lazarovits, G. (2007a). *Azospirillum canadense* sp. nov., a nitrogen-fixing bacterium isolated from corn rhizosphere. *International Journal of Systematic and Evolutionary Microbiology, 57*(3), 620–624.

Mehnaz, S., Weselowski, B., & Lazarovits, G. (2007b). *Azospirillum zeae* sp. nov., a diazotrophic bacterium isolated from rhizosphere soil of *Zea mays*. *International Journal of Systematic and Evolutionary Microbiology, 57*(12), 2805–2809.

Molina, R., Rivera, D., Mora, V., López, G., Rosas, S., Spaepen, S., Vanderleyden, J., & Cassán, F. (2018). Regulation of IAA biosynthesis in Azospirillum brasilense under environmental stress conditions. *Current Microbiology, 75*(10), 1408–1418.

Molina-Favero, C., Creus, C., Lanteri, M., Correa-Aragunde, N., Lombardo, M., Barassi, C., & Lamattina, L. (2007). Nitric oxide and plant growth promoting rhizobacteria: common features influencing root growth and development. *Advances in Botanical Research, 46*, 1–33.

Molina-Favero, C., Creus, C., Simontacchi, M., Puntarulo, S., & Lamattina, L. (2008). Aerobic nitric oxide production by *Azospirillum brasilense* Sp245 and its influence on root architecture in tomato. *Molecular Plant-Microbe Interactions, 21*(7), 1001–1009.

Nelson, L., & Knowles, R. (1978). Effect of oxygen and nitrate on nitrogen fixation and denitrification by *Azospirillum brasilense* growth in continuous culture. *Canadian Journal of Microbiology, 24*, 1395–1403.

Okon, Y., & Kapulnik, Y. (1986). Development and function of *Azospirillum*-inoculated roots. *Plant and Soil, 90*, 3–16.

Okon, Y., & Labandera-Gonzalez, C. (1994). Agronomic applications of *Azospirillum*: an evaluation of 20 years' worldwide field inoculation. *Soil Biology and Biochemistry, 26*, 1591–1606.

Okon, Y., Heytler, P., & Hardy, W. (1983). N_2 fixation by *Azospirillum brasilense* and its incorporation into host *Setaria italica*. *Applied and Environmental Microbiology, 46*, 694–697.

Pagnier, I., Raoult, D., & La Scola, B. (2008). Isolation and identification of amoeba-resisting bacteria from water in human environment by using an *Acanthamoeba polyphaga* co-culture procedure. *Environmental Microbiology, 10*, 1135–1144.

Pedrosa, F., & Yates, G. (1984). Regulation of nitrogen fixation (nif) genes of *Azospirillum brasilense* by nifA and ntrC (glnG) type genes. *FEMS Microbiology Letters, 23*, 95–101.

Peng, G., Wang, H., Zhang, G., Hou, W., Liu, Y., Wang, E. T., & Tan, Z. (2006). *Azospirillum melinis* sp. nov., a group of diazotrophs isolated from tropical molasses grass. *International Journal of Systematic and Evolutionary Microbiology, 56*(6), 1263–1271.

Perrig, D., Boiero, L., Masciarelli, O., Penna, C., Cassán, F., & Luna, V. (2007). Plant growth promoting compounds produced by two agronomically important strains of *Azospirillum brasilense*, and their implications for inoculant formulation. *Applied Microbiology and Biotechnology, 75*, 1143–1150.

Piccoli, P., Masciarelli, O., & Bottini, R. (1996). Metabolism of 17,17 [2H2]-gibberellins A4, A9, and A20 by Azospirillum lipoferum in chemically-defined culture medium. *Symbiosis, 21*, 167–178.

Prinsen, E., Costacurta, A., Michiels, K., Vanderleyden, J., & Van Onckelen, H. (1993). *Azospirillum brasilense* indole-3-acetic acid biosynthesis: evidence for a non-tryptophan dependent pathway. *Molecular Plant-Microbe Interactions Journal, 6*, 609–615.

Priya, R. P., Selastin, A. R., Gopalaswamy, G., & Balachandar, D. (2016). Development of sequence-characterized amplified region (SCAR) markers as a quality standard of inoculants based on *Azospirillum*. *Archives of Microbiology, 198*, 257–267.

Puente, M. L., Gualpa, J. L., López, G. A., Molina, R. M., Carletti, S. M., & Cassán, F. (2017). The benefits of foliar inoculation with *Azospirillum brasilense* in soybean are explained by an auxin signaling model. *Symbiosis*, 1–9.

Puente, M. L., Zawoznik, M., de Sabando, M. L., Pérez, G., Gualpa, J. L., Carletti, S. M., & Cassán, F. (2018). Improvement of soybean grain nutritional quality under foliar inoculation with *Azospirillum brasilense* strain Az39. *Symbiosis*, 1–7.

Pujol, M., Badosa, E., Manceau, C., & Montesinos, E. (2006). Assessment of the environmental fate of the biological control agent of fire blight, *Pseudomonas fluorescens* EPS62e, on apple by culture and real-time PCR methods. *Applied and Environmental Microbiology, 72*(4), 2421–2427.

Reinhold, B., Hurek, T., Fendrik, I., Pot, B., Gillis, M., Kersters, K., Thielemans, S., & De Ley, J. (1987). *Azospirillum halopraeferens* sp. nov., a nitrogen-fixing organism associated with roots of kallar grass (Leptochloa fusca (L.) Kunth). *International Journal of Systematic Bacteriology, 37*(1), 43–51.

Remans, R., Croonenborghs, A., Torres Gutierrez, R., Michiels, J., & Vanderleyden, J. (2007). Effects of plant growth-promoting rhizobacteria on nodulation of *Phaseolus vulgaris* L. are dependent on plant P nutrition. In P. A. H. M. Bakker, J. M. Raaijmakers, G. Bloemberg, M. Höfte, P. Lemanceau, & B. M. Cooke (Eds.), *New perspectives and approaches in plant growth-promoting rhizobacteria research*. Dordrecht: Springer.

Remans, R., Beebe, S., Blair, M., Marique, G., Tovar, E., Rao, I., Croonenbarghs, A., Torres-Gutierrez, R., El-Idoweity, M., Michiels, J., & Vanderlyden, J. (2008a). Physiological and genetic analysis of root responsiveness to auxin-producing plant growth promoting bacteria in common bean (*Phaseolus vulgaris* L.). *Plant and Soil, 302*, 149–161.

Remans, R., Schelkens, S., Hernandez, G., Garcia, A., Luis Reyes, J., Mendez, N., Toscano, V., Mulling, M., Galvez, L., & Vanderleyden, J. (2008b). Effect of *Rhizobium-Azospirillum* coinoculation on nitrogen fixation and yield of two contrasting *Phaseolus vulgaris* L. genotypes cultivated across different environments in cube. *Plant and Soil, 312*, 25–37.

Rivera, D., Revale, S., Molina, R., Gualpa, J., Puente, M., Maroniche, G., Paris, G., Baker, D., Clavijo, B., McLay, K., Spaepen, S., Perticari, A., Vazquez, M., Wisniewski-Dyé, F., Watkins, C., Martínez-Abarca, F., Vanderleyden, J., & Cassan, F. (2014). Complete genome sequence of the model rhizosphere strain *Azospirillum brasilense* Az39, successfully applied in agriculture. *Genome Announcements, 24*(4), pii:e00683-14.

Rivera, D., Mora, V., Lopez, G., Rosas, S., Spaepen, S., Vanderleyden, J., & Cassán, F. (2018). New insights into the indole-3-acetic acid metabolism in Azospirillum brasilense. *Journal of Applied Microbiology., 125*(6), 1774–1785.

Rodriguez-Cáceres, E. A., Gonzalez-Anta, G., Lopez, J. R., Di Ciocco, C., Pacheco-Basurco, J. C., & Parada, J. L. (1994). *Azospirillum brasilense* and *Bacillus polymyxa* inoculation in yield response of field grown wheat in an Argentine semiarid region. *Arid Soil Research and Rehabilitation, 10*, 13–20.

Rodriguez-Cáceres, E., Di Ciocco, C., & Carletti, S. (2008). 25 años de Investigaciones de *Azospirillum brasiliense* AZ39 en Argentina. In F. D. Cassan & I. Garcia de Salamone (Eds.), *Azospirillum sp.: Cell physiology, plant interactions and agronomic research in Argentina* (pp. 179–188). Buenos Aires: Asociación Argentina de Microbiología.

Sahoo, R. K., Ansari, M. W., Pradhan, M., et al. (2014). Phenotypic and molecular characterization of native Azospirillum strains from rice fields to improve crop productivity. *Protoplasma, 251*, 943–953.

Schmidt, W., Martin, P., Omay, H., & Bangerth, F. (1988). Influence of *Azospirillum* brasilense on nodulation of legumes. In W. Klingmüller (Ed.), *Azospirillim IV. Genetics, physiology, ecology* (pp. 92–100). Heidelberg: Springer.

Shime-Hattori, A., Kobayashi, S., Ikeda, S., Asano, R., Shime, H., & Shinano, T. (2011). A rapid and simple PCR method for identifying isolates of the genus *Azospirillum* within populations of rhizosphere bacteria. *Journal of Applied Microbiology, 111*, 915–924.

Somers, E., Ptacek, D., Gysegom, P., Srinivasan, M., & Vanderleyden, J. (2005). *Azospirillum brasilense* produces the auxin-like phenylacetic acid by using the key enzyme for indole-3-acetic acid biosynthesis. *Applied and Environmental Microbiology, 71*, 1803–1810.

Spaepen, S., Vanderleyden, J., & Remans, R. (2007). Indole-3-acetic acid in microbial and microorganism-plant signaling. *FEMS Microbiology Reviews, 31*, 425–448.

Spaepen, S., Dobbelaere, S., Croonenborghs, A., & Vanderleyden, J. (2008). Effects of *Azospirillum brasilense* indole-3-acetic acid production on inoculated wheat plants. *Plant and Soil, 312*, 15.

Star, L., Matan, O., Dardanelli, M., Kapulnik, Y., Burdman, S., & Okon, Y. (2012). The *Vicia sativa* spp. *nigra-Rhizobium leguminosarum* bv. *viciae* symbiotic interaction is improved by *Azospirillum brasilense*. *Plant and Soil, 356*, 165–174.

Stets, M. I., Campbell, M., Souza, M., Pedrosa, D. O., Schmid, M., Hartmann, A., & Cruz, M. (2015). Quantification of *Azospirillum brasilense* FP2 bacteria in wheat roots by strain-specific quantitative PCR. *Applied and Environmental Microbiology, 81*(19), 6700–6709.

Stoffels, M., Castellanos, T., & Hartmann, A. (2001). Design and application of new 16S rRNA-targeted oligonucleotide probes for the *Azospirillum-Skermanella-Rhodocista*-cluster. *Systematic and Applied Microbiology, 24*, 83–97.

Strzelczyk, E., Kamper, M., & Li, C. (1994). Cytocinin-like-substances and ethylene production by *Azospirillum* in media with different carbon sources. *Microbiological Research, 149*, 55–60.

Tarrand, J. J., Krieg, N. R., & Döbereiner, J. (1978). A taxonomic study of the *Spirillum lipoferum* group, with descriptions of a new genus, *Azospirillum* gen. nov. and two species, *Azospirillum lipoferum* (Beijerinck) comb. nov. and *Azospirillum brasilense* sp. nov. *Canadian Journal of Microbiology, 24*(8), 967–980.

Teale, W., Paponov, I., & Palme, K. (2006). Auxin in action: Signalling, transport and the control of plant growth and development. *Nature Reviews. Molecular Cell Biology, 7*, 847–859.

Teyssier, C., Marchandin, H., De Buochberg, M. S., Ramuz, M., & Jumas-Bilak, E. (2003). Atypical 16S rRNA gene copies in *Ochrobactrum intermedium* strains reveal a large genomic rearrangement by recombination between rrn copies. *Journal of Bacteriology, 185*(9), 2901–2909.

Thuler, D., Floh, E., Handro, W., & Barbosa, H. (2003). Plant growth regulators and amino acids released by *Azospirillum* sp. in chemically defined media. *Letters in Applied Microbiology, 37*, 174–178.

Tian, R.-M., Cai, L., Zhang, W.-P., Cao, H.-L., & Qian, P.-Y. (2015). Rare events of intragenus and intraspecies horizontal transfer of the 16S rRNA gene. *Genome Biology and Evolution, 7*(8), 2310–2320.

Tien, T., Gaskins, M., & Hubbell, D. (1979). Plant growth substances produced by *Azospirillum brasilense* and their effect on the growth of pearl millet (*Pennisetum americanum* L.). *Applied and Environmental Microbiology, 37*, 1016–1024.

Tyagi, S., & Singh, D. K. (2014). *Azospirillum himalayense* sp. nov., a *nifH* bacterium isolated from Himalayan valley soil, India. *Annals of Microbiology, 64*(1), 259–266.

Vande Broek, A., Lambrecht, M., Eggermont, K., & Vanderleyden, J. (1999). Auxins upregulate expression of the indole-3-pyruvate decarboxylase gene in *Azospirillum brasilense*. *Journal of Bacteriology, 181*(4), 1338–1342.

Veresoglou, S., & Menexes, G. (2010). Impact of inoculation with *Azospirillum* spp. On growth properties and seed yield of wheat: A meta-analysis of studies in the ISI Web of Science from 1981 to 2008. *Plant and Soil, 337*, 469–480.

Wisniewski-Dyé, F., Borziak, K., Khalsa-Moyers, G., Alexandre, G., Sukharnikov, L. O., Wuichet, K., Hurst, G. B., McDonald, W. H., Robertson, J. S., Barbe, V., Calteau, A., Rouy, Z., Mangenot, S., Prigent-Combaret, C., Normand, P., Boyer, M., Siguier, P., Dessaux, Y., Elmerich, C., Condemine, G., Krishnen, G., Kennedy, I., Paterson, A. H., González, V., Mavingui, P., & Zhulin, I. B. (2011). *Azospirillum* genomes reveal transition of bacteria from aquatic to terrestrial environments. *PLoS Genetics, 7*(2), e1002430.

Wisniewski-Dyé, F., Lozano, L., Acosta-Cruz, E., Borland, S., Drogue, B., Prigent-Combaret, C., Rouy, Z., Barbe, V., Mendoza Herrera, A. M., González, V., & Mavingui, P. (2012). Genome sequence of *Azospirillum brasilense* CBG497 and comparative analyses of *Azospirillum* core and accessory genomes provide insight into niche adaptation. *Genes (Basel), 3*, 576–602.

Xie, C. H., & Yokota, A. (2005). *Azospirillum oryzae* sp. nov., a nitrogen-fixing bacterium isolated from the roots of the rice plant *Oryza sativa*. *International Journal of Systematic and Evolutionary Microbiology, 55*(4), 1435–1438.

Young, C. C., Hupfer, H., Siering, C., Ho, M. J., Arun, A. B., Lai, W. A., et al. (2008). *Azospirillum rugosum* sp. nov., isolated from oil-contaminated soil. *International Journal of Systematic and Evolutionary Microbiology, 58*(4), 959–963.

Young, C. C., Lin, S. Y., Hameed, A., Liu, Y. C., Hsu, Y. H., Huang, H. I., & Lai, W. A. (2016). *Azospirillum agricola* sp. nov., a nitrogen-fixing species isolated from cultivated soil. *International Journal of Systematic and Evolutionary Microbiology, 66*, 1453–1458.

Zhou, Y., Wei, W., Wang, X., Xu, L., & Lai, R. (2009). *Azospirillum palatum* sp. nov., isolated from forest soil in Zhejiang province, China. *The Journal of General and Applied Microbiology, 55*(1), 1–7.

Zhou, S., Han, L., Wang, Y., Yang, G., Li, Z., & Hu, P. (2013). *Azospirillum humicireducens* sp. nov., a nitrogen-fixing bacterium isolated from a microbial fuel cell. *International Journal of Systematic and Evolutionary Microbiology, 63*, 2618–2624.

Zimmer, W., Roeben, K., & Bothe, H. (1988). An alternative explanation for plant growth promotion by bacteria of the genus *Azospirillum*. *Planta, 176*(1988), 333–334.

Chapter 5
The Potential of Nitrogen-Fixing Bacteria in the Sustainability of Agro-Forestry Ecosystems

Isabel Videira e Castro, Márcia de Castro Silva, Concepción Fernandez, Belén Colavolpe, and Helena Machado

Abstract Microorganisms are critical for the maintenance of soil functions in natural and managed agro-forestry ecosystems. This explains the importance of evaluating a particular group of microorganisms, such as legume root nodule bacteria. Legumes and their root nodule bacteria are considered powerful management tools for improving pasture yield in the *montado* ecosystem, which is an agro-forestry system associated with the exploitation of cork and holm oaks. In Portugal, the widespread mortality registered in cork and holm oaks in recent decades has been attributed to infections by *Phytophthora cinnamomi*. In addition to nitrogen fixation, evaluated by the size and symbiotic effectiveness of the rhizobial population, other important functions were also investigated, such as mineral phosphate solubilization and cellulase activity, as well as the antagonistic activity against *P. cinnamomi*. This work showed the important role that root nodule bacteria can play in the sustainability and recovery of these ecosystems by promoting biological nitrogen fixation, especially in low fertility soils, through the establishment of pastures with legumes using pre-selected and characterized rhizobia as biofertilizers. The *Phytophthora* antagonistic activity from some of these bacteria, as well as their ability to degrade cellulose, an important component of *Phytophthora* cell walls, indicate that they may be used as potential biocontrol agents against this disease.

Keywords Cork and holm oaks · Rhizobia · Cellulose degradation · *Phytophthora cinnamomi* · Pastures · Biological nitrogen fixation (BNF)

I. Videira e Castro (✉) · M. de Castro Silva · C. Fernandez · H. Machado
Soil Microbiology Laboratory, National Institute of Agricultural and Veterinary Research (INIAV, I.P.), Oeiras, Portugal
e-mail: isabel.castro@iniav.pt

B. Colavolpe
Soil Microbiology Laboratory, National Institute of Agricultural and Veterinary Research (INIAV, I.P.), Oeiras, Portugal

Instituto Tecnológico de Chascomús (INTECH) (UNSAM-CONICET), Chascomús, Argentina

D. Zúñiga-Dávila et al. (eds.), *Microbial Probiotics for Agricultural Systems*, Sustainability in Plant and Crop Protection,
https://doi.org/10.1007/978-3-030-17597-9_5

5.1 Introduction

Microorganisms play a key role in soil by performing processes that maintain its structure and fertility. In agro-forestry ecosystems a determinant factor for sustainability are natural or introduced biodiverse pastures with legumes. Their associated microbial diversity (mainly nitrogen-fixing bacteria), plays an important role towards increasing and improving soil quality (Crespo 2006). Biological nitrogen fixation (BNF) is a major process providing nitrogen to soil, hence its importance in ecosystems such as the *montado* (or "*dehesa*" in Spain) (Olea and San Miguel-Ayanz 2006; Ferreira et al. 2010; Soares et al. 2014; Castro et al. 2016). In fact, pasture legumes are among the most efficient leguminous plants in terms of nitrogen fixation and, depending on adequate management and on the establishment of effective symbioses with rhizobia, they may contribute to high input rates of fixed nitrogen into the soil (Materon 1988). Therefore, increasing the contribution of BNF represents a significant challenge for more sustainable agro-forestry ecosystems. Different species of rhizobia form nodules with a range of hosts, determined by their nodulation genes. Nodulation and N_2 fixation in these symbioses require host and microorganisms compatibility. Also, soil environment must be appropriate for the exchange of signals that precede infection. Soil abiotic factors, such as temperature, water content, desiccation (drought), salinity and pH (Graham and Vance 2000; Sadowsky 2005; Arrese-Igor et al. 2011) are also critical in the ecology of rhizobia. Symbioses success results in fact from the interaction among legumes, rhizobia, and environmental factors (Fierer and Jackson 2006). Thus, as environmental factors influence all aspects of nodulation and symbiotic N_2 fixation, in some extreme cases they can reduce rhizobial survival and diversity in soil. In other cases, they may affect nodulation, nitrogen fixation and the growth of host plants.

Legumes and root nodule bacteria are amenable to management to improve pasture yield in the *montado* ecosystem. This is an agro-silvo-pastoral ecosystem dominated by cork and holm oaks, that combines the exploitation of cork with livestock husbandry, pastures and other uses. *Quercus suber* L. and *Q. ilex* ssp. *rotundifolia* (syn. *Q. rotundifolia*) are typically endemic to southern European and North African regions, where the climate is characterized by hot and dry summers, and rainy winters with mild temperatures. *Montado* is one of the richest biodiversity ecosystems in Europe and one of the most valuable in Portugal, due to the exploitation (and export) of cork. Portugal is the world's largest producer of cork, accounting for 53% of its production, and cork is one of the Portuguese products with highest impact in the international markets and therefore in the Portuguese economy. Unfortunately, in the last decades there has been a tree decline of Mediterranean *Quercus* species resulting from simultaneous interactions among environmental factors, such as poor soils and climate changes, the action of pests and pathogens, and the negative effects of some anthropic actions. In some cases the pathogenic oomycete *Phytophthora cinnamomi* can be considered as the main factor in the widespread mortality of *Quercus* spp. observed (Brasier et al. 1993; Brasier 1996; Camilo-Alves et al. 2013).

In addition to nitrogen fixation, bacteria root nodules may also directly promote the growth of plants through the solubilization of minerals such as phosphorus or, indirectly, as biological control agents inhibiting the growth of pathogens. Most soil phosphorus (present or applied as fertilizer) is immobilized as insoluble or very poorly soluble forms, thus being unavailable to plants. Phosphate solubilizing bacteria can transform the insoluble mineral phosphates into soluble forms, mainly through the production of organic acids and acidification of the surrounding micro-environment, thus increasing its availability to plants (Dobbelaere et al. 2003). The ability of rhizobia to solubilize inorganic phosphate and to promote plants growth has been extensively reported (Antoun et al. 1998; Alikhani et al. 2006; Sridevi and Mallaiah 2009; Vargas et al. 2009).

The enzymatic activities related to cellulolytic enzymes may also have an important role in controlling the spread and growth of phytopathogenic agents. Downer et al. (2001) showed that the addition of 10–25 U ml^{-1} of cellulase to *Phytophthora* sp. cultures in soil extract impaired zoospore and chlamydospore development. Concentrations higher than 25 U ml^{-1} disrupted the mycelium. Hydrolytic enzymes, including cellulases, were suggested to play an important role as a mechanism of biological control (Schrempf 1995; Tarabily et al. 1996; Richter et al. 2011) since the cell walls of *Phytophthora* spp. consist largely of cellulose (Bartnicki-Garcia and Wang 1983).

This study aimed at characterizing and selecting root nodule bacteria isolated from subterranean clover plants. The latter were grown in soils proceeding from cork and holm oaks from two *montado* areas, in different phytosanitary and vegetative conditions. Main objective was to evaluate the potential of the root nodule bacteria as plant growth promoters and their contribution to the sustainability of these ecosystems. *Trifolium subterraneum* L. was used as host plant because it is an annual legume, commonly used at the onset of upland pastures in the *montado* ecosystems. *Trifolium* is one of the largest genera within the *Fabacea* family and is very specific in its association with rhizobia. It interacts efficiently with *Rhizobium leguminosarum* bv. *trifolii* strains, with a variable symbiotic effectiveness (Ferreira and Marques 1992; Slattery and Coventry 1995; Drew et al. 2011; Carranca et al. 2015). Besides the evaluation of the size and nitrogen fixation capacity of the rhizobial populations, other important functions were also searched, such as the solubilization of mineral phosphate, cellulase activity and antagonism against *P. cinnamomi*.

5.2 Materials and Methods

Soil samples were collected in a cork oak *montado* in Grândola (G) and in an holm oak *montado* in Barrancos (B), both located in the south of Portugal. The soil samples were collected from different plots with dominant trees showing distinct crown defoliation levels. Four classes of crown defoliation were recorded, according to Cadahia et al. (1991): C0 – no defoliation, C1 – slight defoliation (≤ 25%), C2 – moderate defoliation (26–60%) and C3 – severe defoliation (> 60%). Additional symptoms related to the loss of tree vigor, such as wounds and exudations, were noted. From each site, composite soil samples were collected aseptically at 0–20 cm depth. Bulked samples were stored in sealed plastic bags under cooled conditions (6 °C) until microbial analysis and processing.

5.2.1 Size and Nitrogen Fixing Capacity of Rhizobial Populations

For each soil sample, the rhizobial natural population was estimated by the most probable number (MPN) plant infection method using subterranean clover cv. Clare as trap host and a ten-fold dilution series (Somasegaran and Hoben 1994). Clover seeds were rinsed with ethanol, surface sterilized using a freshly prepared 5% solution of calcium hypochlorite for 10 min, washed extensively with sterile distilled water, left in sterile water to hydrate for 2 h and placed in water-agar (0.8%) plates, at 28 °C, in the dark, for 1–2 days. The pre-germinated seeds were transferred to flasks (one seed per flask) containing 50 ml agar slants of Jensen's medium (Jensen 1941). Four replicates were inoculated with 1 ml suspension of each soil dilution. Non-inoculated (T0) and nitrogen (TN) controls were also included, by adding 1 ml liquid Jensen medium (¼ diluted) and 1 ml 1.75% KNO_3, respectively. Plants were grown during 6–8 weeks in a controlled-environment chamber with 16 h light/8 h dark cycle at 23 °C (day)/18 °C (night). After this period, plants were harvested and examined for nodulation. The results were expressed as $\log_{10}$ of rhizobia number per g of dried soil (Somasegraran and Hoben, 1994).

Plants shoots of the first two soil dilutions of each soil sample and also from non-inoculated (T0) and nitrogen (TN) controls, were dried in an oven at 80 °C for 2 days. Shoots dry weight was used to calculate the index of effectiveness (Es) of the rhizobial population according to Ferreira and Marques (1992):

$$\mathrm{Es}(\%) = (X_s - X_{T0}) / (X_{TN} - X_{T0}) \cdot 100$$

where X_s represents the mean dry weight of soil inoculated plant shoots, X_{TN} the mean dry weight of plants from nitrogen control and X_{T0} the mean dry weight of non-inoculated plants.

5.2.2 Isolation of Bacterial Strains

A total of 200 root nodule bacteria (RNB) were isolated from *T. subterraneum* plants nodules inoculated with the soil dilutions of the previous assay, following the methods described in Vincent (1970). Nodules were surface sterilized using a 5% solution of calcium hypochlorite and repeatedly washed with sterilized water. Each nodule was individually crushed and a containing droplet was used for cultivation in yeast-manitol agar media (YMA) with Congo red dye. Isolates were incubated at 27 °C in the dark, until complete growth. Colony purity was checked based on the morphology and Congo red absorption. Each pure isolate was stored at 4 °C.

5.2.3 Other *In Vitro* Activities Related to Plant Growth Promotion

In vitro tests for phosphate-solubilization and cellulase activity were conducted qualitatively. The ability to solubilize mineral phosphate was assayed on yeast extract dextrose (YED) agar medium supplemented with 5 g l^{-1} of $Ca_3(PO_4)_2$ (Peix et al. 2001). Rhizobia strains were inoculated and incubated for 4 days at 27–30 °C. After this period, the colonies showing a surrounding halo were considered as solubilizers.

Cellulase activity was evaluated on TY media supplemented with 0.2% w/v carboxymethylcellulose (CMC), as described by Verma et al. (2001). After inoculation and incubation during 3 days at 27 °C, plates were covered with Congo red solution (1 mg ml^{-1}) for 30 min to facilitate the distinction of an orange halo around the bacterial colonies, associated with the production of the lytic enzyme.

5.2.4 Antagonist Activity Against *Phytophthora cinnamomi*

Phytophthora cinnamomi was isolated from cork oak in Grândola and holm oak in Barrancos, Portugal. Tests for the antagonistic activity against *P. cinnamomi* were performed in Potato Dextrose Agar (PDA) and in YMA media. *Phytophthora cinnamomi* was inoculated in the center of Petri dishes. After 2 days of incubation at 27 °C in the dark, 3 μl of each bacterial culture (grown for 48 h in TY liquid medium) were placed in the periphery of the plate and incubated for 10 days at 27 °C in the dark.

5.3 Results and Discussion

From both cork and holm oak *montado* areas (G and B, respectively) the size of the rhizobial population, estimated by MPN, appeared highly variable among samples (Fig. 5.1). In areas in which trees showed no or slight crown defoliation (C0 and C1) the bacterial densities were close to or above 10^4 cells g^{-1} of soil. These numbers could be considered enough for an efficient nodulation and indicative of a biological nitrogen fixation and an established symbioses contributing to soil fertility. These results agree with those reported by Ferreira et al. (2010) in which the size of the rhizobial population in similar ecosystems was around $5 \cdot 10^4$ bacteria g^{-1} soil.

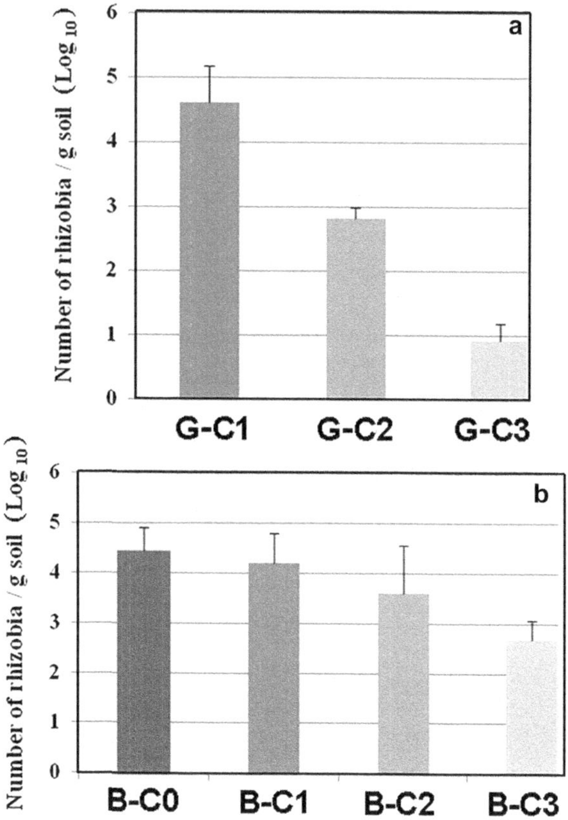

Fig. 5.1 Rhizobial population size (log_{10} of rhizobia bacteria number per g of dried soil) in soils from *montado* ecosystems, using *Trifolium subterraneum* as host plant. Crown defoliation classes are: C0 – no defoliation; C1 – ≤ 25%; C2 – 26–60%; C3 – >60% of crown defoliation. Soils from Grândola (**a**) and Barrancos (**b**)

In soils with trees showing a severe crown defoliation (C3), the rhizobial population had the lowest values with only 10 bacteria g^{-1} of soil (G-C3, Fig. 5.1a). Intermediate values around 10^3 bacteria g^{-1} of soil, were obtained in soils with trees with a moderate crown defoliation (C2).

Data obtained from dry weight evaluation of sub-clover plants inoculated with dilutions of soil samples collected in the *montado* areas showed that nitrogen fixing capacity was variable (Fig. 5.2). In general, values similar to nitrogen controls were observed only in few soil samples, corresponding to soils with holm oak trees in the C0 (data not shown) and C1 defoliation classes (G-C1, Fig. 5.2a and B-C1, Fig. 5.2b). In soils with trees with severe crown defoliation (C3), the dry weight values were the lowest and lower than controls with nitrogen. In some soil samples (G-C3, Fig. 5.2a), the dry weights were even lower than the control (T0), indicative of a failure in nitrogen fixing population for annual clovers.

The effectiveness index data (Fig. 5.2c) confirmed the results mentioned above for dry weight. The effectiveness was higher in soils from trees with slight crown defoliation, reaching 89% in soils from holm oak (B-C1), followed by a decrease in soils from trees showing a moderate defoliation. Lowest values were found in soils proceeding from trees with highest crown defoliation class. Even for soils of cork

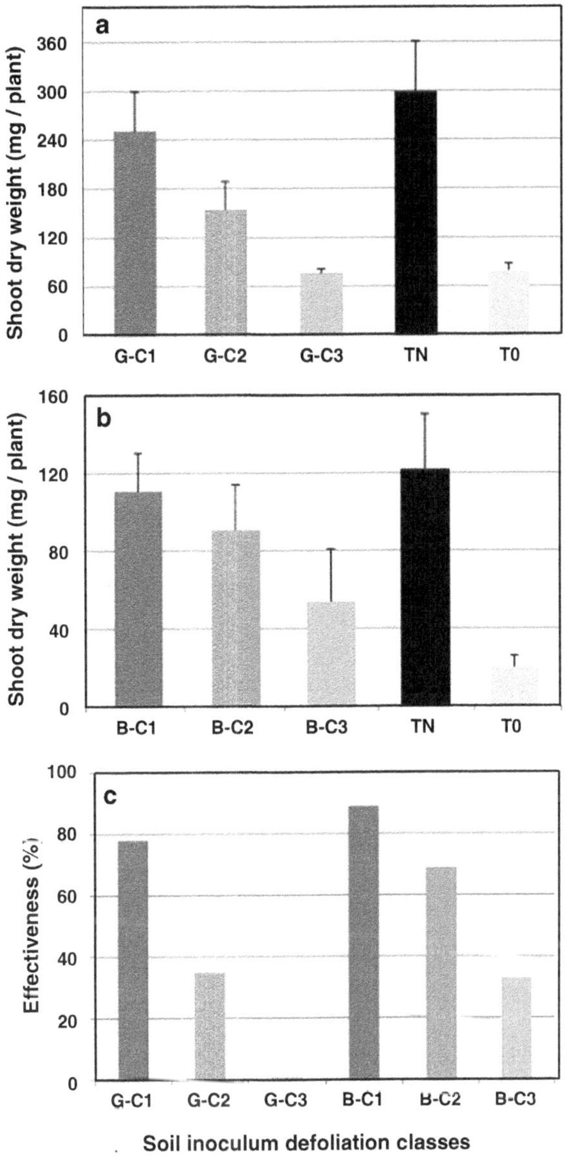

Fig. 5.2 Shoot dry weight values (**a** and **b**) and corresponding effectiveness index (**c**) of *Trifolium subterraneum* plants inoculated with different soil samples from trees with different crown defoliation classes: C1 – ≤ 25%; C2 – 26–60%; C3 – >60% of total crown defoliation. Also included TN (plants only with nitrogen) and T0 (non-inoculated plants and without mineral N) controls. Soils from Grândola (**a**) and Barrancos (**b**). Effectiveness index (%) of rhizobial population in both soils

oak, the effectiveness of rhizobial natural population was about 1%, indicative of a completely ineffective population (G-C3).

The results obtained for both rhizobial counts and nitrogen fixing capacity confirmed a relationship between trees health and rhizobial populations development. The lowest abundance of rhizobia was in fact generally found in soils with trees showing a severe defoliation. These had also a non-effective (cork oak, *montado* G) or less effective (holm oak, *montado* B) rhizobial populations. On the contrary, in soils with the healthiest trees, the number of rhizobia was the highest and these populations were the most effective in nitrogen fixation. These results indicate that these parameters could be good indicators of soil quality in these ecosystems, providing valuable informations about the establishment or maintenance of pasture. As soil microorganisms are critical for the maintenance of soil functions in both natural and managed agricultural ecosystems, it appears of paramount importance to highlight the effect of a particular group of microorganisms, such as the legume root nodule bacteria.

In a second assay, one hundred RNB isolated from each soil were analyzed for *in vitro* activities related to plant growth promoting and antagonism against *P. cinnamomi* (Table 5.1). Results showed that cellulase activity was present in more than half (62%) of the isolates from Grândola (G) with 12% being capable of solubilizing mineral phosphate. On the other hand, 17% of the isolates from Barrancos (B) showed cellulase activity and only 6% were capable of solubilizing mineral phosphate. Moreover, results obtained in the *in vitro* antagonism tests against *P. cinnamomi* showed that 14 and 4% of the isolates from these sites, respectively, could inhibit the growth of the phytopathogen.

Cellulase has a direct impact on *Phytophthora* spp. cell walls, due to the fact that their cell wall is essentially constituted by cellulose (Bartnicki-Garcia and Wang 1983). Consequently, isolates with this activity may contribute to the degradation of the *Phytophthora* propagules, originating by this way a suppressive environment limiting the development of this phytopathogenic agent. The ***in vitro*** results showed that almost all RNB with cellulase activity were isolated from clover plants grown in soils with low crown defoliation (Table 5.2).

Phosphate-solubilizing activity was also found in some isolates from both areas. In the *montado* ecosystem desertification is a common situation due to inconsistent rainfalls, resulting in the progressive degradation of the vegetation cover and the erosion of surface soil. As a consequence, soils are generally poor, deficient in phosphorus and calcium, and contain low levels of organic matter, making arable and intensive farming not sustainable (Tóth et al. 2014). As a possible remedy to this condition, the application of some isolates could solubilize immobilized phosphorus,

Table 5.1 Number (%) of root nodule bacteria (RNB) showing cellulase activity, phosphate solubilization, and antagonistic activity

Isolate group[a]	Celullase activity 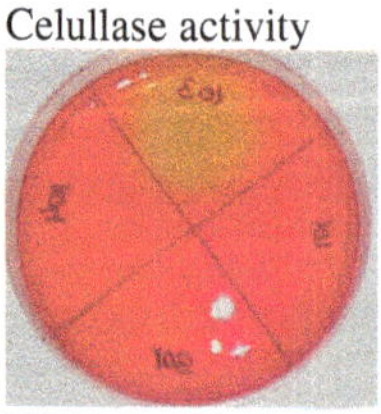	Phosphate solubilization 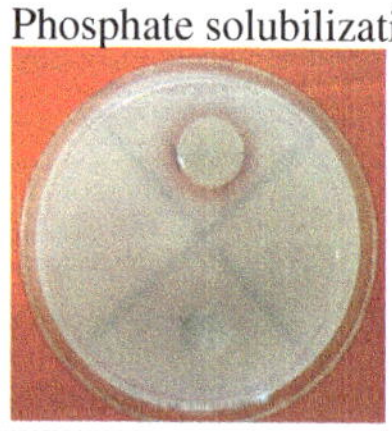	Antagonistic activity 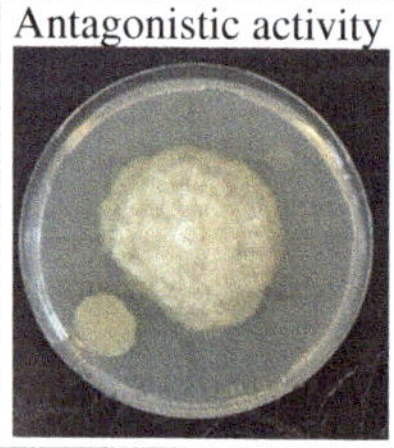
Grândola	62	12	14
Barrancos	17	6	4

[a]One hundred RNB from two sampled *montado* soil

Table 5.2 Frequency of root nodule bacteria (Grândola and Barrancos isolates) with *in vitro* cellulase activity, originating from soils of trees showing different crown defoliation classes

Crown defoliation classes	Cellulase activity (isolates %)
C0 + C1	95.7
C2	69.0
C3	34.6

providing higher amounts of this important macronutrient to plants, contributing by his way to sustain soil fertility in the *montado* ecosystem.

Phytophothora cinnamomi appears to act as a predisposing stress factor in combination with other features, such as soil compaction, shallow soil, drought or water excess events, and other diseases (Camilo-Alves et al. 2013). Effective and long-term rehabilitation of areas infected by *P. cinnamomi* is difficult. Moreover, the use of chemicals to control the pathogen is impractical and environmentally unfriendly. The presence of root nodule bacteria with antagonistic activity against this pathogen may contribute to develop adequate conditions that may help trees to overcome the disease progression. Therefore, an alternative and promising method is the integration of management systems with biological control, through the manipulation of the environment and the introduction of specific microorganisms. The results obtained in this study allowed us to select strains that were highly effective in nitrogen fixation and also capable of solubilizing phosphorus. The production of cellulase also indicates a potential as antagonists against *Phytophothora*. These isolates could be hence considered as biofertilizers/biocontrol agents to be used in the *montado* areas, by inoculating legume seeds before planting.

In summary, and considering the results obtained, the contribution of root nodule bacteria/rhizobia to the sustainability of the *montado* ecosystem, as biofertilizers or/and biocontrol agents of *Phytophothora* is drawn in Fig. 5.3. The conjugation of nitrogen fixation with other important activities, such as phosphate-solubilization and siderophore production (not evaluated in this work) should be considered, as well the production of the lytic enzyme cellulase and the antagonistic activity, which seem also essential for this process.

5.4 Conclusion

Rhizobial population size and nitrogen fixation capacity could be used as a possible bioindicator of the soil quality/fertility in the *montado* ecosystems due to their relationship with the vegetative status of cork and holm oaks, evaluated by crown defoliation. The protection of more biodiverse pastures may help to overcome diseases and, together with biological nitrogen fixation, should contribute to the recovery of soil fertility in the *montado* ecosystem.

Tests in controlled and field conditions should be performed using several plant species naturally occurring in *montado* undercover, inoculated with selected strains

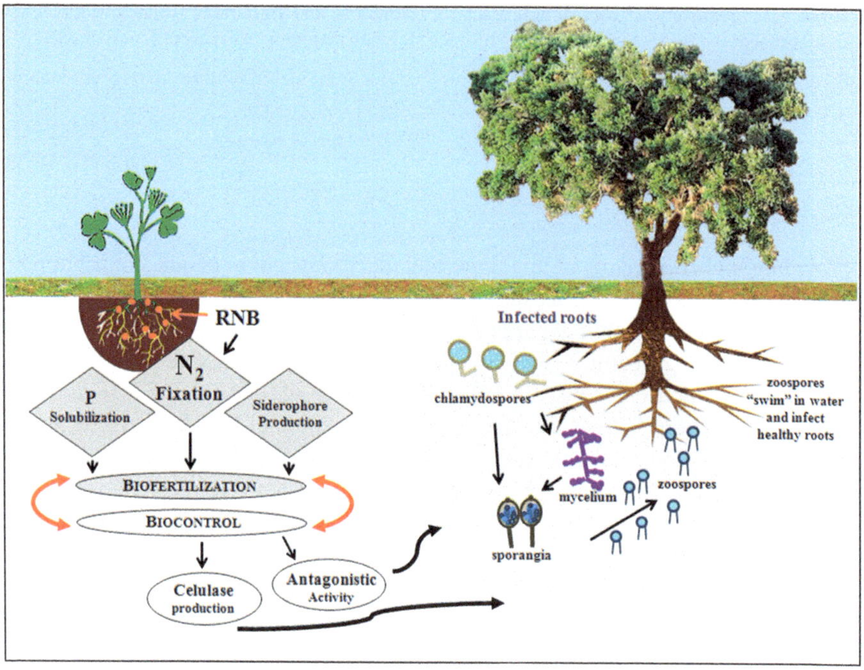

Fig. 5.3 Contribution of root nodule bacteria (RNB) to the sustainability of the *montado* ecosystem, and their interactions with *Phytophothora cinnamomi* life-cycle

with antagonistic activity against *Phythophora* to identify the best, highly-performant "bacteria-plant" associations.

Acknowledgements Financial support was provided by the project "Life Montado & Climate – A Need to Adapt" 15 CCA/PT/000043 (EU) and also by the project from "Fundo Florestal Permanente" (Portugal). Márcia de Castro Silva is granted by the above mentioned EU project and Concepcion Fernandez was granted by Quercus V- Leonardo da Vinci program (Fundecyt), Spain. The authors are grateful to Augusta Costa and Maria João Barrento for all the assistance in collecting soil samples.

References

Alikhani, H. A., Saleh-Rastin, N., & Antoun, H. (2006). Phosphate solubilization activity of rhizobia native to Iranian soils. *Plant and Soil, 287*, 35–41.

Antoun, H., Beauchamp, C. J., Goussard, N., Chabot, R., & Lalande, R. (1998). Potential of *Rhizobium* and *Bradyrhizobium* species as plant growth promoting rhizobacteria on non-legumes: Effect on radishes (*Raphanus sativus* L.). *Plant and Soil, 204*, 57–67.

Arresc-Igor, C., González, E. M., Marino, D., Ladrera, R., Larrainzar, E., & Gil-Quintana, E. (2011). Physiological response of legumes nodules to drought. *Plant Stress, 5*, 24–31.

Bartnicki-Garcia, S., & Wang, M. C. (1983). Biochemical aspects of morphogenesis in *Phytophthora*. In D. C. Erwin, S. Bartnicki-Garcia, & P. H. Tsao (Eds.), *Phytophthora: Its biol-*

ogy, taxonomy, ecology, and pathology (pp. 121–137). St. Paul: American Phytopathological Society.
Brasier, C. M. (1996). *Phytophthora cinnamomi* and oak decline in southern Europe. Environmental constraints including climate change. *Annals of Forest Science, 53*(2–3), 347–358.
Brasier, C. M., Robredo, F., & Ferraz, J. F. P. (1993). Evidence for *Phytophthora cinnamomi* involvement in Iberian oak decline. *Plant Pathology, 42*, 140–145.
Cadahía, D., Cobos, J. M., Soria, S., Clauser, F., Gellini, R., Grossoni, P., & Ferreira, M. C. (1991). *Observação de Danos em Espécies Florestais Mediterrâneas*. MAPA. Secretaría General Técnica, Madrid. Comissão das Comunidades Europeias, Bruxels, 97 pp. ISBN 84-7479-880-9.
Camilo-Alves, C. S. P., Clara, M. I. E., & Ribeiro, N. M. C. A. (2013). Decline of Mediterranean oak trees and its association with *Phytophthora cinnamomi*: A review. *European Journal of Forest Research, 132*, 411–432.
Carranca, C., Castro, I. V., Figueiredo, N., Redondo, R., Rodrigues, A. R. F., Saraiva, I., Maricato, R., & Madeira, M. A. V. (2015). Influence of tree canopy on N_2 fixation by pasture legumes and soil rhizobial abundance in Mediterranean oak woodlands. *Science of the Total Environment, 506–507*, 86–94.
Castro, I. V., Fareleira, P., & Ferreira, E. (2016). Nitrogen fixing symbioses in a sustainable agriculture. In M. Akhtar, K. R. Hakeem, M. S. Akhtar, et al. (Eds.), *Plant, soil and microbes. Volume 1: Implications in crop science* (pp. 55–91). Cham: Springer. ISBN 978-3-319-27453-9.
Crespo, D. G. (2006). The role of pasture improvement on the rehabilitation of the *montado*/dehesa system and in developing its traditional products. In J. M. C. Ramalho Ribeiro, A. E. M. Horta, C. Mosconi, et al. (Eds.), *Animal products from the Mediterranean area* (pp. 185–197). Wageningen: The Netherlands Academic Publishers.
Dobbelaere, S., Vanderleyden, J., & Okon, Y. (2003). Plant growth-promoting effects of diazotrophs in the rhizosphere. *Critical Reviews in Plant Sciences, 22*, 107–149.
Downer, A. J., Menge, J. A., & Pond, E. (2001). Effects of cellulytic enzymes on *Phytophthora cinnamomi*. *Phytopathology, 91*, 839–846.
Drew, E. A., Charman, N., Dingemanse, R., Hall, E., & Ballard, R. A. (2011). Symbiotic performance of Mediterranean *Trifolium* spp. with naturalised soil rhizobia. *Crop & Pasture Science, 62*(10), 903–913.
El-Tarabily, K. A., Sykes, M. L., Kurtboke, I. D., Hardy, G. E. S. J., Barbosa, A. M., & Dekker, R. F. H. (1996). Synergistic effects of a cellulase-producing *Micromonospora carbonacea* and an antibiotic-producing *Streptomyces violascens* on the suppression of *Phytophthora cinnamomi* root rot of *Banksia grandis*. *Canadian Journal of Botany, 74*, 618–624.
Ferreira, E. M., & Marques, J. F. (1992). Selection of Portuguese *Rhizobium leguminosarum* bv. *trifolii* strains for production of legume inoculants. *Plant and Soil, 147*(1), 151–158.
Ferreira, E. M., Simões, N., Castro, I. V., & Carneiro, L. C. (2010). Relationships of selected soil parameters and natural pastures yield in the Montado ecosystem of the Mediterranean area using multivariate analysis. *Silva Lusitana, 18*(2), 151–166.
Fierer, N., & Jackson, R. B. (2006). The diversity and biogeography of soil bacterial communities. *Proceedings of the National Academy of Sciences of the United States of America, 103*, 626–631.
Graham, P. H., & Vance, C. P. (2000). Nitrogen fixation in perspective, an overview of research and extension needs. *Field Crops Research, 65*, 93–106.
Jensen, H. L. (1941). Nitrogen fixation in leguminous plants. I. General characters of root-nodule bacteria isolated from species of *Medicago* and *Trifolium* in Australia. *Proceedings of the Linnean Society of New South Wales, 67*, 98–108.
Materon, L. A. (1988). Maximizing biological nitrogen fixation by forage and pasture legumes in semi-arid areas. In D. P. Beck & L. A. Materon (Eds.), *Nitrogen fixation by legumes in Mediterranean agriculture* (pp. 33–40). Dordrecht: ICARDA, Martinus Nijhoff Publishers.
Olea, L., & San Miguel-Ayanz, A. (2006). The Spanish dehesa. A traditional Mediterranean silvopastoral system linking production and nature conservation. In J. Lloveras, A. González-Rodríguez, O. Vázquez-Yañez, et al. (Eds.), *Sustainable grassland productivity: Proceedings of the 21st general meeting of the European Grassland federation* (pp. 3–15). Spain: Badajoz.

Peix, A., Rivas-Boyero, A. A., Mateos, P. F., Rodriguez-Barrueco, C., Martínez-Molina, E., & Velazquez, E. (2001). Growth promotion of chickpea and barley by a phosphate solubili-zing strain of *Mesorhizobium mediterraneum* under growth chamber conditions. *Soil Biology and Biochemistry, 33*(1), 103–110.

Richter, B. S., Ivors, K., Shi, W., & Benson, D. M. (2011). Cellulase activity as a mechanism for suppression of *Phytophthora* root rot in mulches. *Phytopathology, 101*, 223–230.

Sadowsky, M. J. (2005). Soil stress factors influencing symbiotic nitrogen fixation. In D. Werner & W. E. Newton (Eds.), *Nitrogen fixation research in agriculture, forestry, ecology, and the environment* (pp. 89–102). Dordrecht: Springer.

Schrempf, H. (1995). Degradation of crystalline cellulose and chitin by streptomycetes. In V. N. Danilenko, D. YuV, & V. G. Debabov (Eds.), *Biology of actinomycetes, ISBA'94. Proceedings of the 9th international symposium on the biology of actinomycetes* (pp. 165–168). Moscow: Russia Academy of Science.

Slattery, J. F., & Coventry, D. R. (1995). Acid-tolerance and symbiotic effectiveness of *Rhizobium leguminosarum* bv. *trifolii* isolated from subterranean clover growing in permanent pastures. *Soil Biology and Biochemistry, 27*(1), 111–115.

Soares, R., Fernandéz, C., Carranca, C., Madeira, M., & Castro, I. V. (2014). Does tree environment in agro-forestry ecosystems influence the population of N_2 fixing soil bacteria? *Revista de Ciências Agrárias, 37*(4), 472–481.

Somasegaran, P., & Hoben, H. J. (1994). *Handbook for rhizobia*. Berlin: Springer.

Sridevi, M., & Mallaiah, K. V. (2009). Phosphate solubilisation by *Rhizobium* strains. *Indian Journal of Microbiology, 49*, 98–102.

Tóth, G., Guicharnauda, R. A., Tóth, B., & Hermann, T. (2014). Phosphorus levels in croplands of the European Union with implications for P fertilizer use. *European Journal of Agronomy, 55*, 42–52.

Vargas, L. K., Lisboa, B. B., Schlindwein, G., Granada, C. E., Giongo, A., Beneduzi, A., & Passaglia, L. M. P. (2009). Occurrence of plant growth-promoting traits in clover-nodulating rhizobia strains isolated from different soils in Rio Grande do Sul state. *Revista Brasileira de Ciência do Solo, 33*, 1227–1235.

Verma, S. C., Ladha, J. K., & Tripathi, A. K. (2001). Evaluation of plant growth promoting and colo-nization ability of endophytic diazotrophs from deep water rice. *Journal of Biotechnology, 91*, 127–141.

Vincent, J. M. (1970). *A manual for practical study of root nodule bacteria* (IBPHandbook No. 15). Oxford: Blackwell Scientific Publishers.

Chapter 6
Challenges, Regulations and Future Actions in Biofertilizers in the European Agriculture: From the Lab to the Field

Marcia Barquero, Raquel Pastor-Buies, Beatriz Urbano, and Fernando González-Andrés

Abstract Microorganisms have been used in agriculture for more than a century, beginning with the rhizobia inoculants and, more recently, the so-called plant growth-promoting rhizobacteria (PGPR). Generally, bacteria have proven to be a valid and useful biotechnology for crop production. In spite of the existing knowledge about functional aspects of the interaction between microorganisms and plants and their effects on plants growth, adoption of such products by farmers is still incipient in some regions of the world, especially in industrialised areas While in Asia and Latin America they are widespread, in Europe they are still emerging. This chapter analyses the challenges of the European sector, including: (i) avoiding inconsistences in field performance, and (ii) informing and training farmers about this technology. Emerging regulation in Europe are also examined. Last, it discusses the prospective actions to help overcome challenges while also staying within the current regulation guidelines, including: (i) searching for autochthonous strains, (ii) optimisation of the industrial production and formulation, (iii) development of techniques for precise strain identification in products, especially for non-sterile carriers, (iv) field experiments at the "farmers scale," and (v) screening action mechanisms from a genetic viewpoint. This chapter reviews the scientific information about field trials from a critical standpoint.

Keywords Biofertiliser · PGPR · PGPB · Microbial biostimulant · Field trial

---

Authors Marcia Barquero and Raquel Pastor-Buies have been equally contributed to this chapter

M. Barquero · R. Pastor-Buies · F. González-Andrés (✉)
Research Group IQUIMAB, Institute of Environment Natural Resources and Biodiversity, University of León, León, Spain
e-mail: fgona@unileon.es

B. Urbano
Area of Economy, Sociology and Agrarian Policy, University of Valladolid, Palencia Campus, Palencia, Spain

D. Zúñiga-Dávila et al. (eds.), *Microbial Probiotics for Agricultural Systems*, Sustainability in Plant and Crop Protection,
https://doi.org/10.1007/978-3-030-17597-9_6

6.1 The Challenges for Microorganism-Based Products in Agriculture

Products based on plant growth-promoting rhizobacteria (PGPR), or plant growth-promoting bacteria (PGPB) for use in agriculture have received widespread attention in recent years (Pastor-Bueis et al. 2017). It has been demonstrated that these kind of products lead to an increase in crop yields when used properly, which results in a reduced need for chemicals (Bhardwaj et al. 2014). This technology is compatible with, and may be complementary to, conventional technologies based on mineral, synthetic or organic products. Eventually, microorganism-based products could partially, or even totally replace conventional agricultural products. However, microbial products face several challenges, which pose a threat towards their more generalised use in agriculture (Fig. 6.1).

Avoiding the well-known inconsistences in the performance of microorganisms on the field scale is one of the most important challenges (Morrissey et al. 2004; Vejan et al. 2016). The success of microorganisms in the field depends on the

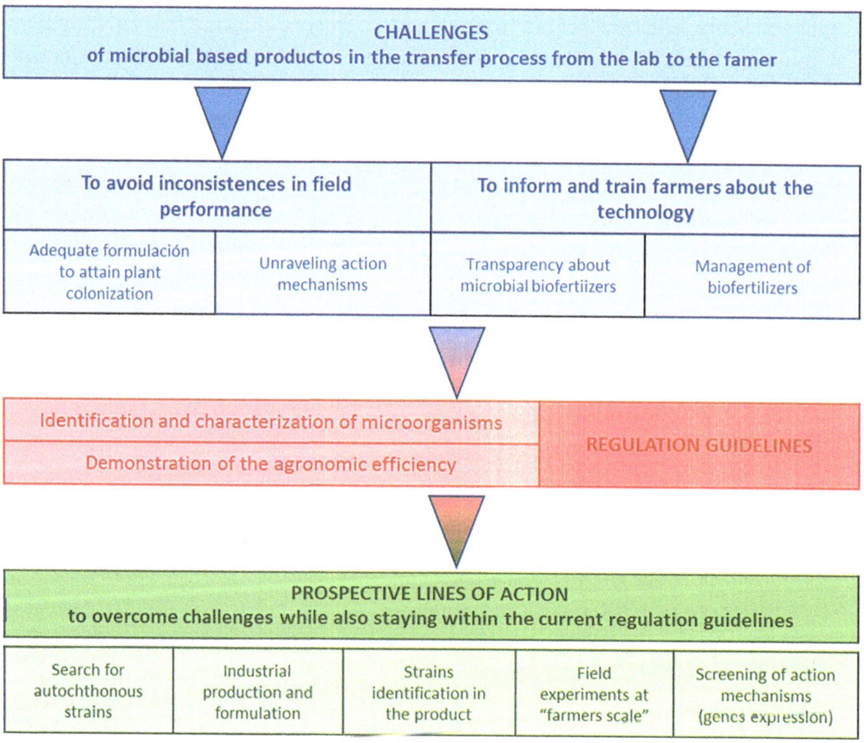

Fig. 6.1 Addressing the challenges of using microorganisms in agriculture, the emerging regulations in Europe, and prospective actions to overcome challenges while also staying within the current regulation guidelines

effective plant colonisation of bacteria, which is influenced by the intrinsic bacterial properties, as well as the physical, chemical and biological nature of the environment. Among the factors affecting bacterial colonisation, some of the most significant are soil particle aggregation, quantity and quality of available carbon, temperature and pH (Timmusk et al. 2017), as well as the ability to interact with indigenous soil microflora (Martínez-Viveros et al. 2010). Usually, populations of inoculated bacteria decline rapidly after initial inoculation. Consequently, they do not attain a sufficient number of viable cells to successfully colonise the root and thus cannot trigger a plant response. Therefore, many efforts are being concentrated in overcoming this issue (Herrmann and Lesueur 2013). Keeping a sufficient number of viable cells is imperative, and preparing the correct bio-formulation is a key challenge. The microorganisms must be prepared methodically in order to provide an appropriate micro-environment, including physical protection for a sustained period of time to avoid decline (Bashan et al. 2016; Timmusk et al. 2017). Unfortunately, there is a lack of available scientific knowledge about the formulation of biofertilisers. In fact, most of this information is either patented or declared an industrial secret. However, even taking into account the significant efforts of private companies in formulation development, the existing information is far from being optimised.

Another contribution to the apparently inconsistent effects of these kind of products from the farmer's viewpoint is given by the diversity of the modes of action (Choudhary et al. 2011; Vejan et al. 2016). For example, a single microorganism can have multiple simultaneous actions in a crop, and can also exhibit multiple mechanisms for a given action (Etesami and Maheshwari 2018). This effect can make it sometimes difficult to identify the action of a given product in the crop, and ultimately confuses the farmer.

A third challenge involves the information that farmers receive and their training in this technology. The rhizobia for legumes, and the PGPR or PGPB, either directly or indirectly facilitate or promote plant growth under nutritional, abiotic or biotic stress conditions. In the last case they are called biocontrol-PGPB (Cassán et al. 2014). When the primary action mode is nutritional or relates to abiotic stress, the microorganism-based products are generally called biofertilisers or microbial-biofertilisers (Pastor-Bueis et al. 2017). However, there is general confusion about what a biofertiliser is. Frequently, anaerobic digestates and their derivatives (Mekki et al. 2017; Du et al. 2018), along with several kinds of composts, are wrongly considered biofertilisers, solely due to the fact that they have a high microbial load (Mulas et al. 2013). Instead, microbial biofertilisers are products that contain specific bacteria strains, which have been carefully selected after the isolation and the biochemical and/or genomic processes of identification and characterization. Afterwards, the products are tested in plants, including crop testing in field situations. Such confusion has been detrimental to the image of microbial biofertilisers and may preclude the use of appropriate products. Therefore, another challenge is to provide a clear explaination to farmers about the differences between organic fertilisers and microbial biofertilisers, along with what they can expect from these products and how they must manage them. Such products consist in fact of living organisms, and their management requires specific care. Moreover, it is unlikely that farming practices will significantly change to accommodate biofertiliser requirements. Therefore, more effort must be put towards developing farmer-friendly products (Bashan et al. 2014). Private companies play an important role in transferring knowledge to farmers. Thus they need to train technicians and sales agents in such products. Regardless, a lack of responsibility in the commercial distribution of microorganism-based products in the past also contributed to a general distrust among farmers.

Finally, the "Nagoya Protocol on Access to Genetic Resources" pursues a fair and equitable sharing of the benefits arising from the utilization of genetic resources, including access to these resources, technology transfer and funding (IEEP, Ecologic and GHK 2012). However, it poses a threat regarding the bureaucratic procedures, which can become a hindrance for the development of new and more effective products, if based on new isolates.

6.2 Regulations on Microorganism-Based Products in European Agriculture

Products based on PGPR burst into the market in the decade 1980–1990, but their presence was dramatically reduced shortly thereafter. Among other reasons, this was due to the lack of formal and standard regulation of the sector, which resulted in situations of poor quality and low efficiency in the field. Currently, the quality is still far from being adequate, and in some cases it is considered poor. In order to increase the agricultural use of microorganism-based products, the desired quality and stability should be maintained (Bashan et al. 2014, Stamenković et al. 2018).

Nevertheless, during the last decade, microorganism-based products have made a strong comeback in certain regions, such as Latin America and southern Asia (Bashan et al. 2014). Conversely, Europe has always been reluctant to use microbial products in agriculture because of the strength of the chemical industry, which until a few years ago did not show interest in microorganism-based products because they were outside of their scope. Still, the most important companies in agricultural enterprise are currently creating production lines for microbial products, and several small or medium sized companies are entering into the business as well.

A key aspect of the safe commercialisation of products based on microorganisms, as well as for the safeguard of the farmers and the consumers' rights, is the development of standard regulations. Hence, Europe has been conscious of the interest of small and large companies in this business and has started creating a regulation to define rules for the availability of microorganism-based products on the market of what is called "CE-marked fertilising products". This new regulation will replace (EC) No 2003/2003 (the existing "Fertilisers Regulation"), and it will amend two other regulations, including (EC) No 1107/2009 concerning the placement of plant protection products on the market. The proposal of the European

Parliament and the Council of the European Union is still in draft form (No 2016/0084, COD), but it already allows for an understanding of the forthcoming regulation. On the one hand, it considers the biocontrol-PGPB agents, which will be out of the range of the new regulation on marked fertilising products. Moreover, it recognises that substances, mixtures and microorganisms commonly referred to as plant biostimulants are not as such nutrients, but nevertheless stimulate the plants nutrition processes. The draft indicates that since such products are aimed solely at improving the plants nutrient use efficiency, tolerance to abiotic stress, or crop quality traits, they are by nature more similar to fertilising products than to other categories of plant protection products. These products should therefore be eligible for CE-marking under the regulation on CE-marked fertilising products, and excluded from the scope of Regulation (EC) No 1107/2009 of the European Parliament and of the Council on plant protection products.

Some European Union countries already regulate the sector, such as Spain for example, with the RD 999/2017 about fertilising products. This regulation includes a special section for "special products with micro-organisms," which includes (Spain 2017): (i) mycorrhizal fungi, (ii) fertiliser with mycorrhizal fungi, (iii) non-mycorrhizal microorganisms, (iv) fertiliser with non-mycorrhizal microorganisms, (v) a mix of mycorrhizae and non-mycorrhizal microorganisms, and (vi) a mix of fertilisers with mycorrhizae and non-mycorrhizal microorganisms. According to the European regulations, the key aspects required to register a microorganism-based product by manufactures are: (i) the identification and characterisation of the microorganisms, and (ii) the demonstration of its agronomic efficiency.

6.2.1 Identification and Characterisation of Microorganisms

At the moment, no specific list of accepted microorganisms taxa exists. The European regulation draft tentatively includes nitrogen fixing bacterial (*Azospirillum*, *Azotobacter* and *Rhizobium*) and mycorrhizal fungi. In any case, any addition to the component material category (CMC) will include the following data on the new microorganism: (i) name, (ii) taxonomic classification, (iii) historical data of safe production and use, (iv) taxonomic relation to micro-organism species, which fulfills the requirements for the Qualified Presumption of Safety as established by the European Food Safety Agency, (v) information on the residue levels of toxins, (vi) information on the production process, and (vii) information on the identity of residual intermediates or microbial metabolites in the component material. The identification of the microorganisms included in a registered product must be based on molecular sequences, such as the 16S rRNA ribosomal gene in bacteria and the ITS-18S rRNA in the case of mycorrhizal fungi.

The minimum microorganism concentration has been tackled by the Spanish regulation, and for bacteria it has been set at 10^7 CFU/ml or 10^7 CFU/g, depending on the product formulation. The regulation does not take into account the estimated final number of microorganisms per plant in the field, as other regulations do

(Herrmann and Lesueur 2013), but only the concentration in the product. However, the Spanish regulation accepts products with a lower concentration of bacteria, provided that their effectiveness is proven with statistical significance in two different microcosm experiments (one experiment with one crop, and another with a second crop, or two different experiments with the same crop).

6.2.2 Demonstration of the Agronomic Efficiency

A vital aspect of the successful registration of a microbial biostimulant is proving its agronomic proficiency in field experiments, using the necessary controls and an adequate experimental design, with a statistical evaluation of results. According to the Spanish regulation, a different experiment is necessary for each group of crops (i.e. horticultural crops, open field crops, trees, products for plants nurseries, etc.), and the registration of products in different groups will follow parallel processes.

6.3 Field Evidences in Scientific and Academic Literature of Effective Microorganisms for Agricultural Use

Multiple advanced "-omics" technologies have enabled us to gain insights into the structure and function of plant-associated microbes (Quin et al. 2016), as the number of scientific and academic studies in such disciplines does not stop growing. The evaluation of biofertilisers and strain selection still chiefly remains in controlled environments rather than under field conditions, whereas scientifically sound field experiments are a necessary step in the development of innovative products based on microorganisms (Herrmann and Lesueur 2013). Table 6.1 gathers recent existing worldwide information about field experiments on microorganism-based products on a medium or large scale. As can be observed in the Table 6.1, there is a broad range of microorganisms used in inoculants, including PGPR, PGP, rhizobia as N-fixing with legumes and mycorrhizal fungi. Likewise, the field assays include inoculations with only one microorganism and those with cocktails containing two or more microorganisms. Although a few of the experiments tested commercial inoculants, most of them reflected the results of the initial stage of strains testing, which are applied directly to the seed (not formulated). In other cases, the microorganisms have been mixed (formulated) with a carrier, such as peat or compost, but a microbial protectant has not been added, and the survival of the inoculum has not been evaluated. Even though the product is tested in the field, rarely its expiry date has been appraised, nor the shelf-life of the product in which its efficiency and quality can be assured. What this means is that such products cannot be released from the commercial viewpoint.

Table 6.1 Medium or large scale field experiments performed with microorganism-based products, published in scientific and academic literature during the last 4 years. The intended action of the microorganisms refers to the main action reported

Type of microorganism	Intended action of the microorganism in the crop	Formulation	Product development phase	Crop	Geographical area of testing	Number of different environments covered by the experiment	Size of an individual field experiment (where available)	Brief summary of the results obtained	References
PGPR: Several species from *Pseudomonas*, *Azotobacter* and *Bacillus*	Not specified	Formulated	Commercial: Phylazonit MC®	Tomato (*Solanum lycopersicum* L.)	Gödöllő, (Hungary)	2	Not available	Positive effect on yield with irrigation.	Le et al. (2018)
PGPR: *Azospirillum brasilense*, *Pseudomonas fluorescens*	Not specified	Formulated (commercial) and not formulated (experimental)	Commercial: Rhizoflo premium Maíz ™ (mix of *A. brasilense* and *P. fluorescens* and experimental (strains of *A. brasilense*)	Maize (*Zea mays* L.)	Buenos Aires province (Argentina)	1	3.150 m^2	PGPR + nitrogen fertilisation increased grain yield and modified rhizosphere microbial communities.	Di Salvo et al. (2018)
PGPR: *Paenibacillus mucilaginosus*	N fixation, P and K solubilization	Not formulated	Experimental	Soybean (*Glycine max* L.)	Shandong Province (China)	1	90 m^2	Positive effects on soybean growth, nodulation and yields. Improved soil bacterial community.	Ma et al. (2018)

(continued)

Table 6.1 (continued)

Type of microorganism	Intended action of the microorganism in the crop	Formulation	Product development phase	Crop	Geographical area of testing	Number of different environments covered by the experiment	Size of an individual field experiment (where available)	Brief summary of the results obtained	References
PGP: *P. fluorescens*, *Pseudomonas* sp., *Serratia* sp., *Enterobacter* sp.	P solubilization, ACC-deaminase activity, siderophore production	Not formulated	Experimental	Oilseed rape (*Brassica napus* L.)	Carlow (Ireland)	1	3.312 m^2	Increased crop height and aerial/pods biomass. Best results with consortium. Not a statistically significant increase of seeds or oil yields.	Lally et al. (2017)
PGPR: *Bacillus siamensis*	Not specified	Formulated	Experimental	Sweet pepper (*Capsicum annuum* L.)	León (Spain)	2	45 m^2	PGPR + decreased mineral N fertilisation (80%) produced significantly better yields than the N–80% and full N (100%) controls. Improved N use efficiency.	Pastor-Bueis et al. (2017)
PGPR: *Pseudomonas oryzihabitans, Bradyrhizobium japonicum*	ACC-deaminase activity, auxin production	Formulated (commercial) and not formulated (experimental)	Commercial: Rhizotorfin (*B. japonicum*) and experimental (*P. oryzihabitans*)	Soybean	Village Lavrovo, Orel, (Russia)	1	160 m^2	Increased plant growth driven by soybean genotype, explained by the interaction of PGPR and root exudates.	Kuzmicheva et al. (2017)

PGPR: *Arthrobacter scleromae*	Not specified	Formulated	Experimental	*Lactuca sativa* L., *Raphanus raphanistrum*, *Brassica pekinensis* L.	Paltan-myeon, Hwaseong-si, Gyeonggi-do, (South Korea)	1		250 m^2	Increased shoot lengths in three crops and increased leaf number in lettuce. Especially effective in salinised environments.	Hong and Lee (2017)
PGPR: *Azorhizobium* spp., *Azoarcus* spp., *Azospirillum* spp.	Not specified	Formulated	Commercial: TripleN®	Wheat (*Triticum aestivum* L.)	Padua province (Italy)	2		540 m^2	PGPR + N-fixing bacteria improved root growth and increased plant resilience to environmental stressors.	Dal Cortivo et al. (2017)
PGPR: *Serratia. marcescens*, *Microbacterium arborescens*, *Enterobacter* sp.	N fixation	Not formulated	Experimental	Wheat (*T. aestivum* L.)	Uttar Pradesh (India)	2		280 m^2	Increased growth and yield; best results with the treble consortium.	Kumar et al. (2017)
PGPR: *Azospirillum lipoferum*	Not specified	Formulated	Commercial: *Azospirillum* strain	Maize	Sérézin-de-la-tour, Chatonnay, saint Savin and Corg, (France)	8		5.750 m^2, 3.080 m^2, 4.600 m^{2}and 2.300 m^2	Improved yield through effects on sugar metabolism and securing mature plant density; no suggested impact on N and P assimilation.	Rozier et al. (2017)
PGPR: Several species of *Bacillus* and one of *Virgibacillus*	P solubilization	Formulated	Experimental	Wheat (*T. aestivum* L.)	Not specified	1		445 m^2	Increased several plant growth parameters.	Mukhtar et al. (2017)

(continued)

Table 6.1 (continued)

Type of microorganism	Intended action of the microorganism in the crop	Formulation	Product development phase	Crop	Geographical area of testing	Number of different environments covered by the experiment	Size of an individual field experiment (where available)	Brief summary of the results obtained	References
PGPR: *Pseudomonas aeruginosa*	Not specified	Formulated	Experimental	Sunflower (*Helianthus annuus* L.)	Faisalabad (Pakistan)	2	2.250 m^2	PGPR + N-enriched compost optimised N uptake efficiency, reduces N fertiliser losses.	Arif et al. (2017)
PGPR: *P. aeruginosa*	Not specified	Formulated	Experimental	Sunflower (*H. annuus*)	Faisalabad (Pakistan)	2	2.700 m^2	PGPR + N-enriched compost improved yield and soil fertility in nutrient-poor agrosystems in drylands.	Arif et al. (2016)
PGPR: *Pseudomonas rhodesiae*, *Paenibacillus polymyxa*, *Rahnella* sp., *Serratia* sp.	N fixation, P solubilization	Formulated	Experimental	Switchgrass (*Panicum virgatum*)	Quebec (Canada)	9	Not available	Inoculation with biochar + consortium improved crop height.	Shanta et al. 2016
PGPR: *Rhizobium* sp., *Burkholderia* sp. AMF: *Claroideoglomus etunicatum*, *Acaulospora* sp.	Not specified	PGPR: Not formulated AMF: Formulated	Experimental	*Schizolobium parahyba* var. *Amazonicum*	Pará state, (Brazil)	1	5.670 m^2	AMF + PGPR + fertiliser increased wood yield by 20% compared to fertiliser alone.	Cely et al. (2016)

PGPR: *Bacillus amyloliquefaciens*	Indole-3-acetic acid (IAA) production	Formulated	Experimental	Banana (*Musa* AAA cv. Dwarf Cavendish)	Hainan Province (China)	2	1.410 m^2	Application of a bio-organic fertiliser significantly promoted banana growth/fruit yield while suppressing *Fusarium* wilt disease.	Wang et al. (2016)
PGPR: *Pseudomonas plecoglossicida, Pseudomonas mosselii, Pseudomonas taiwanensis*	IAA production, siderophore production,	Not formulated	Experimental	Banana (*Musa* AAA Cavendish cv. Brazil)	Azua, (Dominican Republic)	1	432 m2	Improved banana fruit yield and controlled the incidence of black Sigatoka disease.	Marcano et al. (2016)
PGPR: *Enterobacter cloacae, Bacillus drentensis*	IAA production, P solubilization, ACC-deaminase activity, siderophore production.	Formulated	Experimental	Mung bean (*Vigna radiata* L.)	Jeddah, (Saudi Arabia)	2	648 m^2	PGPR + silicon enhanced salinity tolerance.	Mahmood et al. (2016)
PGPR: *Exiguobacterium oxidotolerans, AMF: Glomus fasciculatum*	PGPR halotolerant	PGPR: Not formulated, AMF: Formulated	Experimental	Mentha (*Mentha arvensis* L.)	Uttar Pradesh (India)	1	150 m^2	PGPR + AMF improved plant growth and AMF colonisation under salt stress conditions; better if combined with vermicompost.	Bharti et al. (2016a)

(continued)

Table 6.1 (continued)

Type of microorganism	Intended action of the microorganism in the crop	Formulation	Product development phase	Crop	Geographical area of testing	Number of different environments covered by the experiment	Size of an individual field experiment (where available)	Brief summary of the results obtained	References
PGPR: *Dietzia natronolimnaea*, AMF: *Glomus intraradices*	Not specified	PGPR: Not formulated, AMF: Formulated	Experimental	Indian basil (*Ocimum basilicum* L.)	Uttar Pradesh (India)	1	72 m^2	PGPR + AMF + vermicompos improved plant growth under salt stressed, improved indigenous microbial community structure.	Bharti et al. (2016b)
PGPR: *Azospirillum* sp., *Azotobacter* sp.	Not specified	Not formulated	Experimental	Safflower (*Carthamus tinctorius* L.)	Islamabad (Pakistan)	2	36 m^2	PGPR + significantly reduced use of NP fertilisers (up to 75%) improved quality/ quantity of seed protein.	Nosheen et al. (2016)
PGPR: *Bacillus* sp., *Pseudomonas* sp.	Not specified	Formulated	Experimental	Sunflower (*H. annuus*)	Faisalabad (Pakistan)	2	Not available	PGPR + different P rates increased yield and P use efficiency.	Sarwar et al. (2016)
PGPR: *P. polymyxa*, *Pantoea agglomerans* AMF: *Glomus mosseae*	Not specified	Formulated	Experimental	French beans (*Phaseolus vulgaris* L.)	Bangalore (India)	1	216 m2	Microbial consortium saved 25% of the recommended NPK fertiliser.	Chauhan and Bagyaraj (2015)

PGPR cocktail: 13 different *Bacillus* species. AMF cocktail.	Not specified	Formulated	Commercial PGPR cocktail: Symbio (*Bacillus* sp. on bran) Commercial AMF cocktail: (micronised Endo mycorrhizae; Symbio)	Durum wheat (*Triticum durum* Desf.)	Sicily (Italy)	1	270 m^2	Soil inoculation with AMF consortium and PGPR consortium (alone or in combination) improved nutrient uptake.	Saia et al. (2015a)
PGPR cocktail: 13 different *Bacillus* species. AMF cocktail.	Not specified	Formulated	Commercial PGPR cocktail: Symbio (*Bacillus* sp. on bran) Commercial AMF cocktail: (micronised Endo mycorrhizae; Symbio)	Durum wheat (*T. durum* Desf)	A typical semiarid Mediterranean area	2	445 m^2	AMF negatively affected amination activity in the root. Combination of AMF + PGPR increased concentrations of amino acids in roots. Exotic AMF reprogrammed primary metabolism in plants.	Saia et al. (2015b)
PGPR: *Bacillus megaterium, Bacillus subtilis*	N fixation, IAA production, ACC-deaminase activity	Formulated	Experimental	Lettuce (*L. sativa* L)	Erzurum (Turkey)	2	110 m^2	Alleviated the deleterious effects of irrigation shortage in plant growth and yield.	Sahin et al. (2015)

(continued)

Table 6.1 (continued)

Type of microorganism	Intended action of the microorganism in the crop	Formulation	Product development phase	Crop	Geographical area of testing	Number of different environments covered by the experiment	Size of an individual field experiment (where available)	Brief summary of the results obtained	References
PGPR: *Achromobacter xylosoxidans*, *P. oryzihabitans*, *Variovorax paradoxus*	ACC-deaminase activity	Not formulated	Experimental	Potato (*Solanum tuberosum* L.)	St. Petersburg (Russia)	2	160 m^2	Increased tuber number and crop yield.	Belimov et al. (2015)
PGPR: *P. fluorescens*, *Bacillus lecheniformis*, *Bacillus* sp., *Mesorhizobium ciceri*. FUNGUS: *Trichoderma harzianum*	Not specified	Formulated	Mix of commercial and experimental	Chickpea (*Cicer arietinum* L.)	New Delhi (India)	2	128 m^2	Combination of fungal and bacterial bio-agents had a synergistic effect, increasing grain yield.	Dubey et al. (2015)
PGPR: *B. subtilis*, *Bacillus mucilaginosus*	N fixation, P and K solubilization	Formulated	Experimental	Tomato, spinach (*Spinacia oleracea* L.)	Suzhou (China)	1	121 m^2	Vermicompost + PGPR enhanced soil's nutrient availability, microbial biomass, and crop yield and quality.	Song et al. (2015)

PGPR: *Enterobacter* sp., *Bacillus* sp., *Klebsiella pneumoniae*, *Serratia* sp., *Staphylococcus saprophyticus*, *Klebsiella* sp.	N fixation, P, K and Zn solubilization, IAA production, ACC-deaminase activity, siderophore production	Formulated	Experimental	Rice (*Oryza sativa* L.)	Coimbatore (India)	1	Not specified	Improved growth parameters and crop yield.	Sarathambal et al. (2015)
PGPR: *Azotobacter chroococcum, A. brasilense*	N fixation	Not formulated	Experimental	Onion (*Allium cepa* L.)	Kafr El-Sheikh (Egypt)	2	94,5 m^2	PGPR + compost extract + reduced chemical fertilisation improved yield and soil characteristics.	Mahmoud et al. (2015)
PGPR: *Burkholderia* sp.	N fixation, IAA production, P solubilization, ACC-deaminase activity, biocontrol activity	Not formulated	Experimental	Tomato (*S. lycopersicum*)	Beijing (China)	1	48 m^2	Improved crop growth/yield and enhanced soil enzymatic activity.	Gao et al. (2015)
PGPR: *Pseudomonas moraviensis, Bacillus cereus*	Not specified	Not formulated	Experimental	Wheat (*T. aestivum* L.)	Islamabad (Pakistan)	2	100 m^2	PGPR improved growth and physiological activity. In combination with tryptophan there was a synergic effect.	Hassan and Bano (2015)

(continued)

Table 6.1 (continued)

Type of microorganism	Intended action of the microorganism in the crop	Formulation	Product development phase	Crop	Geographical area of testing	Number of different environments covered by the experiment	Size of an individual field experiment (where available)	Brief summary of the results obtained	References
PGPR: *B. amyloliquefaciens, S. marcescens*	Not specified	Not formulated	Experimental	Ginger (*Zingiber officinale* Rosc.)	Calicut (India)	2	Not available	Growth promotion with two bacteria; only B. Amyloliquefaciens was recommended for safety reasons.	Dinesh et al. (2015)
PGPR: *Azospirillum* sp. *Bacillus pumilus*	High transfer factor of cesium (Cs)	Not formulated	Experimental	*Brassica rapa* L. var. *Perviridis, B. juncea* (L.) Czern., *Fagopyrum esculentum*	Nihonmatsu, (Japan)	1	160 m^2	Increased biomass production and 137Cs final content (not statistically significant). Despite a positive effect of inoculation, the removal of 137Cs from soil was very low.	Djedidi et al. (2015)
PGPR: *Bacillus aquimaris, B. subtilis*	Not specified	Formulated	Experimental	Wheat (*T. aestivum* L.)	Not specified	1	480 m^2	Induction of proline and sugar accumulation in plants. Increased N, P, and K content in leaves, reduction of Na.	Upadhyay and Singh (2015)

PGPR: *Bradyrhizobium. diazoefficiens*, *B. japonicum*, *B. subtilis*, *Staphylococcus* sp.	N fixation, IAA production, ACC-deaminase activity, siderophore production	Not formulated	Experimental	Soybean (*G. max* L.)	Ratchasima province, Buriram province (Thailand)	2	7296,75 m^2, 2923 m^2	Coinoculation increased number of active nodules, plant yield and nitrogen fixation.	Prakamhang et al. (2015)

Of the 37 studies, published between 2015 and the present moment, in the first or second quartile of the Journal Citation Index, and which performed PGPR or mycorrhizal fungi field tests, only seven were tested in the EU. Thus, it is necessary for European scientists to increase the efforts in formulating products based on native microbes, in order to design successful products that are ready-to-market under the EU regulations.

6.4 Prospective Actions for a Microorganism-Based Agriculture in Europe

The EU is emerging regarding the use of microorganisms for agriculture. The European regulations on "plant biostimulants", which, independently of the products nutrient content, aim to improve the efficiency of nutrients uptake, the tolerance to abiotic stress, and/or the crop quality traits, is the starting point for a promising future of this kind of products. In light of the challenges posed in Sect. 6.1 and the new regulation, the prospective lines of action are discussed below and shown in Fig. 6.1.

One important line of action is the use of autochthonous strains. The advantages of using autochthonous microorganisms is still a controversial aspect, even though there is enough scientific evidence to verify the better adaptation and field performance of autochthonous microorganisms (Mulas et al. 2013). For instance, in *Phaseolus vulgaris* L. the use of allochthonous rhizobia usually does not produce a response to inoculation (Rodríguez-Navarro et al. 2000; Daza et al. 2000), while the use of autochthonous strains have produced a good response in field studies in South America (Motasso et al. 2002; Hungría et al. 2003; Díaz-Alcántara et al. 2014), Africa (Mrabet et al. 2005), and Europe (Mulas et al. 2011, 2015).

The genes encoding an adaptation to a given environment are generally located in the bacterial chromosome (García-Fraile et al. 2010; Mulas et al. 2011; Cao et al. 2017). It has been demonstrated that native rhizobia strains, which are well adapted to environmental conditions, incorporated the plasmid containing the nodC gene typical of biovar phaseoli into their genome (García-Fraile et al. 2010; Mulas et al. 2011, 2015; Díaz Alcántara et al. 2014). This plasmid comes from America, the centre of origin of the common bean, and was persistently transferred to the native rhizobia species, up to date. Such a plasmid confers to the native rhizobia the ability to successfully fix nitrogen. According to manufacturing companies, the main drawback of using autochthonous strains is the increase of the portfolio, as it is necessary to design several products for the same crop, depending on the geographic region. An alternative solution is to use multi-strain inoculants; however, even though there are many laboratory-based studies describing the advantages of strain combinations, there is still a lack of information about the performance of these kind of formulations, on the field scale (Bashan et al. 2014).

Another aspect is to define the range of autochthony, that is the adaptation ranges of each strain. There is a lack of information about this issue. Unraveling this question would involve testing each strain in a set of field experiments in the same and different agroclimatic regions. A study by Marcano et al. (2016) showed that within the same agroclimatic region in a transect of 150 km, the biodiversity of cultivable soil bacteria was mainly located within populations, indicating homogeneity between populations in the agroclimatic region. However, Marasco et al. (2013) observed a broader transect of 1000 km across the agroclimatic region known as the "Mediterranean basin" and found greater diversity between distantly located populations in the transect. Hence, the bacterial diversity across different regions depends not only on the agroclimatic region, but also on the transect length and plant genotype. This must be taken into account when designing biofertilisers based on autochthonous bacteria because locations that are very far apart, but which belong to the same agroclimatic region, could need different strains in order for them to be considered autochthonous.

Another action line is the optimisation of the fermentative process for microorganism production and the formulation at an industrial scale (Bashan et al. 2014). The cost of the growth media for microorganisms needs to be feasible, and for this reason the use of residues has been proposed as a cheap option (Pastor-Bueis et al. 2017). More basic research is needed to develop formulations that maximise the shelf life of the microorganisms, while also optimising plant colonisation.

One important bottleneck in research is identifying the inoculated strain or strains in a product in order to count the CFU per g or ml, as requested by the regulation. This is especially important when the carrier is not sterile, for instance in the mix of a fertiliser with microorganisms. In addition, another major point for microorganism tracking in the field is the development of specific strain markers. Even if the regulations do not require this assessment at this moment, it is of high interest to control the populations change in soils and survival across the different growth stages of the crop. In such cases, it is necessary to have a strain-specific marker to precisely identify the inoculated strain and to distinguish it from other resident microorganisms, even from the same species. Moreover, the identification system must be cheap and effective. For this reason, it has been proposed to design a Sequence Characterized Amplified Region (SCAR) marker for each strain (Reddy Priya et al. 2016), although due to the decrease of the price in genome sequencing, in the future it will be more feasible to find distinctive sequences, based on the analysis of the full genome.

Finally, future research has to be carried out under field situations in order to concentrate the efforts only on those strains which are consistently effective in field conditions. Theoretically, it is possible to achieve a very large range of responses in plants using adequate microorganisms (Etesami and Maheshwari 2018). However, at this moment it is sometimes difficult to see the effects of some actions at the field scale, even if it has been observed in crop tests, at the lab scale or in controlled conditions. Moreover, it is necessary to gain a better understanding of the mechanisms of action of the microorganisms in the plant. Frequently, several mechanisms are working simultaneously. Presently, several studies have sequenced and characterised the plant genes whose expression is affected by interactions with PGPRs, resulting in an improved plant performance in stress situations. For example, studies by Kaushal and Wani (2016), Jatan et al. (2018) and Tiwari et al. (2017) identified several drought stress-related genes that were up-regulated in plants inoculated with PGPR, which resulted in a growth promotion under drought and salinity stress. Similarly, nitrate, ammonium and phosphorus transporter genes in wheat were up-regulated after inoculation with PGPR and mycorrhizal fungi (Saia et al. 2015a). The list of known genes is continuously increasing, and therefore, in the near future it will be relatively easy to screen for the molecular mechanisms of action for a given strain.

6.5 Conclusion

This chapter reviews the challenges facing microorganism-based products in the market of agricultural inputs, such as products for improving the plants efficiency of nutrient uptake, the tolerance to abiotic stress, and the quality traits of crops. Main challenges of using microbial stimulants reside in overcoming the inconsistences in field response and in adequately informing and training farmers. The response of the EU regulation to tackle such challenges was discussed. Such regulations aim to guarantee the quality of the product and their effectiveness in the field, as well as defining what has to be demonstrated in field experiments. In this scenario, the following prospective lines of action are discussed: (i) searching for autochthonous strains, (ii) optimising industrial production and formulation, (iii) developing techniques for precise strain identification in the product, (iv) performing field experiments at the "farmer's scale", and (v) screening action mechanisms from a genetic viewpoint.

Acknowledgements The research line "Biofertilisers for Sustainable Agriculture" belonging to the research group IQUIMAB (University of León), has been financially supported since 2014 to date by the Spanish Ministry of Science, Innovation and Universities (project RTC2014-1793-2), by the Spanish Ministry of Foreign Affairs and Cooperation (project 2015/ACDE/1339), and by the European Comission call H2020 (project NEWFERT).

References

Arif, M. S., Riaz, M., Shahzad, S. M., Yasmeen, T., Akhtar, M. J., Riaz, M. A., Jassey, V. E. J., Bragazza, L., & Buttler, A. (2016). Associative interplay of plant growth promoting rhizobacteria (*Pseudomonas aeruginosa* QS40) with nitrogen fertilizers improves sunflower (*Helianthus annuus* L.) productivity and fertility of aridisol. *Applied Soil Ecology, 108*, 238–247. https://doi.org/10.1016/j.apsoil.2016.08.016.

Arif, M. S., Shahzad, S. M., Riaz, M., Yasmeen, T., Shahzad, T., Akhtar, M. J., Bragazza, L., & Buttler, A. (2017). Nitrogen-enriched compost application combined with plant growth-promoting rhizobacteria (PGPR) improves seed quality and nutrient use efficiency of sunflower. *Journal of Plant Nutrition and Soil Science, 180*, 464–473. https://doi.org/10.1002/jpln.201600615.

Bashan, Y., de-Bashan, L. E., Prabhu, S. R., & Hernandez, J.-P. (2014). Advances in plant growth-promoting bacterial inoculant technology: Formulations and practical perspectives (1998–2013). *Plant and Soil, 378*, 1–33. https://doi.org/10.1007/s11104-013-1956-x.

Bashan, Y., de-Bashan, L. E., & Prabhu, S. R. (2016). Superior polymeric formulations and emerging innovative products of bacterial inoculants for sustainable agriculture and the environment. In H. B. Singh, B. K. Sarma, & C. Keswani (Eds.), *Agriculturally important microorganisms: Commercialization and regulatory requirements in Asia* (pp. 15–46). Singapore: Springer.

Belimov, A. A., Dodd, I. C., Safronova, V. I., Shaposhnikov, A. I., Azarova, T. S., Makarova, N. M., Davies, W. J., & Tikhonovich, I. A. (2015). Rhizobacteria that produce auxins and contain 1-amino-cyclopropane-1-carboxylic acid deaminase decrease amino acid concentrations in the rhizosphere and improve growth and yield of well-watered and water-limited potato (*Solanum tuberosum*). *The Annals of Applied Biology, 167*, 11–25. https://doi.org/10.1111/aab.12203.

Bhardwaj, D., Wahid Ansari, M., Kumar Sahoo, R., & Tuteja, N. (2014). Biofertilizers function as key player in sustainable agriculture by improving soil fertility, plant tolerance and crop productivity. *Microbial Cell Factories, 13*(2014), 66. https://doi.org/10.1186/1475-2859-13-66.

Bharti, N., Barnawal, D., Shukla, S., Tewari, S. K., Katiyar, R. S., & Kalra, A. (2016a). Integrated application of Exiguobacterium oxidotolerans, *Glomus fasciculatum*, and vermicompost improves growth, yield and quality of *Mentha arvensis* in salt-stressed soils. *Industrial Crops and Products, 83*, 717–728. https://doi.org/10.1016/j.indcrop.2015.12.021.

Bharti, N., Barnawal, D., Wasnik, K., Tewari, S. K., & Kalra, A. (2016b). Co-inoculation of *Dietzia natronolimnaea* and *Glomus intraradices* with vermicompost positively *influences Ocimum basilicum* growth and resident microbial community structure in salt affected low fertility soils. *Applied Soil Ecology, 100*, 211–225. https://doi.org/10.1016/J.APSOIL.2016.01.003.

Cao, Y., Wang, E., Tong, W., Qiao, Y., Zhao, L., Chen, W., & Wei, G. (2017). Population structure of *Rhizobium etli*-like strains nodulated with *Phaseolus vulgaris* in two ecoregions of China. *Soil Biology and Biochemistry, 112*, 14–23. https://doi.org/10.1016/j.soilbio.2017.04.017.

Cassán, F., Vanderleyden, J., & Spaepen, S. (2014). Physiological and agronomical aspects of phytohormone production by model plant-growth-promoting rhizobacteria (PGPR) belonging to the genus *Azospirillum*. *Journal of Plant Growth Regulation, 33*, 440–459. https://doi.org/10.1007/s00344-013-9362-4.

Cely, M. V. T., Siviero, M. A., Emiliano, J., Spago, F. R., Freitas, V. F., Barazetti, A. R., Goya, E. T., Lamberti G de, S., dos Santos, I. M. O., De Oliveira, A. G., & Andrade, G. (2016). Inoculation of *Schizolobium parahyba* with mycorrhizal fungi and plant growth-promoting rhizobacteria increases wood yield under field conditions. *Frontiers in Plant Science, 22*. https://doi.org/10.3389/fpls.2016.01708.

Chauhan, H., & Bagyaraj, D. J. (2015). Inoculation with selected microbial consortia not only enhances growth and yield of french bean but also reduces fertilizer application under field condition. *Scientia Horticulturae (Amsterdam), 197*, 441–446. https://doi.org/10.1016/j.scienta.2015.10.001.

Choudhary, D. K., Sharma, K. P., & Gaur, R. K. (2011). Biotechnological perspectives of microbes in agro-ecosystems. *Biotechnology Letters, 33*, 1905–1910. https://doi.org/10.1007/s10529-011-0662-0.

Dal Cortivo, C., Barion, G., Visioli, G., Mattarozzi, M., Mosca, G., & Vamerali, T. (2017). Increased root growth and nitrogen accumulation in common wheat following PGPR inoculation: Assessment of plant-microbe interactions by ESEM. *Agriculture, Ecosystems and Environment, 247*, 396–408. https://doi.org/10.1016/j.agee.2017.07.006.

Daza, A., Santamarıa, C., Rodrıguez-Navarro, D., Camacho, M., Orive, R., & Temprano, F. (2000). Perlite as a carrier for bacterial inoculants. *Soil Biology and Biochemistry, 32*, 567–572. https://doi.org/10.1016/S0038-0717(99)00185-6.

Di Salvo, L. P., Cellucci, G. C., Carlino, M. E., & García de Salamone, I. E. (2018). Plant growth-promoting rhizobacteria inoculation and nitrogen fertilization increase maize (*Zea mays* L.)

grain yield and modified rhizosphere microbial communities. *Applied Soil Ecology, 126*, 113–120. https://doi.org/10.1016/j.apsoil.2018.02.010.

Díaz-Alcántara, C. A., Ramírez-Bahena, M. H., Mulas, D., García-Fraile, P., Gómez-Moriano, A., Peix, A., Velázquez, E., & González-Andrés, F. (2014). Analysis of rhizobial strains nodulating *Phaseolus vulgaris* from Hispaniola island, a geographic bridge between Meso and South America and the first historical link with Europe. *Systematic and Applied Microbiology, 67*, 113–124. https://doi.org/10.1016/j.syapm.2013.09.005.

Dinesh, R., Anandaraj, M., Kumar, A., Bini, Y. K., Subila, K. P., & Aravind, R. (2015). Isolation, characterization, and evaluation of multi-trait plant growth promoting rhizobacteria for their growth promoting and disease suppressing effects on ginger. *Microbiological Research, 173*, 34–43. https://doi.org/10.1016/j.micres.2015.01.014.

Djedidi, S., Terasaki, A., Aung, H. P., Kojima, K., Yamaya, H., Ohkama-Ohtsu, N., Bellingrath-Kimura, S. D., Meunchang, P., & Yokoyama, T. (2015). Evaluation of the possibility to use the plant–microbe interaction to stimulate radioactive 137Cs accumulation by plants in a contaminated farm field in Fukushima, Japan. *Journal of Plant Research, 128*, 147–159. https://doi.org/10.1007/s10265-014-0678-3.

Du, C., Abdullah, J. J., Greetham, D., Fu, D., Yu, M., Ren, L., Li, S., & Lu, D. (2018). Valorization of food waste into biofertiliser and its field application. *Journal of Cleaner Production, 187*, 273–284. https://doi.org/10.1016/J.JCLEPRO.2018.03.211.

Dubey, S. C., Singh, V., Priyanka, K., Upadhyay, B. K., & Singh, B. (2015). Combined application of fungal and bacterial bio-agents, together with fungicide and *Mesorhizobium* for integrated management of *Fusarium* wilt of chickpea. *BioControl, 60*, 413–424. https://doi.org/10.1007/s10526-015-9653-8.

Etesami, H., & Maheshwari, D. K. (2018). Use of plant growth promoting rhizobacteria (PGPRs) with multiple plant growth promoting traits in stress agriculture: Action mechanisms and future prospects. *Ecotoxicology and Environmental Safety, 156*, 225–246. https://doi.org/10.1016/J.ECOENV.2018.03.013.

Gao, M., Zhou, J. J., Wang, E. T., Chen, Q., Xu, J., & Sun, J. G. (2015). Multiphasic characterization of a plant growth promoting bacterial strain, *Burkholderia* sp. 7016 and its effect on tomato growth in the field. *Journal of Integrative Agriculture, 14*, 1855–1863. https://doi.org/10.1016/S2095-3119(14)60932-1.

García-Fraile, P., Mulas-García, D., Peix, A., Rivas, R., González-Andrés, F., & Velázquez, E. (2010). *Phaseolus vulgaris* is nodulated in northern Spain by *Rhizobium leguminosarum* strains harboring two *nodC* alleles present in american *Rhizobium etli* strains: Biogeographical and evolutionary implications. *Canadian Journal of Microbiology, 56*, 657–666. https://doi.org/10.1139/w10-048.

Hassan, T., & Bano, A. (2015). The stimulatory effects of L-tryptophan and plant growth promoting rhizobacteria (PGPR) on soil health and physiology of wheat. *Journal of Soil Science and Plant Nutrition, 15*(1), 190–201. https://doi.org/10.4067/S0718-95162015005000016.

Herrmann, L., & Lesueur, D. (2013). Challenges of formulation and quality of biofertilizers for successful inoculation. *Applied Microbiology and Biotechnology, 97*, 8859–8873. https://doi.org/10.1007/s00253-013-5228-8.

Hong, S. H., & Lee, E. Y. (2017). Phytostabilization of salt accumulated soil using plant and biofertilizers: Field application. *International Biodeterioration and Biodegradation, 124*, 188–195. https://doi.org/10.1016/j.ibiod.2017.05.001.

Hungria, M., Campo, R. J., & Mendes, I. C. (2003). Benefits of inoculation of the common bean (*Phaseolus vulgaris*) crop with efficient and competitive *Rhizobium tropici* strains. *Biology and Fertility of Soils, 39*, 88–93. https://doi.org/10.1007/s00374-003-0682-6.

IEEP, Ecologic and GHK. (2012). *Study to analyse legal and economic aspects of implementing the Nagoya Protocol on ABS in the European Union*. Final report for the European Commission, DG Environment, Institute for European Environmental Policy, Brussels and London.

Jatan, R., Chauhan, P. S., & Lata, C. (2018). *Pseudomonas putida* modulates the expression of miRNAs and their target genes in response to drought and salt stresses in chickpea (*Cicer arietinum* L.). *Genomics*. (in press).

Kaushal, M., & Wani, S. P. (2016). Rhizobacterial-plant interactions: Strategies ensuring plant growth promotion under drought and salinity stress. *Agriculture, Ecosystems and Environment, 231*, 68–78.

Kumar, A., Maurya, B. R., Raghuwanshi, R., Meena, V. S., & Tofazzal Islam, M. (2017). Co-inoculation with *Enterobacter* and *Rhizobacteria* on yield and nutrient uptake by wheat (*Triticum aestivum L.*) in the alluvial soil under indo-Gangetic plain of India. *Journal of Plant Growth Regulation, 36*, 608–617. https://doi.org/10.1007/s00344-016-9663-5.

Kuzmicheva, Y. V., Shaposhnikov, A. I., Petrova, S. N., Makarova, N. M., Tychinskaya, I. L., Puhalsky, J. V., Parahin, N. V., Tikhonovich, I. A., & Belimov, A. A. (2017). Variety specific relationships between effects of rhizobacteria on root exudation, growth and nutrient uptake of soybean. *Plant and Soil, 419*, 83–96. https://doi.org/10.1007/s11104-017-3320-z.

Lally, R. D., Galbally, P., Moreira, A. S., Spink, J., Ryan, D., Germaine, K. J., & Dowling, D. N. (2017). Application of endophytic *Pseudomonas fluorescens* and a bacterial consortium to *Brassica napus* can increase plant height and biomass under greenhouse and field conditions. *Frontiers in Plant Science, 8*. https://doi.org/10.3389/fpls.2017.02193.

Le, T. A., Pék, Z., Takács, S., Neményi, A., Daood, H. G., & Helyes, L. (2018). The effect of plant growth promoting rhizobacteria on the water-yield relationship and carotenoid production of processing tomatoes. *Hortscience, 53*, 816–822. https://doi.org/10.21273/HORTSCI13048-18.

Ma, M., Jiang, X., Wang, Q., Guan, D., Li, L., Ongena, M., & Li, J. (2018). Isolation and identification of PGPR strain and its effect on soybean growth and soil bacterial community composition. *International Journal of Agriculture and Biology, 20*, 1289–1297. https://doi.org/10.17957/IJAB/15.0627.

Mahmood, S., Daur, I., Al-Solaimani, S. G., Ahmad, S., Madkour, M. H., Yasir, M., Hirt, H., Ali, S., & Ali, Z. (2016). Plant growth promoting rhizobacteria and silicon synergistically enhance salinity tolerance of mung bean. *Frontiers in Plant Science, 7*. https://doi.org/10.3389/fpls.2016.00876.

Mahmoud, E., El-Gizawy, E., & Geries, L. (2015). Effect of compost extract, N2-fixing bacteria and nitrogen levels applications on soil properties and onion crop. *Archives of Agronomy and Soil Science, 61*(2), 185–201. https://doi.org/10.1080/03650340.2014.928409.

Marasco, R., Rolli, E., Fusi, M., Cherif, A., Abou-Hadid, A., El-Bahairy, U., Borin, S., Sorlini, C., & Daffonchio, D. (2013). Plant growth promotion potential is equally represented in diverse grapevine root-associated bacterial communities from different biopedoclimatic environments. *BioMed Research International, 2013*, 1–17. https://doi.org/10.1155/2013/491091.

Marcano, I. E., Díaz-Alcántara, C. A., Urbano, B., & González-Andrés, F. (2016). Assessment of bacterial populations associated with banana tree roots and development of successful plant probiotics for banana crop. *Soil Biology and Biochemistry, 99*, 1–20. https://doi.org/10.1016/j.soilbio.2016.04.013.

Martínez-Viveros, O., Jorquera, M., Crowley, D., Gajardo, G., & Mora, M. (2010). Mechanisms and practical considerations involved in plant growth promotion by rhizobacteria. *Journal of Soil Science and Plant Nutrition, 10*, 293–319. https://doi.org/10.4067/S0718-95162010000100006.

Mekki, A., Arous, F., Aloui, F., & Sayadi, S. (2017). Treatment and valorization of agro-wastes as biofertilizers. *Waste and Biomass Valorization, 8*, 611–619. https://doi.org/10.1007/s12649 016 9620 3.

Morrissey JP, Dow JM, Louise G, Fergal M&, Gara O' (2004) Are microbes at the root of a solution to world food production? Rational exploitation of interactions between microbes and plants can help to transform agriculture. EMBO Reports 5:922–926. doi: https://doi.org/10.1038/sj.embor.7400263.

Mostasso, L., Mostasso, F. L., Dias, B. G., Vargas, M. A., & Hungria, M. (2002). Selection of bean (*Phaseolus vulgaris* L.) rhizobial strains for the Brazilian Cerrados. *Field Crops Research, 73*, 121–132. https://doi.org/10.1016/S0378-4290(01)00186-1.

Mrabet, M., Mhamdi, R., Tajini, F., Tiwari, R., Trabelsi, M., & Aouani, M. E. (2005). Competitiveness and symbiotic effectiveness of a *R. gallicum* strain isolated from root nodules of *Phaseolus vulgaris*. *European Journal of Agronomy, 22*, 209–216. https://doi.org/10.1016/J.EJA.2004.02.006.

Mukhtar, S., Shahid, I., Mehnaz, S., & Malik, K. A. (2017). Assessment of two carrier materials for phosphate solubilizing biofertilizers and their effect on growth of wheat (*Triticum aestivum* L.). *Microbiological Research, 205*, 107–117. https://doi.org/10.1016/j.micres.2017.08.011.

Mulas, D., García-Fraile, P., Carro, L., Ramírez-Bahena, M. H., Casquero, P., Velázquez, E., & González-Andrés, F. (2011). Distribution and efficiency of *Rhizobium leguminosarum* strains nodulating *Phaseolus vulgaris* in northern spanish soils: Selection of native strains that replace conventional N fertilization. *Soil Biology and Biochemistry, 43*, 2283–2293. https://doi.org/10.1016/j.soilbio.2011.07.018.

Mulas, D., Díaz-Alcántara, C. A., Mulas, R., Marcano, I., Barquero, M., Serrano, P., & González-Andrés, F. (2013). Inoculants based in autochthonous microorganisms, a strategy to optimize agronomic performance of biofertilizers. In M. B. Rodelas González & J. Gonzalez-López (Eds.), *Beneficial plant-microbial interactions: Ecology and applications*. Boca Raton: CRS Press.

Mulas, D., Seco, V., Casquero, P. A., Velázquez, E., & González-Andrés, F. (2015). Inoculation with indigenous rhizobium strains increases yields of common bean (*Phaseolus vulgaris* L.) in northern Spain, although its efficiency is affected by the tillage system. *Symbiosis, 67*, 113–124. https://doi.org/10.1007/s13199-015-0359-6.

Nosheen, A., Bano, A., Yasmin, H., Keyani, R., Habib, R., Shah, S. T. A., & Naz, R. (2016). Protein quantity and quality of safflower seed improved by NP fertilizer and rhizobacteria (*Azospirillum* and *Azotobacter* spp.). *Frontiers in Plant Science, 7*. https://doi.org/10.3389/fpls.2016.00104.

Pastor-Bueis, R., Mulas, R., Gómez, X., & González-Andrés, F. (2017). Innovative liquid formulation of digestates for producing a biofertilizer based on *Bacillus siamensis*: Field testing on sweet pepper. *Journal of Plant Nutrition and Soil Science, 180*, 748–758. https://doi.org/10.1002/jpln.201700200.

Prakamhang, J., Tittabutr, P., Boonkerd, N., Teamtisong, K., Uchiumi, T., Abe, M., & Teaumroong, N. (2015). Proposed some interactions at molecular level of PGPR coinoculated with *Bradyrhizobium diazoefficiens* USDA110 and *B. japonicum* THA6 on soybean symbiosis and its potential of field application. *Applied Soil Ecology, 85*, 38–49. https://doi.org/10.1016/j.apsoil.2014.08.009.

Qin, Y., Druzhinina, I. S., Pan, X., & Yuan, Z. (2016). Microbially mediated plant salt tolerance and microbiome-based solutions for saline agriculture. *Biotechnology Advances, 34*, 1245–1259. https://doi.org/10.1016/J.BIOTECHADV.2016.08.005.

Reddy Priya, P., Selastin Antony, R., Gopalaswamy, G., & Balachandar, D. (2016). Development of sequence-characterized amplified region (SCAR) markers as a quality standard of inoculants based on *Azospirillum*. *Archives of Microbiology*. https://doi.org/10.1007/s00203-016-1187-7.

Rodriguez-Navarro, D. N., Buendia, A. M., Camacho, M., Lucas, M. M., & Santamaria, C. (2000). Characterization of *Rhizobium* spp. bean isolates from south-West Spain. *Soil Biology and Biochemistry, 32*, 1601–1613. https://doi.org/10.1016/S0038-0717(00)00074-2.

Rozier, C., Hamzaoui, J., Lemoine, D., Czarnes, S., & Legendre, L. (2017). Field-based assessment of the mechanism of maize yield enhancement by *Azospirillum lipoferum* CRT1. *Scientific Reports, 7*, 1–12. https://doi.org/10.1038/s41598-017-07929-8.

Sahin, U., Ekinci, M., Kiziloglu, F. M., Yildirim, E., Turan, M., & Kotan, R. (2015). Ameliorative effects of plant growth on water-yield relationships, growth, and nutrient uptake of lettuce plants under different irrigation levels. *Hort Science, 50*(9), 1379–1386.

Saia, S., Rappa, V., Ruisi, P., Abenavoli, M. R., Sunseri, F., Giambalvo, D., Frenda, A. S., & Martinelli, F. (2015a). Soil inoculation with symbiotic microorganisms promotes plant growth and nutrient transporter genes expression in durum wheat. *Frontiers in Plant Science*. https://doi.org/10.3389/fpls.2015.00815.

Saia, S., Ruisi, P., Fileccia, V., Di Miceli, G., Amato, G., & Martinelli, F. (2015b). Metabolomics suggests that soil inoculation with arbuscular mycorrhizal fungi decreased free amino acid content in roots of durum wheat grown under N-limited, P-rich field conditions. *PLoS One*. https://doi.org/10.1371/journal.pone.0129591.

Sarathambal, C., Ilamurugu, K., Balachandar, D., Chinnadurai, C., & Gharde, Y. (2015). Characterization and crop production efficiency of diazotrophic isolates from the rhizosphere of semi-arid tropical grasses of India. *Applied Soil Ecology, 87*, 1–10. https://doi.org/10.1016/j.apsoil.2014.11.004.

Sarwar, M. A., Tahir, M., Tanveer, A., & Yaseen, M. (2016). Evaluating role of plant growth promoting rhizobacteria for improving phosphorus use efficiency and productivity in sunflower (*Helianthus annuus*). *International Journal of Agriculture and Biology, 18*, 881–888. https://doi.org/10.17957/IJAB/15.0110.

Shanta, N., Schwinghamer, T., Backer, R., Allaire, S. E., Teshler, I., Vanasse, A., Whalen, J., Baril, B., Lange, S., MacKay, J., Zhou, X., & Smith, D. L. (2016). Biochar and plant growth promoting rhizobacteria effects on switchgrass (*Panicum virgatum* cv. Cave-in-rock) for biomass production in southern Québec depend on soil type and location. *Biomass and Bioenergy, 95*, 167–173. https://doi.org/10.1016/j.biombioe.2016.10.005.

Song, X., Liu, M., Wu, D., Griffiths, B. S., Jiao, J., Li, H., & Hu, F. (2015). Interaction matters: Synergy between vermicompost and PGPR agents improves soil quality, crop quality and crop yield in the field. *Applied Soil Ecology, 89*, 25–34. https://doi.org/10.1016/j.apsoil.2015.01.005.

Spain. (2017). Royal Decree 999/2017, of November 24th, modifying the Royal Decree 506/2013, of June 28th, about fertilizer products. *Official State Gazette (BOE), 296*, 119396–119450.

Stamenković, S., Beškoski, V., Karabegović, I., Lazić, M., & Nikolić, N. (2018). Microbial fertilizers: A comprehensive review of current findings and future perspectives. *Spanish Journal of Agricultural Research, 16*. https://doi.org/10.5424/sjar/2018161-12117.

Timmusk, S., Behers, L., Muthoni, J., Muraya, A., & Aronsson, A.-C. (2017). Perspectives and challenges of microbial application for crop improvement. *Frontiers in Plant Science, 8*. https://doi.org/10.3389/fpls.2017.00049.

Tiwari, S., Prasad, V., Chauhan, P. S., & Lata, C. (2017). *Bacillus amyloliquefaciens* confers tolerance to various abiotic stresses and modulates plant response to phytohormones through osmoprotection and gene expression regulation in rice. *Frontiers in Plant Science, 8*, 1510. https://doi.org/10.3389/fpls.2017.01510.

Upadhyay, S. K., & Singh, D. P. (2015). Effect of salt-tolerant plant growth-promoting rhizobacteria on wheat plants and soil health in a saline environment. *Plant Biology*. https://doi.org/10.1111/plb.12173.

Vejan, P., Abdullah, R., Khadiran, T., Ismail, S., Nasrulhaq Boyce, A., Vejan, P., Abdullah, R., Khadiran, T., Ismail, S., & Nasrulhaq Boyce, A. (2016). Role of plant growth promoting rhizobacteria in agricultural sustainability—A review. *Molecules, 21*. https://doi.org/10.3390/molecules21050573.

Wang, B., Shen, Z., Zhang, F., Raza, W., Yuan, J., Huang, R., Ruan, Y., Li, R., & Shen, Q. (2016). *Bacillus amyloliquefaciens* strain W19 can promote growth and yield and suppress *Fusarium* wilt in banana under greenhouse and field conditions. *Pedosphere, 26*, 733–744. https://doi.org/10.1016/S1002-0160(15)60083-2.

Part II
Original Research Papers

Chapter 7
PGPR Characterization of Non-Nodulating Bacterial Endophytes from Root Nodules of *Vigna unguiculata* (L.) Walp.

Renzo A. Valdez-Nuñez, Rony Castro-Tuanama, Antonio Castellano-Hinojosa, Eulogio J. Bedmar, and Winston F. Ríos-Ruiz

Abstract Seventeen morphological groups of Non-Nodulating Endophytic Bacteria (NNEB) were isolated from the interior of healthy cowpea root nodules (*Vigna unguiculata* L. Walp.) growing in seven soils from three regions of Peru. The amplification and partial sequencing of the 16S rRNA ribosomal gene for representatives of each morphological group showed that they were closely related to members of genera *Rhizobium*, *Agrobacterium*, *Phyllobacterium*, *Mesorhizobium*, *Bosea*, *Ochrobactrum*, *Bradyrhizobium*, *Labrys*, *Ensifer*, *Starkeya* and *Nordella* (Proteobacteria), and of *Mycobacterium* (Actinobacteria). The plant growth promotion capacity of representative NNEB strains was examined. *Agrobacterium radiobacter* 5620I and 5722H isolates showed high IAA production (> 400 μg ml^{-1}). *Ochrobactrum haematophilum* 5410F and *Starkeya novella* 5740O were the best isolates solubilizing tricalcium phosphate (> 300%). *Agrobacterium radiobacter* 5722H was the best isolate for production of siderophores (272.59%), whereas *A. radiobacter* 5620I showed the greatest antagonistic activity against *Fusarium oxysporum* (88.52%). The NNEB strains identified in this study showed that cowpea root nodules are a potential source of Plant Growth Promoting Bacteria (PGPB) which may be used for the development of new inoculants.

Keywords Cowpea · Peru · Endophytes · Indole-3-acetic-acid · *Fusarium oxysporum*

R. A. Valdez-Nuñez (✉) · R. Castro-Tuanama · W. F. Ríos-Ruiz
Universidad Nacional de San Martín, Tarapoto, Perú
e-mail: revanu@unsm.edu.pe

A. Castellano-Hinojosa · E. J. Bedmar
Department of Soil Microbiology and Symbiotic Systems, Estación Experimental Del Zaidín, CSIC, Granada, Spain

D. Zúñiga-Dávila et al. (eds.), *Microbial Probiotics for Agricultural Systems*, Sustainability in Plant and Crop Protection,
https://doi.org/10.1007/978-3-030-17597-9_7

7.1 Introduction

Cowpea, *Vigna unguiculata* (L.) Walp., is a grain legume native to Africa, which belongs to the tribe Phaseoleae of subfamily Papilionaceae (LPWG 2013). Since ancient times, cowpea is an important crop both in developing and developed countries (Carvalho et al. 2017). In 2014, 11 x 10^6 ha were cultivated with cowpea worldwide, accounting for a total of 5.59 x 10^6 tonnes of dry grain (FAOSTAT 2017). In Peru, cowpea cultivation is an important economic activity in the North coast and in the amazonic forest area of the country.

Cowpea grows in acid soils and tolerates drought, salinity and high temperatures (Soares et al. 2014). Moreover, its cultivation improves soil quality when used as green manure or in crop rotation systems (Martins et al. 2015). It establishes symbiotic associations with soil bacteria such as rhizobia, which results in the formation of root nodules where N_2 fixation takes place. Cowpea is a promiscuous legume capable to nodulate with members of genera *Rhizobium*, *Mesorhizobium*, *Ensifer*, *Methylobacterium*, *Ralstonia*, *Cupriavidus*, *Microvirga* and *Bradyrhizobium*, the latter being the most frequent (Peix et al. 2015; Shamsheldin et al. 2016).

In addition to the rhizobial strains, many other bacteria which lack the ability to induce legume nodulation have been isolated from root nodules. These nodule-isolated bacteria have been called non-rhizobial endophytes (De Meyer et al. 2015), nodule endophytes (Velázquez et al. 2013) or nodules associated-bacteria (Rajendran et al. 2012). Here we will use the term non-nodulating bacterial endophytes (NNBE) which is broader in scope than other designations.

The diversity of the NNBE depends on the plant genotype and the edaphoclimatic conditions (Leite et al. 2017). However, their function within the root nodules is largely unknown. NNBE include the Alphaproteobacteria *Azospirillum*, *Gluconacetobacter*, *Ochrobactrum*, *Methylobacterium*, *Phyllobacterium*, *Endobacter*, *Bosea*, *Tardiphaga*, *Herbaspirillum*, *Rhizobium* and *Labrys*, the Betaproteobacteria *Burkholderia* and *Variovorax*, the Gammaproteobacteria *Klebsiella*, *Pseudomonas* and *Pantoea*, the Actinobacteria *Arthrobacter*, *Brevibacterium*, *Micromonospora*, *Mycobacterium*, *Streptomyces* and *Xiangella* and the Firmicutes *Bacillus*, *Cohnella*, *Fontibacillus* and *Paenibacillus* (Martínez-Hidalgo and Hirsch 2017; Velázquez et al. 2013, 2017).

Many NNBE reported to date have been characterized as plant growth promoting bacteria (PGPB), and express the capability of nitrogen fixation, phytohormone production, solubilization of mineral phosphate, production of siderophores and control of soil pathogens (Egamberdieva et al. 2017; Zhao et al. 2017; Velázquez et al. 2017 and references therein). Information on the isolation and characterization of NNBE from cowpea nodules is scarce (Da Costa et al. 2013, 2016; Leite et al. 2017) and its role as possible PGPB is not well known. Accordingly, the aim of this study was: a) the isolation of endophytes from root nodules of agriculturally cultivated cowpea plants, b) the molecular identification of the isolates and c) the evaluation of their PGPB properties.

7.2 Materials and Methods

7.2.1 Soil Collection and Plant Inoculation

Soil samples (100 g) were collected from seven different locations within the departments of Lambayeque, Piura and San Martín. The soils without previous history of legume-bacteria inoculation were sampled from September to November 2014, and their physicochemical characteristics were determined (Table 7.1) following the methodology proposed by Soil Survey Staff (2009). Cowpea seeds (*Vigna unguiculata* L. Walp var. CAR-3005) were surface-sterilized and germinated as described by Valdez et al. (2016). Seedlings were planted in Leonard jars (2 per jar) containing sterile vermiculite and independently inoculated with 1 ml of soil solutions, prepared by homogenization of 1 g of soil in 9 mL of sterile physiological saline solution. The bottom part of the jars contained the sterile N-free mineral solution described by Rigaud and Puppo (1975). The jars were placed under controlled environmental conditions (16/8 h light/dark photoperiod, 26/18 °C day/night temperature and 70–75% relative humidity) for 40 days.

7.2.2 Isolation of Endophytes from Nodules, Bacterial Culture Conditions and Reinfection Test

Nodules collected from roots of cowpea from the same plant and location were pooled and sterilized according to Somasegaran and Hoben (1994). Twelve nodules from each pool were placed independently in Petri dishes and crushed in a drop of sterile water with a sterile glass rod. The resulting suspension was streaked onto Petri dishes containing yeast extract mannitol (YEM) medium (Vincent 1970) or peptone salts yeast (PSY) medium (Regensburger and Hennecke 1983). After incubation of the plates at 30 °C for 12 days, colony forming units (CFU), which represented all colony types that could be distinguished by microscopic observation of living cells, were chosen. Tests to validate surface sterilization of plant tissues were performed by touching the disinfected nodules several times on the surface of the corresponding solid medium prior to isolation of the internal microbiota. Single colonies were picked and checked for purity by repeated streaking on growth media. To test the nodulation capacity of the isolated strains, seeds of cowpea were surface-sterilized with 70% ethanol for 1 min, washed with 5% sodium hypochlorite for 5 min, rinsed thoroughly with sterile distilled water, embedded in sterile distilled water for 1 h, and then allowed to germinate at 30 °C in the dark. Seedlings were separately inoculated at planting with 1 mL (about 10^8 cells) of each one of the strains isolated from the cowpea nodules and placed under controlled environmental conditions as above. The non-nodulating strains were selected as NNBE.

Table 7.1 Location and physicochemical properties of the cowpea-cultivated soils used in this study

Soil	Department, District, Locality	Altitude m.a.s.l [a]	Climate[b]	Climatic region	Latitude/ Altitude	Mechanical analysis			Textural class	Soil pH		Organic carbon %	Total nitrogen %	Relation C/N
						Sand	Silt	Clay						
L1	Lambayeque, Pacora, La curva.	57	Semi-warm and desert climate. Rain deficiency all year round, humid environment.	Coast	06°25′33″ S 79°49′51″ w	55	30	15	Sandy loamy	7.07	Neutral	0.12	0.04	3
L2	Lambayeque, Pacora, El puente	57			06°25′33″ S 79°49′51″ w	85	5	10	Sandy	8.14	Moderately alkaline	0.09	0.04	2
L3	Lambayeque, Illimo, Compuerta Marcelo	46			06°32′51″ S 79°51′49″ w	45	15	40	Sandy loamy	7.20	Neutral	0.32	0.08	4
P4	Piura, Morropón, La Entrada	1475	Warm and desert climate. With deficiency of rains all year round, humid environment.	Interandine	05°01′39″ S 79°52′27″ w	30	15	55	Clay loamy	7.80	Slightly alkaline	0.84	0.14	6
P5	Piura, Piura, Cura Mori	25	Desert climate and deficiency of rains, dry environment.	Coast	05°19′32″ S 80°39′51″ w	43	28	30	Clay loamy	7.69	Slightly alkaline	0.32	0.08	4
S6	San Martín, Cacatachi, Cuñumbuque	504	Semi-dry and warm weather. Deficiency of rains in autumn, winter and spring, tropical zone.	High jungle	06°27′0″ S 76°30′0″ w	53	41	6	Sandy loamy	6.71	Neutral	2.38	0.12	20
S7	San Martín, Juanjui, La Victoria	565			07°10′49″ S 76°43′35″ w	59	32	9	Sandy loamy	6.49	Slightly acid	1.63	0.08	20

[a]m.a.s.l, meters above sea level; [b]Climatic classification according to Werren Thornthwaite

7.2.3 Morphocolonial Characterization and Diversity Analysis

All NNBE were grouped based on their morphocolonial characteristics (Table 7.2) as suggested by Somasegaran and Hoben (1994). The Shannon-Wiener diversity index and the Simpson dominance and diversity indices were used to study the relative abundance and diversity of NNBE within each group (Leite et al. 2009).

7.2.4 DNA Extraction and Amplification of the 16S rRNA Gene

Genomic DNA was isolated from the bacterial cells using the Real Pure Genomic DNA Extraction Kit (Durviz, Spain) and quantified using a NanoDrop spectrophotometer, model ND1000 (Thermo Fisher Scientific, USA). PCR amplifications of 16S rRNA gene were carried out using primers and thermal conditions described by Weisburg et al. (1991). Amplification products were purified using the Qiagen PCR product purification system and subjected to cycle sequencing using the same primers as for PCR amplification, with ABI Prism dye chemistry and analyzed with a 3130 XL automatic sequencer at the sequencing facilities of the Estación Experimental del Zaidín, CSIC, Granada, Spain. The 16S rRNA gene sequences were compared to those deposited in the EZBioCloud database (Yoon et al. 2017). Phylogenetic trees were inferred based on the neighbor-joining algorithm (NJ) (Saitou and Nei 1987) using MEGA7 (Kumar et al. 2016). The accession numbers of the obtained nucleotide sequences are shown in the phylogenetic trees.

7.2.5 Determination of PGPR Properties

The selected representative strains from each morphological group were used to test the production of indole-3-acetic acid (IAA) and siderophores, the solubilization of tricalcium phosphate and production of the hydrolytic enzymes cellulose, chitinase, glucanase and protease. The antagonism against *Fusarium oxysporum* was also determined. The methods used have been reported earlier (Valdez et al. 2016; Castellano-Hinojosa and Bedmar 2017; Egamberdieva et al. 2017).

7.2.6 Statistical Analysis

The measured variables were first explored using the Shapiro-Wilk test and the Bartlett's test to check whether they met the normality and homoscedasticity assumptions, respectively. Since most data fitted the normal distribution, the Duncan test (ANOVA, $p \leq 0.05$) was chosen to search for significant differences using the InfoStat program.

Table 7.2 Identification of NNBE from cowpea nodules

Isolate	Isolation culture medium	Morphocolonial group	Cultural characteristics[a]					Phylogenetic affiliation	
			GR (h)[b]	Appearance	Color	Elevation	Mu[c]	Closest relative species (according to Ez-taxon-e)	Similarity (%)
5101B, **5102B**	PSY	I	24	HM	Yellowish	Pulvinate	+++	*Mycobacterium cosmeticum*	99.86
5110C, 5111C, **5112C**	YEM	II	24	HM	Aqueous white	Elevated	+++	*Rhizobium yanglingense*	98.31
5120D, 5601D, **5602D**	YEM	III	24	HM	Opaque cream	Flat	+	*Phyllobacterium brassicacearum*	97.69
5401E, **5402E**	YEM	IV	36	HM	Opaque cream	Flat	+	*Mesorhizobium huakuii*	98.75
5610E, 5701E, 5702E	YEM	V	36	HM	Yellowish	Flat	+	*Bosea vestrisii*	98.27
5410F	YEM	VI	96	HM	Aqueous cream	Flat	+	*Ochrobactrum haematophilum*	98.92
5420G, 5421G, 5422G, 5501G, **5502G,** 5710G, 5711G, 5301G, 5210G, 5211G	YEM	VII	24	HT	Transparent	Flat	+++	*Rhizobium miluonense*	98.48
5721H, **5722H,** 5221H	YEM	VIII	96	HM	Opaque white	Elevated	++	*Rhizobium radiobacter*	99.56
5230I, **5620I**	YEM	IX	48	HT	Aqueous white	Flat	+	*Rhizobium radiobacter*	99.56
5730K	PSY	X	42	HM	Pink	Elevated	++	*Labrys neptuniae*	99.20
5630L	YEM	XI	72	HT	Opaque cream	Flat	++	*Bradyrhizobium jicamae*	100.00
5430M, 5640M	YEM	XII	48	HM	Transparent	Elevated	+++	*Mesorhizobium acaciae*	99.56
5440N	YEM	XIII	24	HT	Opaque cream	Flat	+	*Ensifer adhaerens*	99.71
5740O, 5741O	PSY	XIV	42	HM	Transparent	Flat	+++	*Starkeya novella*	98.70
5650P	YEM	XV	24	HM	Opaque white	Elevated	++	*Rhizobium miluonense*	99.86
5660Q, 5661Q	YEM	XVI	24	HT	Transparent	Elevated	++	*Rhizobium miluonense*	99.93
5130S	YEM	XVII	96	HT	Watery orange	Flat	+++	*Nordella oligomobilis*	91.58

Strains in bold were chosen as the representative strain of each morphocolonial group
[a]Cultural characteristics in growth media after incubation at 30 °C, [b] GR: Time for an Colony forming unit to reach 1 mm diameter, [c]Mucosity: +++, abundant; ++, regular; +, scarce; –, absent

7.3 Results

7.3.1 *Isolation of Bacteria, Morphocolonial Groups and Diversity Indices*

Out of the 108 bacterial strains isolated from cowpea nodules, 40 (37%) were unable to reinfect their original host *V. unguiculata*. Analysis of the morphocolonial traits of the NNBE allowed their distribution into 17 clusters (Table 7.2) and showed that most of the strains were isolated from soil S7 (9 strains), followed by S6 and P4 (8 strains each), L1 (7 strains), L2 (4 strains), P5 (3 strains) and L3 (1 strain) (Table 7.3). According to the Shannon-Wiener biodiversity index, soils S6 and S7 were the most diverse and soils L3 and P5 showed the lowest biodiversity indices (Table 7.3). No dominance of a species was found in soil S6 as the corresponding Simpson indices were below 0.25 (Table 7.3).

7.3.2 *16S rRNA Gene Phylogenetic Analysis*

The partial sequence of the 16S rRNA gene from a representative strain of each morphological group showed that the isolated NNBE were members of genera: *Rhizobium* (52.5%), *Mesorhizobium* (10%), *Phyllobacterium* (7.5%), *Bosea* (7.5%), *Mycobacterium* (5%), *Starkeya* (5%), and *Bradyrhizobium*, *Ensifer*, *Labrys*, *Nordella* and *Ochrobactrum* (2.5% each, respectively) (Table 7.2). Most (95%) of the strains belonged to class Alphaproteobacteria and the remaining 5% were members of class Actinobacteria. According to our results (Table 7.2), within genus *Rhizobium*, the closest relative species of strain 5112C (II) was *R. yanglingense* SH22623^T with 98.31% similarity, strain 5502G (VII) showed 98.48% similarity with *R. miluonense* HAMBI 2971^T, strains 5722H (VIII) and 5620I (IX) showed 99.56% similarity with *A. radiobacter* ATCC 19358^T and strains 5650P (XV) and 5660Q (XVI) were 99.86 and 99.93% similar to *R. miluonense* HAMBI 2971^T, respectively. Regarding genus *Ensifer*, strain 5440N (XIII) had 99.71% similarity with *E. adhaerens* Casida A^T. Belonging to genus *Mesorhizobium*, strain 5402E (IV) was 98.75% similar to *M. huakuii* IAM 14158^T and strain 5430M (XII) had 99.56% similarity with *M. acaciae* RITF 741^T. Finally, strain 5630L (XI) showed 100% similarity with *B. jicamae* PAC68^T. A NJ phylogenetic tree inferred from the 16S rRNA gene sequences of the NNBE included in the rhizobia together with those of their corresponding type strains is shown in Fig. 7.1.

On the other hand, a 98.92% similarity was found between strains 5410F (VI) and *Ochrobactrum haematophilum* DD234^T, and a 97.69% between strains 5602D (III) and *Phyllobacterium brassicacearum* STM 196^T. Strain 5130S (XVII) showed 99.43% similarity with *Nordella oligomobilis* N21, and strain 5740O (XIV) had 98.70% similarity with *Starkeya novella* DSM 506^T. Strains 5610E (V), 5730K (X) and 5102B (I) had 98.27%, 99.20% and 99.86% similarity with *Bosea vestrisii*

Table 7.3 Number of strains from each morphological group in Northern Peruvian soils cultivated with *V. unguiculata* var. CAR-3005, and Shannon-Wiener and Simpson diversity indices

Soil	Number of NNEB	Morphological group (number of strains)	Shannon-Wiener	Simpson	
				Dominance	Diversity
L1	7	I (2), II (3), III (1), XVII (1)	1.28	0.31	0.69
L2	4	VII (2), VIII (1), IX (1)	1.04	0.56	0.44
L3	1	VII (1)	0	1	0
P4	8	V (2), VII (1), VIII (3), XIV (1), XV (1)	1.49	0.25	0.75
P5	3	VII (2), IX (1)	0.64	1	0
S6	8	III (2), V (1), XI (1), XII (1), XV (1), XVI (2)	1.73	0.19	0.81
S7	9	V (2), VII (2), VIII (2), X (1), XIV (2)	1.58	0.25	0.75

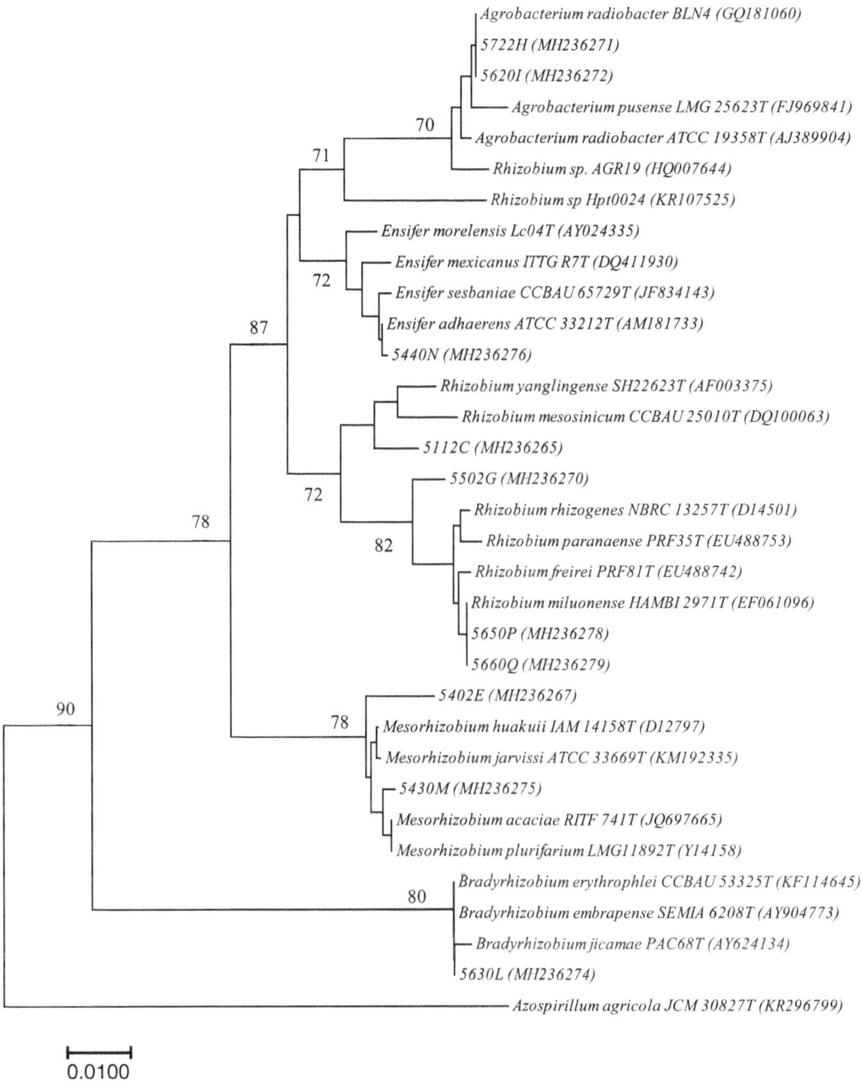

Fig. 7.1 NJ-phylogenetic tree based on partial 16S rRNA gene sequences of the NNEB included in the rhizobia and phylogenetically related type species. The significance of each branch is indicated by a bootstrap value calculated for 1000 subsets; values lower than 70% are not indicated. Bar, 1 substitution per 100 nucleotide position. The tree is rooted on *Azospirillum agricola* JCM 30827^{T}

34635^{T}, *Labrys neptuniae* Liujia-146^{T} and *Mycobacterium cosmeticum* LTA 388^{T}, respectively. A NJ phylogenetic tree built from the 16S rRNA gene sequences of the NNBE not included in the rhizobia together with those of their corresponding type strains are shown in Fig. 7.2.

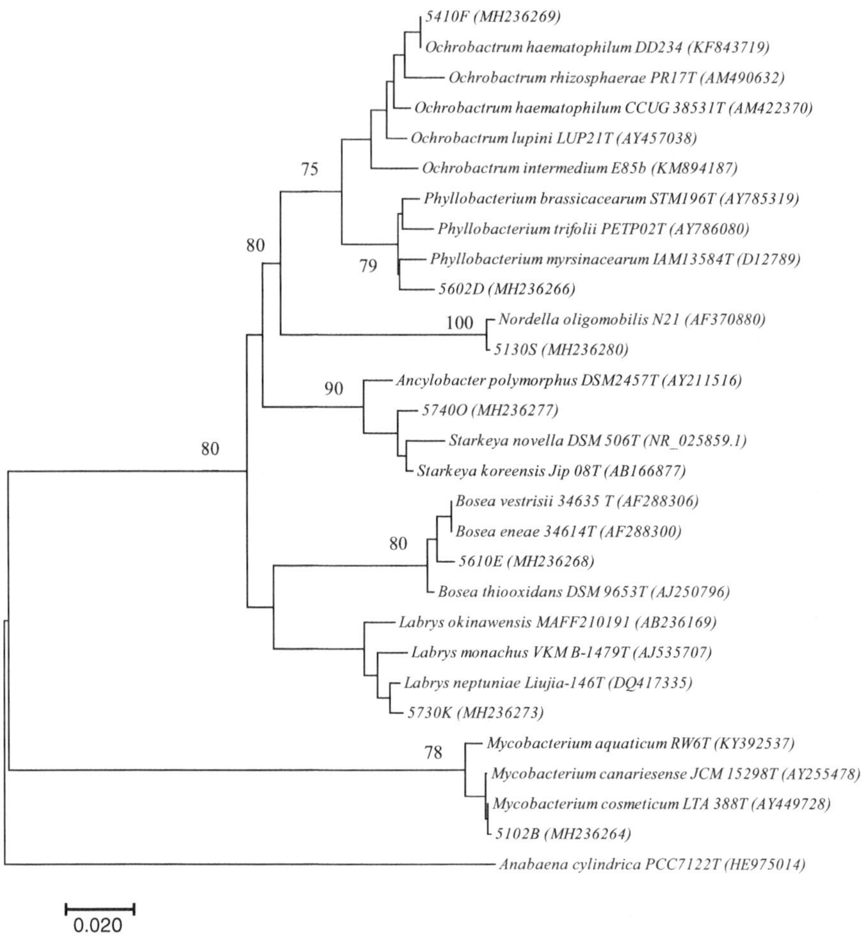

Fig. 7.2 NJ-phylogenetic tree based on partial 16S rRNA gene sequences of NNEB not included in the rhizobia and phylogenetically related type species. The significance of each branch is indicated by a bootstrap value calculated for 1000 subsets; values lower than 70% are not indicated. Bar, 2 substitution per 100 nucleotide position. The tree is rooted on *Anabaena cylindrica* PCC7122^T

7.3.3 Plant Growth-Promoting Traits

The plant growth promotion characteristics of each one of the 17 representative strains are presented in Table 7.4. In the absence of tryptophan most strains (82.35%) produced IAA or IAA-related compounds with values ranging from 0.59 to 85.65 μg ml^{-1}, and the presence of tryptophan in the culture medium increased the production of IAA. The strains 5610E, 5410F and 5130S did not produce IAA. The solubilization of tricalcium phosphate was performed by 70.5% of the strains with

Table 7.4 Plant growth promotion traits of NNBE from cowpea nodules

Isolate	IAA Production (μg ml^{-1})		Solubilization index (%)[c]	Siderophore production (%)[d]	Antagonism against *Fusarium oxysporum*[e]
	−Tryp[a]	+Tryp[b]			
5102B	0.00 ^{E}	0.00 ^{L}	NS	0.00 ^{D}	0.00 ^{E}
5112C	16.83 (± 3.34) ^{BC}	279.84 (± 1.46) ^{D}	152.09 (± 11.67) ^{D}	149.46 (± 1.49) ^{BC}	42.04 (± 2.14) ^{D}
5602D	7.37 (± 1,52) ^{CDE}	19.19 (± 0.75) ^{K}	207.43 (± 10.32) ^{BCD}	142.25 (± 18.21) ^{C}	0.00 ^{E}
5402E	8.01 (± 0.57) ^{CDE}	164.46 (± 2.10) ^{G}	187.43 (± 29.12) ^{BCD}	158.27 (± 5.42) ^{BC}	40.56 (± 0.56) ^{D}
5610E	0.0 ^{DE}	0.00 ^{L}	NS	142.73 (± 5.40) ^{C}	0.00 ^{E}
5410F	0.0 ^{E}	259.41 (± 0.39) ^{E}	327.46 (± 27.78) ^{A}	164.73 (± 10.19) ^{BC}	73.52 (± 1.03) ^{B}
5502G	85.65 (± 5.66) ^{A}	246.61 (± 4.73) ^{F}	286.67 (± 6.67) ^{AB}	175.54 (± 7.97) ^{B}	75.74 (± 0.67) ^{B}
5722H	10.16 (± 0.97) ^{CD}	440.27 (± 1.76) ^{B}	177.38 (± 35.90) ^{CD}	272.59 (± 6.34) ^{A}	0.00 ^{E}
5620I	10.16 (± 0.93) ^{CD}	454.68 (± 2.80) ^{A}	NS	0.00 ^{D}	88.52 (± 0.49) ^{A}
5730K	5.22 (± 1.32) ^{DE}	172.63 (± 0.60) ^{G}	164.58 (± 14.92) ^{D}	161.55 (± 3.23) ^{BC}	73.89 (± 0.85) ^{B}
5630L	0.59 (± 0.28) ^{DE}	5.75 (± 1.63) ^{L}	128.60 (± 0.07) ^{D}	0.00 ^{D}	0.00 ^{E}
5430M	3.92 (± 0.28) ^{DE}	124.57 (± 0.22) ^{H}	278.18 (± 31.23) ^{ABC}	0.00 ^{D}	47.78 (± 0.01) ^{C}
5440N	6.40 (± 1.06) ^{DE}	50.38 (± 0.65) ^{J}	163.64 (± 31.06) ^{D}	167.07 (± 7.36) ^{BC}	46.67 (± 2.32) ^{C}
5740O	4.78 (± 1.61) ^{DE}	96.61 (± 0.75) ^{I}	325.92 (± 45.11) ^{A}	0.00 ^{D}	46.67 (± 11.67) ^{C}
5650P	21.77 (± 3.72) ^{B}	115.43 (± 1.03) ^{H}	159.52 (± 21.19) ^{D}	161.58 (± 2.61) ^{BC}	0.00 ^{E}
5660Q	23.71 (± 0.67) ^{B}	58.44 (± 5.97) ^{J}	NS	0.00 ^{D}	0.00 ^{E}
5130S	0.0 ^{E}	0.00 ^{L}	NS	163.89 (± 2.78) ^{BC}	0.00 ^{E}
[f]Control strain	4.78 (± 0.28) ^{DE}	379.19 (± 3.23) ^{C}	250.00 (± 0.07) ^{ABC}	ND	ND

Values represent the mean followed by the standard error (n = 5). Values in a column followed by the same letter are not statistically different according to the Duncan test ($p \leq 0.05$). [a]– Tryp, production of IAA in culture medium without tryptophan. [b] + Tryp, production of IAA in culture medium supplemented with 600 mg L^{-1} tryptophan. [c] The solubilization index (SI) was calculated according to the formula SI = halo diameter (mm)/colony diameter (mm) (Castellano-Hinojosa and Bedmar 2017). [d] Siderophore production was estimated following the formula: % siderophore units = [(Ar-As)/Ar] × 100, where Ar = absorbance of reference (minimal media + CAS assay solution) and As = absorbance of sample (Castellano-Hinojosa and Bedmar 2017). [e] Inhibition of mycelial growth was estimated following the formula: % inhibition = [(Gc-Gs)/Gc]/100, where, Gc = diameter of control mycelial growth and Gs = diameter of the bacterial growth (Castellano-Hinojosa and Bedmar 2017). [f] Control strains were *Brevibacillus brevis* BEA1 for production of IAA and *Azospirillum brasilense* C16 for solubilization of tricalcium phosphate efficiency values varying between 128.6 and 327.4%. Strains 5410F and 5740O were statistically more efficient than the remaining ones. The production of siderophores was observed in 64.7% of the strains with values of efficiency from 142.2 to 272.5%, the highest production corresponding to strain 5722H. None of the strains produced chitinases and glucanases, and only strains 5502G, 5730K, 5740 and 5660Q showed cellulase activity (data not shown). Many isolates (52.94%) inhibited significantly the mycelial growth of *F. oxysporum*, and strain 5620I showed the highest percentage of inhibiton (88.5%).

7.4 Discussion

Members of the family Rhizobiaceae have been considered for decades to be the only inhabitants of legume nodules. However, soil bacteria other than the classical nodulating rhizobia, *Burkholderia* and *Cupriavidus* are often found inside nodules. Many of those non-rhizobial bacteria can be N_2-fixers and some also induce nitrogen-fixing nodules on legume roots. More surprising, however, is the vast diverse population of NNBE residing within the nodules that neither elicit nodulation nor nitrogen fixation (Martínez-Hidalgo and Hirsh 2017; Velázquez et al. 2017). The function of the NNBE is not well characterized and whether or not their presence within the nodules is biologically important is under debate. Synergistic effects between rhizobia and NNBE have been reported due to the expression of plant growth promotion mechanisms by the endophytes (Velázquez et al. 2013, 2017). NNBE may also use the nodule as a niche without any obvious benefit for the plant, albeit its lodging has been reported to be under host genetic control (Zgadzaj et al. 2015).

In this study, members of genera *Rhizobium*, *Mesorhizobium*, *Bradyrhizobium* and *Ensifer* that were unable to reinfect their host plant have been isolated from nodules of cowpea (Table 7.2), and therefore can be considered as true NNBE. This is not surprising because bacterial strains with non-specific symbiotic genes for legumes or those which lost their symbiotic genes can be found within the nodules as NNBE (Peix et al. 2015; Velázquez et al. 2013, 2017 and references therein). Beghalem et al. (2017) also recovered *Neorhizobium galegae* and *Sinorhizobium meliloti* from nodules of *Sulla* that were not capable to nodulate the host plants. Moreover, a group of bacteria affiliated to genus *Bradyrhizobium* isolated from different ecozones across North America were incapable of nodulating the usual host legume of the species (Van Insberghe et al. 2015).

In addition to those non-nodulating members of the classic rhizobial species, the presence of genera *Bacillus*, *Paenibacillus*, *Brevibacillus*, *Blastobacter*, *Enterobacter*, *Pseudomonas*, *Klebsiella*, *Dyadobacter*, *Chitinophaga*, *Chryseobacterium* and *Sphingobacterium* have been reported inside the root nodules of cowpea (e.g. Da Costa et al. 2013, 2016; Pandya et al. 2013, 2015; Leite et al. 2017). Isolation, however, of NNBE from interior of nodules of cowpea plants grown in the coast and amazonic area of Northern Peru has not been reported.

Differences in the number and composition of the isolates among the seven soils used for plant-tramp experiments were found, suggesting that the soil type could influence the NNBE inside the nodules. Our results agree with data from a study of the nodule microbiome of two different cowpea genotypes growing in Brazil, which showed that bacterial cells inside the nodules were more influenced by the soil type than the plant genome. Bacterial cells of genera *Mycobacterium*, *Bosea*, *Ochrobactrum*, *Labrys*, *Starkeya* and *Phyllobacterium*, have never been reported as NNBE in cowpea nodules, and *Nordella* has never been reported as NNBE in legume nodules. Although *Mycobacterium* is considered a human pathogen, the species *M. frederiksbergensis* was found within the nodules of *Astragalus armatus*, but was unable to reinfect its hos plant after isolation (Zakhia et al. 2006). Strain 5102B, closely related to *M. cosmeticum*, suggests that members of genus *Mycobacterium* could be true NNBE.

The genus *Phyllobacterium* has frequently been reported as a NNBE within legume nodules (Zakhia et al. 2006; Flores-Félix et al. 2013; De Meyer et al. 2015; Benghalem et al. 2017). In our study, strain 5602D showed a 97.69% identity with the type strain *P. brassicacearum* (Table 7.2). Accordingly, whether or not strain 5602D is a true NNBE in cowpea nodules should be taken with precaution.

The species *Bosea lupini*, *B. lathyri*, *B. robiniae* and *B. vaviloviae* were obtained from legume nodules and later identified as NNEB (Zakhia et al., 2006; De Meyer and Willems 2012, De Meyer et al. 2015). The 98.27% identity of strain 5610E isolated in this study with the type species *B. vestrisii* confirms its status as a true NNBE in cowpea nodules.

The species *O. ciceri* was isolated from *Cicer arietinum* which was unable to reinfect chickpea (Imran et al. 2010). Pea nodules were reported as colonized by non-nodulating *Ochrobactrum* (Tariq et al. 2014), whereas *O. intermedium* and *O. anthropi* were isolated from *Acacia ehrenbergiana*, *A. laeta* and *A. seyal* (Boukhatem et al. 2016). Considering that strain 5410F (*O. haematophilum*) isolated from cowpea does not form nodules on cowpea, our results support the endophytic nature of strain 5410F.

So far, members of genus *Labrys* have been reported to behave as NNBE when isolated from nodules of the aquatic legume *Neptunia oleracea* (Chou et al. 2007). We can extend those findings by showing that strain 5730K isolated from nodules of cowpea is also an NNBE.

Although bacteria of genus *Starkeya* were isolated from *Retama raetam*, they could not induce nodulation neither on *Retama* nor on the legume species *Macroptillium atropurpureum*, albeit *nifH* sequences showed 90% similarity with those of *Sinorhizobium meliloti* (Zakhia et al. 2006).

The NNBE isolated in this study showed characteristics involved in plant growth promotion. Many of them solubilized tricalcium phosphate, and the isolates 5410F and 5740O showed the highest capacity. Ability of *Ochrobactrum* sp. to solublize phosphate was previously reported in NNBE from nodules of *Pisum sativum* (Tariq et al. 2014). The highest rates of siderophores production was shown for 5722H, most likely an *A. radiobacter* affiliate, whose ability to produce siderophores was reported by Palaniappan et al. (2010) after isolation from nodules of *Lespedeza* sp.

Although the production of hydrolytic enzymes is common among NNBE, cellulase activity was determined only in isolates 5502G and 5660Q (*R. miluonense*), 5730K (*L. neptuniae*) and 5740 (*S. novella*). Cellulase, protease and chitinase activities have not been detected in NNBE from cowpea nodules. In contrast to cowpea, cellulase, lipase, protease and chitinase activities have been detected in NNBE from nodules of *C. arietinum* (Egamberdieva et al. 2017). Antagonistic activity of the bacteria isolated in this study against the pathogenic soil fungus *F. oxysporum* was detected in strains 5620I (*A. radiobacter*), 5410F (*O. haematophilum*), 5502G (*R. miluonense*) and 5730K (*L. neptuniae*).

7.5 Conclusions

Studies on the microbiome indicate that bacteria inside the nodules show a complex structure and composition, including very different phylogenetic groups, albeit the role of the inhabitants within the nodules is poorly known. Nevertheless, results suggest that cowpea nodules are an important source of microorganisms that could be of interest for the development of inoculants to be used in agronomic practices, in a sustainable agriculture context.

Acknowledgements This study was supported by the ERDF-cofinanced grant PEAGR2012-1968 from Consejería de Economía, Innovación y Ciencia (Junta de Andalucía, Spain) and funds from the National University of San Martín-Tarapoto-Peru.

References

Beghalem, H., Aliliche, K., Chriki, A., et al. (2017). Molecular and phenotypic characterization of endophytic bacteria isolated from *Sulla* nodules. *Microbial Pathogenesis, 111*, 225–231.

Boukhatem, Z. F., Merabet, C., Bekki, A., et al. (2016). Nodular bacterial endophyte diversity associated with native *Acacia* spp. in desert region of Algeria. *African Journal of Microbiology Research, 10*, 634–645.

Carvalho, M., Muñoz-Amatriaín, M., Castro, I., et al. (2017). Genetic diversity and structure of Iberian Peninsula cowpeas compared to world-wide cowpea accessions using high density SNP markers. *BMC Genomics, 18*, 891.

Castellano-Hinojosa, A., & Bedmar, E. J. (2017). Methods for evaluating plant growth-promoting rhizobacteria traits. In H. B. Singh, B. K. Sarma, & C. Keswani (Eds.), *Advances in PGPR research* (pp. 255–274). Oxford: CABI international.

Chou, Y. J., Elliott, G. N., James, E. K., et al. (2007). *Labrys neptuniae* sp. nov., isolated from root nodules of the aquatic legume *Neptunia oleracea*. *International Journal of Systematic and Evolutionary Microbiology, 57*, 577–581.

Da Costa, E. M., Nóbrega, R. S. A., de Carvalho, F., et al. (2013). Promoção do crescimento vegetal e diversidade genética de bactérias isoladas de nódulos de feijão-caupi. *Pesquisa Agropecuária Brasileira, 48*, 1275–1284.

Da Costa, E. M. D., Carvalho, F. D., Nóbrega, R. S. A., et al. (2016). Bacterial strains from floodplain soils perform different plant-growth promoting processes and enhance cowpea growth. *Scientia Agricola, 73*, 301–310.

De Meyer, S. E., & Willems, A. (2012). Multilocus sequence analysis of *Bosea* species and description of *Bosea lupini* sp. nov., *Bosea lathyri* sp. nov. and *Bosea robiniae* sp. nov., isolated from legumes. *International Journal of Systematic and Evolutionary Microbiology, 62*, 2505–2510.
De Meyer, S. E., de Beuf, K., Vekeman, B., et al. (2015). A large diversity of non-rhizobial endophytes found in legume root nodules in Flanders (Belgium). *Soil Biology and Biochemistry, 83*, 1–11.
Egamberdieva, D., Wirth, S. J., Shurigin, V. V., et al. (2017). Endophytic bacteria improve plant growth, symbiotic performance of chickpea (*Cicer arietinum* L.) and induce suppression of root rot caused by *Fusarium solani* under salt stress. *Frontiers in Microbiology, 8*, 1887.
FAOSTAT. (2017). *Agricultural database*. Available in: http://faostat.fao.org/site/567/default.aspx#ancor
Flores-Félix, J. D., Carro, L., Velázquez, E., et al. (2013). *Phyllobacterium endophyticum* sp. nov., isolated from nodules of *Phaseolus vulgaris*. *International Journal of Systematic and Evolutionary Microbiology, 63*, 821–826.
Imran, A., Hafeez, F. Y., Frühling, A., et al. (2010). *Ochrobactrum ciceri* sp. nov., isolated from nodules of *Cicer arietinum*. *International Journal of Systematic and Evolutionary Microbiology, 60*, 1548–1553.
Kumar, S., Stecher, G., & Tamura, K. (2016). MEGA7: Molecular evolutionary genetics analysis version 7.0 for bigger datasets. *Molecular Biology and Evolution, 33*, 1870–1874.
Legume Phylogeny Working Group. (2013). Towards a new classification system for legumes: Progress report from the 6th international legume conference. *South African Journal of Botany, 89*, 3–9.
Leite, J., Seido, S. L., & Passos, S. R. (2009). Biodiversity of rhizobia associated with cowpea cultivars in soils of the lower half of the São Francisco River valley. *Revista Brasileira de Ciência do Solo, 33*, 1215–1226.
Leite, J., Fischer, D., & Rouws, L. F. (2017). Cowpea nodules harbor non-rhizobial bacterial communities that are shaped by soil type rather than plant genotype. *Frontiers in Plant Science, 7*, 2064.
Martínez-Hidalgo, P., & Hirsch, A. M. (2017). The nodule microbiome: N_2-fixing rhizobia do not live alone. *Phytobiomes, 1*, 70–82.
Martins, J. C. R., Freitas, A. D. S. D., & Menezes, R. S. C. (2015). Nitrogen symbiotically fixed by cowpea and *Gliricidia* in traditional and agroforestry systems under semiarid conditions. *Pesquisa Agropecuaria Brasileira, 50*, 178–184.
Palaniappan, P., Chauhan, P. S., & Saravanan, V. S. (2010). Isolation and characterization of plant growth promoting endophytic bacterial isolates from root nodule of *Lespedeza* sp. *Biology and Fertility of Soils, 46*, 807–816.
Pandya, M., Naresh Kumar, G., & Rajkumar, S. (2013). Invasion of rhizobial infection thread by non-rhizobia for colonization of *Vigna radiata* root nodules. *FEMS Microbiology Letters, 348*, 58–65.
Pandya, M., Rajput, M., & Rajkumar, S. (2015). Exploring plant growth promoting potential of non rhizobial root nodules endophytes of *Vigna radiata*. *Microbiology, 84*, 80–89.
Peix, A., Ramírez-Bahena, M. H., & Velázquez, E. (2015). Bacterial associations with legumes. *Critical Reviews in Plant Sciences, 34*, 17–42.
Rajendran, G., Patel, M. H., & Joshi, S. J. (2012). Isolation and characterization of nodule-associated Exiguobacterium sp. from the root nodules of fenugreek (*Trigonella foenum-graecum*) and their possible role in plant growth promotion. *International Journal of Microbiology, 2012*, 1–7.
Regensburger, B., & Hennecke, H. (1983). RNA polymerase from *Rhizobium japonicum*. *Archives of Microbiology, 135*, 103–109.
Rigaud, J., & Puppo, A. (1975). Indole-3-acetic acid catabolism by soybean bacteroids. *Journal of General Microbiology, 88*, 223–228.
Saitou, N., & Nei, M. A. (1987). Neighbour-joining method: A new method for reconstructing phylogenetic trees. *Molecular Biology and Evolution, 44*, 406–425.

Shamsheldin, A., Abdelkhalek, A., & Sadowsky, M. J. (2016). Recent changes to the classification of symbiotic, nitrogen-fixing, legume-associating bacteria: A review. *Symbiosis, 71*, 91–109.

Soares, B. L., Ferreira, P. A. A., & de Oliveira-Longatti, S. M. (2014). Cowpea symbiotic efficiency, pH and aluminum tolerance in nitrogen-fixing bacteria. *Scientia Agricola (Piracicaba, Braz.), 71*, 171–180.

Soil Survey Staff. (2009). *Soil survey field and laboratory methods manual.* Soil survey investigations Report No. 51, Version 1.0. In R. Burt (Ed.). U.S. Department of Agriculture, Natural Resources Conservation Service.

Somasegaran, P., & Hoben, H. J. (1994). *Handbook for rhizobia: Methods in legume-rhizobium technology*. New York: Springer.

Tariq, M., Hameed, S., & Yasmeen, T. (2014). Molecular characterization and identification of plant growth promoting endophytic bacteria isolated from the root nodules of pea (*Pisum sativum* L.). *World Journal of Microbiology and Biotechnology, 30*, 719–725.

Valdez, R. A., Soriano, B., Prado, G., et al. (2016). Symbiotic and agronomic characterization of bradyrhizobial strains nodulating cowpea in Northern Peru. In F. González-Andrés & E. James (Eds.), *Biological nitrogen fixation and beneficial plant-microbe interaction* (pp. 195–212). Cham: Springer.

Van Insberghe, D., Maas, K. R., Cárdenas, E., et al. (2015). Non-symbiotic *Bradyrhizobium* ecotypes dominate North American forest soils. *Int Soc Microb Ecol*, 1–7.

Velázquez, E., Martínez-Hidalgo, P., Carro, L., et al. (2013). Nodular endophytes: An untapped diversity. In M. B. Rodelas-González & J. González-López (Eds.), *Beneficial plant-microbe interactions: Ecology and applications* (pp. 214–236). Boca Raton: CRC Press.

Velázquez, E., Carro, L., Flores-Félix, J. D., et al. (2017). The legume nodule microbiome: A source of plant growth-promoting bacteria. In V. Kumar et al. (Eds.), *Probiotics and plant health* (pp. 41–70). Singapore: Springer.

Vincent, J. (1970). *A manual for the practical study of the root nodule bacteria international biological programme handbook* (Vol. 15). Oxford: Blackwell.

Weisburg, W. G., Barns, S. M., Pelletier, D. A., et al. (1991). 16S ribosomal DNA amplification for phylogenetic study. *Journal of Bacteriology, 173*, 697–703.

Yoon, S. H., Ha, S. M., Kwon, S., et al. (2017). Introducing EzBioCloud: A taxonomically united database of 16S rRNA and whole genome assemblies. *International Journal of Systematic and Evolutionary Microbiology, 67*, 1613–1617.

Zakhia, F., Jeder, H., Willem, A., et al. (2006). Diverse bacteria associated with root nodules of spontaneous legumes in Tunisia and first report for *nifH*-like gene within the genera *Microbacterium* and *Starkeya*. *Microbiology Ecology, 51*, 375–393.

Zgadzaj, R., James, E. K., Kelly, S., et al. (2015). A legume genetic framework controls infection of nodules by symbiotic and endophytic bacteria. *PLoS Genetics*, 1–21.

Zhao, L., Xu, Y., & Lai, X. (2017). Antagonistic endophytic bacteria associated with nodules of soybean (*Glycine max* L.) and plant growth-promoting properties. *Brazilian Journal of Microbiology, 49*, 269–278.

Chapter 8
Endophytic Bacteria from *Passiflora incarnata* L. Leaves with Genetic Potential for Flavonoid Biosynthesis

Marcela Cristina Goulart, Luis Gabriel Cueva-Yesquén, Derlene Attili-Angelis, and Fabiana Fantinatti-Garboggini

Abstract *Passiflora incarnata*, a member of Passifloraceae family, is a traditional herb widely used as medicine since ancient times. It has extensive medicinal uses because of its high flavonoid content. Due to the high degree of interactions between endophytic bacteria and plants, it is believed that these microorganisms can produce metabolites initially produced by their host. Based on this hypothesis, the objective of this work was to analyze the genetic potential for flavonoid production of endophytic bacteria isolated from passion fruit plants, applying a PCR-based approach with genus-specific primers. Twenty strains from ten species within the genus *Sphingomonas* were tested for presence of the flavonol synthase (FLS) gene. The primers used effectively detected the FLS gene in all strains, but in two species, nonspecific bands appeared. A phylogenetic tree was constructed to evaluate the distribution of the gene among the strains used in this work. Phylogenetic analyses suggests that horizontal gene transfer (HGT) events occurred among various strains. This work may contribute to further research efforts aiming at the production of flavonoids by endophytic bacteria, based on the ability of these microorganisms to acquire genes from their hosts.

Keywords PCR · Flavonoids · Endophytic bacteria · Passionflower

---

M. C. Goulart · L. G. Cueva-Yesquén (✉) · F. Fantinatti-Garboggini
Graduate Program in Genetics and Molecular Biology, Institute of Biology, University of Campinas, São Paulo, Brazil

Division of Microbial Resources, Research Center for Agricultural, Biological and Chemical, University of Campinas, São Paulo, Brazil
e-mail: luisg_cueva@yahoo.es

D. Attili-Angelis
Division of Microbial Resources, Research Center for Agricultural, Biological and Chemical, University of Campinas, São Paulo, Brazil

D. Zúñiga-Dávila et al. (eds.), *Microbial Probiotics for Agricultural Systems*, Sustainability in Plant and Crop Protection,
https://doi.org/10.1007/978-3-030-17597-9_8

8.1 Introduction

Flavonoids are plant-specific secondary metabolites synthesized by the phenylpropanoid pathway. They consist of a large group of polyphenolic compounds with a benzo-γ-pyrone as central structure. Flavonoids are ubiquitous in the plant kingdom (Kumar and Pandey 2013; Weston and Mathesius 2013). Over 9000 compounds of this group have been described, and due to their physical and biochemical properties, they can interact with several plant targets, at different subcellular levels (Mierziak et al. 2014).

Flavonoids play various biological functions, including protection against ultraviolet (UV) radiation and phytopathogens, signaling during nodulation, root and shoot development, auxin transport and providing pigmentation to attract pollinators. Due to their antioxidative properties they also maintain a redox state in the plant cells (Falcone Ferreyra et al. 2012; Samanta et al. 2011). Because they are metabolites with many beneficial functions, investigators have genetically manipulated the molecular mechanisms of their synthesis for exploitation in human health and nutrition, pharmaceutical industry, crops production and protection (Trantas et al. 2015; Ververidis et al. 2007).

Currently, flavonoids are obtained by extraction from plant systems. However, this is a time-consuming process that requires the use of non-environmentally safe solvents (Trantas et al. 2015). In addition, separating pure flavonoid compounds from plants material results an unprofitable and laborious activity, due to the low concentrations of certain flavonoids and the numerous phytoconstituents with similar molecular structure (Du et al. 2010). Therefore, these molecules have been the target of metabolic engineering for the last 10–15 years, as various approaches have been developed and used to produce flavonoids and their derivatives (Pandey et al. 2016).

Characterization of genes encoding both structural enzymes and regulatory proteins, involved in the biosynthesis of flavonoids, has allowed the synthesis of flavonoids in heterologous hosts. Furthermore, some studies have also reported the natural production of several flavonoids in endophytic fungi (Ebada et al. 2016; Pan et al. 2017; Qiu et al. 2010; Seetharaman et al. 2017). In bacteria, heterologous expression and engineering studies have only been reported in the cases of *E. coli* (Fowler et al. 2009; Horinouchi 2008; Kaneko et al. 2003; Katsuyama et al. 2007; Leonard et al. 2007), *Streptomyces venezuelae* (Park et al. 2010) and *Streptomyces albus* (Marín et al. 2017); and also in yeasts such as *Saccharomyces cerevisiae* (Yan et al. 2005).

The genus *Passiflora* L. comprises about 520 species of dicotyledonous plants from the family Passifloraceae. Most species are vines and have been widely found throughout Central and South America. Some species also occur in North America, Southeast Asia and Australia (Dhawan et al. 2004). Several species have a long history of use as traditional herbal medicines. *Passiflora incarnata* is the most popular species of this genus. It contains several phytoconstituents including alkaloids, phenols, glycosyl flavonoids and cyanogenic compounds (Patel et al. 2009). Therefore, it has been widely used as analgesic, anti-spasmodic, anti-asthmatic, de-worming medicine and for the treatment of insomnia and anxiety (Miroddi et al. 2013).

The main chemical constituents of the passionflower are the C-glycosyl flavonoids (2.5%) such as vitexin, isovitexin, orientin, isoorientin, apigenin, kaempferol and quercetin, and β-carbolinic alkaloids (harman, harmin, harmalin, harmol, and harmalol) (Marchart et al. 2003). At the moment of publication of this chapter, there is only one study that aims at the isolation of endophytic microorganisms from *P. incarnata* and the prospection of flavonoids from them. This study reported production of the flavonoid chrysin (5,7-dihydroxyflavone) in three endophytic fungi *Alternaria alternata*, *Colletotrichum capsici*, and *C. taiwanense* (Seetharaman et al. 2017). Taking into account the evolutionary hypothesis that endophytic microorganisms are able to acquire host genes through horizontal gene transfer (HGT) (Taghavi et al. 2005) and, as a consequence, become able to produce plant-derived compounds (Caruso et al. 2000; Flores-Bustamante et al. 2010; Stierle et al. 1993; Yang et al. 2003; Zhao et al. 2010); endophytic bacteria emerge as an unexplored reserve for obtaining high value-added metabolites. These include flavonoids for use in agriculture, where they can be applied to improve crop production and protection. Thus, to make progress toward these ends, the present study aimed at evaluating the genetic potential for flavonoid production of *P. incarnata* endophytic bacteria, by using a PCR-based approach.

8.2 Material and Methods

8.2.1 Bacterial Strains

A total of 20 endophytic bacterial strains isolated from *Passiflora incarnata* leaves were used in this study. All endophytic strains belonged to the genus *Sphingomonas* (Table 8.1) and were provided by the Microbial Resources Division of the Research Center for Agriculture, Biology and Chemistry, University of Campinas. They were isolated from passionflower leaf tissue samples collected at the Centroflora Group's agricultural fields located in the city of Botucatu, São Paulo, Brazil. Endophytic isolates were taxonomically assigned using their partial 16S rRNA ribosomal gene sequences (Goulart, M.C., Cueva-Yesquén, L.G., Attili-Angelis, D., Fantinatti-Garboggini, F., unpublished data) in Identify server from the EzBioCloud platform (Yoon et al. 2017).

8.2.2 Design of Generic Genus-Specific Primers

Genus-specific primers were designed for this study. From the Gene portal of the National Center for Biotechnology Information (NCBI) all putative flavonol synthase gene sequences found in *Sphingomonas* were recovered. These sequences

Table 8.1 Endophytic *Sphingomonas* strains from *Passiflora incarnata* L. leaves used in this study

Strains	Candidate Identification[a]	Accession number[b]
EP37, EP41.2	*Sphingomonas yabuuchiae*	AB071955
EP42, EP76	*Sphingomonas sanguinis*	BCTY01000091
EP3, EP14, EP45, EP16	*Sphingomonas zeae*	KP999966
EP165, EP120, EP124	*Sphingomonas parapaucimobilis*	BBPI01000114
EP84	*Sphingomonas panni*	AJ575818
EP106, EP143	*Sphingomonas pseudosanguinis*	AM412238
EP61.2	*Sphingomonas leidyi*	AJ227812
EP1, EP167	*Sphingomonas melonis*	KB900605
EP30.1	*Sphingomonas paucimobilis*	BBJS01000072
EP15, EP20	*Sphingomonas sp.*	AORY01000018

[a]Taxonomic identification based on partial 16S rRNA gene identity (>99%) to closest species type strains

[b]Accession number of the type strains 16S rRNA gene sequences

Fig. 8.1 Partial scheme of multiple sequence alignment of putative *FLS* gene sequences recovered from GenBank. Red boxes showing conserved regions were selected for primer design. The multiple sequence alignment was visualized using UGENE software

were used to carry out a BLASTN analysis to increase the coverage of homologous *Sphingomonas*-specific sequences to the target gene. The *Sphingomonas*-specific sequences producing significant alignments (>80% identity) were included in this analysis. All recovered sequences were aligned online with CLUSTALW (http://www.genome.jp/tools-bin/clustalw) by searching for conserved regions (Fig. 8.1). Primers were designed in Primer3 (http://research.dfci.harvard.edu/cgi-bin/primer3/primer3_www.cgi).

Table 8.2 PCR primers used in PCR based screening of a putative *FLS* gene

Target	Gene size (bp)	Primer	Primer Sequence (5′ → 3′)	°Tm	Product size (bp)
Flavonol synthase(*FLS*)	957 bp	F-*Sphi*	CTSAAGGAATTCTGGCAYGTC	54°	~659 bp
		R-*Sphi*	CTTGATCARCCCRATCTCGAC	55°	

8.2.3 *PCR-Based Screening of Putative Flavonol Synthase Gene*

Bacterial genomic DNA was extracted using the methods described by van Soolingen et al. (1993), with modifications. PCR was performed in vials containing 25 μL total volume, 0.2 mM of dNTP, 1X reaction buffer (20 mM Tris, pH 8.4), 1.5 mM $MgCl_2$, 0.5 μM of each primer (Table 8.2), 1 U of Taq DNA polymerase, and 25 ng of template DNA. The PCR cycling protocol consisted of an initial denaturation at 95 °C for 3 min, followed by 30 cycles of 94 °C for 1 min, 55 °C for 1 min and 72 °C for 3 min, followed by a final extension of one cycle at 72 °C for 5 min. The PCR products were separated by 1% (v/w) agarose gel electrophoresis, searching for bands equivalent to 659 bp.

8.2.4 *Putative Flavonol Synthase Gene Sequencing and Phylogenetic Analysis*

PCR products showing unique bands with the expected size were purified directly using the GFX™ PCR DNA Purification kit (GE Healthcare Life Sciences, Germany). PCR products with non-specific fragments were not included in the sequencing analyses. Amplicons were sequenced by the dideoxy terminator method to confirm that the employed PCR strategy amplified the target gene. The sequencing reaction was performed with BigDye Terminator v3.1 Cycle Sequencing Kit (Applied Biosystems, Life Technologies), using the same *Sphingomonas*-specific primers. The sequencing cycling protocol consisted in an initial denaturation at 96 °C for 1 min, followed by 30 cycles of 96 °C for 15 s, 50 °C for 15 s and 60 °C for 4 min. The above reaction products were sequenced in an ABI3500XL Series (Applied Biosystems) sequencer. The forward and reverse sequences of each strain were assembled in BioEdit 7.2.6.1 software (Hall 1999). The putative flavonol synthase gene sequences were deposited in GenBank database under accession numbers MK106372 to MK106379.

A phylogeny analysis was carried out to establish the evolutionary history of the putative flavonol synthase gene among strains. A representative strain for each *Sphingomonas* species was selected for further analysis. The selected strain sequences were translated *in silico* through the EMBOSS Transeq tool (https://www.ebi.ac.uk/Tools/st/emboss_transeq/) from the EMBL-EBI platform. The rendered deduced amino acid sequences were analyzed with BLAST algorithm (https://www.uniprot.org/blast/) from UniProt, to find homologous sequences with significant alignments. Hits of at least 90% amino acid sequence identity were included in this analysis. The multiple-sequence alignment of deduced amino acid sequences and most closely related sequences was generated by the Muscle algorithm. The phylogenetic tree was inferred by Maximum Likelihood (Guindon et al. 2003) based on the Poisson substitution model with 1000 bootstrap replications. Gaps and positions with ambiguities were excluded from the phylogenetic analysis. Evolutionary analyses were conducted in MEGA6 (Molecular Evolutionary Genetic Analysis) software. The putative FLS gene sequence of *Sphingobium japonicum* UT26S was used as the outgroup.

8.3 Results

8.3.1 Detection of a Putative FLS Gene in Endophytic Sphingomonas *Strains*

Genus-specific primers (F-Sphi and R-Sphi) were used to amplify putative flavonol synthase gene sequences in endophytic *Sphingomonas* strains. The PCR based screening was successful in targeting flavonol synthase gene sequences. All PCR screens obtained unique ~659 bp fragments, except for *S. sanguinis* (EP42, EP76) and *S. panni* (EP84) species, from which additionally amplified a nonspecific fragment (~500 pb) (Fig. 8.2) and thus were not included in further analyses. Amplification of the target gene was confirmed via sequencing and BLAST analysis (against protein and nucleotide databases from NCBI). Table 8.3 shows BLASTX and BLASTN analysis of representative sequences for each *Sphingomonas* species from which the target fragment was obtained. Based on BLASTX matches to non-redundant protein database, all sequences were homologous to bacterial FLSs, except for the strain EP106 sequence, which shares 97% amino acid identity with a putative oxidoreductase from *S. paucimobilis* NBRC 13935. Only the second-best match for strain EP106 sequence showed 93% identity with an FLS. Also, the BLASTX analysis detected a protein (isopenicillin N synthase family oxygenase) identical to putative flavonol synthase in this fragment. Moreover, BLASTN analyses revealed less significant matches (some sequences only reached 86% identity) than in BLASTX, indicating that the sequences were relatively novel at the nucleotide level. The search in the nucleotide database revealed that all sequences aligned to flavonol synthase gene homologs.

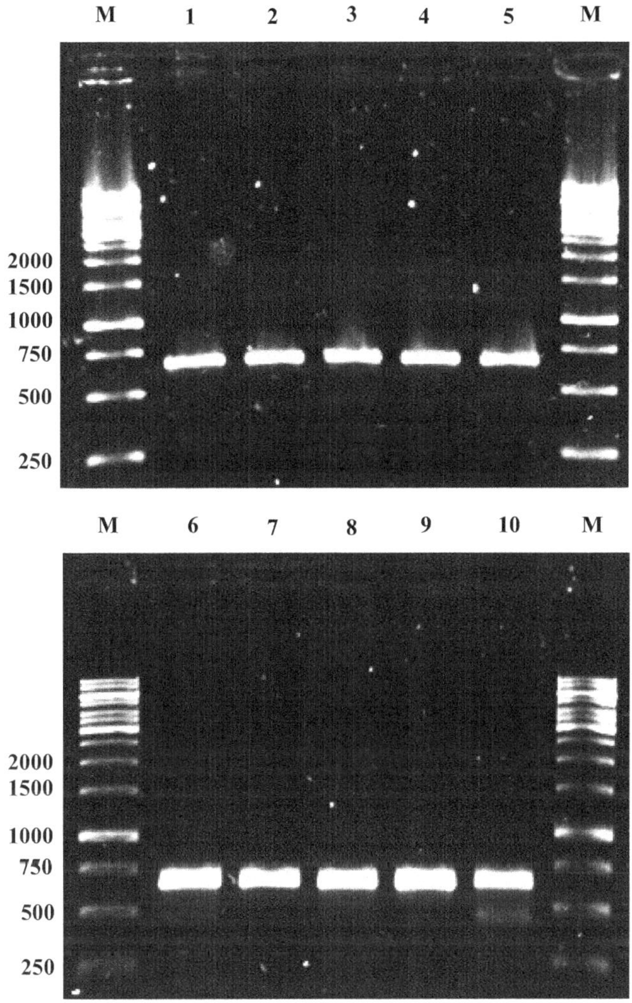

Fig. 8.2 Detection of FLS gene in endophytic *Sphingomonas* strains by PCR screening. Lane 1: *S. yabuuchiae* EP37, Lane 2: *S. zeae* EP3, Lane 3: *S. parapaucimobilis* EP120, Lane 4: *S. pseudosanguinis* EP106, Lane 5: *S. panni* EP84, Lane 6: *S. sanguinis* EP42, Lane 7: *S. leidyi* EP61.2, Lane 8: *S. melonis* EP167, Lane 9: *S. paucimobilis* EP30.1, Lane 10: *S. sanguinis* EP76, and Lane M: 1 Kb DNA Ladder

8.3.2 Phylogenetic Distribution of Putative Flavonol Synthase in Sphingomonas *Strains*

In an attempt to discover the distribution and evolution of flavonol synthase among *Sphingomonas* strains, we used the deduced amino acid sequences of flavonol synthase gene fragments, compared them in multiple sequence alignments, and reconstructed their phylogenetic relationships. From the topology of the phylogenetic

Table 8.3 Putative flavonol synthase gene identified with genus-specific primers

Test strain	BLASTX Macht	Identity (%)	Accession number[P]	BLASTN Macht	Identity (%)	Accession number[N]
EP37	Flavonol synthase, *Sphingomonas* sp. Leaf257	199/203 (98%)	KQO51024.1	Flavonol synthase, *Sphingomonas* sp. LK11	585/609 (96%)	CP013916.1
EP3	Flavonol synthase, *S. sanguinis* NS319	208/213 (98%)	KTT67831.1	Flavonol synthase, *Sphingomonas* sp. LK11	581/639 (91%)	CP013916.1
EP120	Flavonol synthase, *Sphingomonas* sp. Sph1(2015)	190/191 (99%)	OMJ31146.1	Flavonol synthase, *Sphingomonas* sp. LK11	526/573 (92%)	CP013916.1
EP106	Oxidoreductase, *S. paucimobilis NBRC 13935*	188/193 (97%)	GAN13320.1	Flavonol synthase, *Sphingomonas* sp. LK11	518/581 (89%)	CP013916.1
EP61.2	Flavonol synthase, *Sphingomonas* sp. IBVSS2	189/202 (94%)	OSZ70063.1	Flavonol synthase, *S. koreensis* ABOJV	527/602 (88%)	CP018820.1
EP167	Flavonol synthase, *Sphingomonas* sp. NIC1	183/190 (96%)	ANC88394.1	Flavonol synthase, *Sphingomonas* sp. NIC1	526/571 (92%)	CP015521.1
EP30.1	Flavonol synthase, *Sphingomonas* sp. Sph1(2015)	182/192 (95%)	OMJ31146.1	Flavonol synthase, *Sphingomonas* sp. LK11	507/578 (88%)	CP013916.1
EP15	Flavonol synthase, *S. sanguinis* NS319	202/205 (99%)	KTT67831.1	Flavonol synthase, *Sphingomonas* sp. LK11	561/615 (91%)	CP013916.1

[P]Protein database from NCBI
[N]Nucleotide database from NCBI

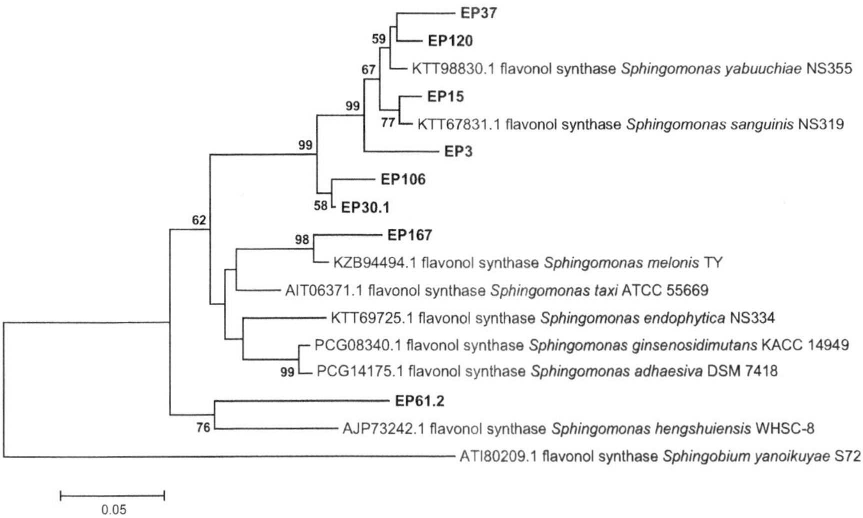

Fig. 8.3 Maximum likelihood tree of the deduced amino acid sequences of representative strains and closely related sequences available in the UniProtKB database. Bootstrap values are indicated as percentages derived from 1000 replications. Values lower than 50 are not shown. The horizontal bar represents 0.05 substitutions per amino acid site

tree (Fig. 8.3), we deduced the presence of three main clades. The first clade is constituted by strain sequences EP120, EP37, EP15, EP3, EP106, and EP30.1, suggesting that *S. sanguinis* and *S. yabuuchiae* could be more closely related to each other than to other *Sphingomonas* species. This clade represented the larger cluster on the phylogenetic tree where, specifically, the strain EP15 and *S. sanguinis* sequences formed a distinct monophyletic cluster. In the second main clade, strain EP167 and *S. melonis* YT formed a highly supported clade, which coincides with candidate identification (Table 8.1) by 16S rRNA gene sequence identity. Strain EP61.2 and *S. hengshuiensis* WHS-8 formed the third main clade, despite strain EP61.2 16S rRNA sequence also showed a higher identity with *S. leidyi* (Table 8.1).

8.4 Discussion

The importance of flavonoids for all living organisms has been well documented (Mierziak et al. 2014). These compounds are found throughout the plant kingdom. Because of their diverse chemical structure and variety resulting from the attached substituents, they play various biological roles with relevance to plant physiology and development. *Passiflora incarnata* shows the highest production of flavonoids, and their leaves are the main source of these compounds. However, extracting flavonoids from plant systems is economically expensive and may negatively affect the sustainability of agroecosystems. In addition, within the specific and intimate endophyte-plant relationship, these microorganisms have evolutionarily acquired the capability to synthesize phytochemicals naturally characteristic of the host (Tan and Zou 2001). The flavonoid synthesis by endophytic fungi *Alternaria alternata* PI-1 is one of the most outstanding evidence of the hypothesis mentioned above (Seetharaman et al. 2017).

The present study shows the first evidence of the potential of endophytic bacteria for flavonoid synthesis. This potential was confirmed by PCR screening of the flavonol synthase gene, encoding the enzyme that catalyzes the formation of flavonols from dihydroflavonols. A study on *Arabidopsis thaliana* characterized the flavonol synthase gene and revealed its central role in flavonoid metabolism (Owens et al. 2008). Enzymatic activity of flavonol synthase requires the presence of 2-oxoglutarate, ascorbate, and Fe^{2+} (Britsch 1990). These components are commonly found in bacterial metabolic pathways (Sugisawa et al. 2005), indicating a likely production of flavonols by the bacterial flavonol synthase. Nevertheless, many enzymes, intermediaries and precursors of the flavonoid biosynthesis pathway are rare or may be produced in low concentrations in bacteria (Du et al. 2010; Hwang et al. 2003). In fact, the presence of a flavonol synthase gene does not necessarily mean that these strains possess the complete flavonoid biosynthesis pathway. Thereby, it is likely that other enzymes and regulatory proteins of this biosynthetic pathway may be supported by the plant. The detected flavonol synthase gene in studied strains might express proteins with the same function of oxidoreductase, but with distinct substrates and products.

Endophytic *Sphingomonas* used in this study were selected because the flavonol synthase gene was predicted by sequence homology in the genome of some *Sphingomonas* species. Genus-specific primers were designed from multiple sequence alignment of conserved regions of this gene (~957 bp). The aligned sequences were selected by BLAST analysis and all belonged to *Sphingomonas* related species. The primers pair amplified an internal region (~650 bp) of the target gene.

The analysis of sequences generated by BLASTN and BLASTX confirmed the amplification of the target gene, although for the sequence of strain EP106, the top match in the BLASTX analysis was an oxidoreductase. The same analysis showed the existence of a close protein (isopenicillin N synthase family oxygenase), indicating that the functionality of the flavonol synthase genes in the studied strains remains still putative. The BLASTN analysis showed higher percentages of sequences variation than BLASTX analyses, though the majority of nucleotide sequences aligned exclusively with *Sphingomonas* sp. LK11. In contrast, BLASTX analysis revealed a higher diversity of matches, suggesting that reconstruction of phylogenetic relations might be more complex and representative at the amino acid level.

The phylogenetic analysis of deduced amino acid sequences showed that sequences clustered in three main groups (Fig. 8.3). The larger cluster contained a clade, with the higher bootstrap value (99%), formed by sequences of the strains EP37 (*S. yabuuchiae*), EP120 (*S. parapaucimobilis*), EP15 (*Sphingomonas* sp.), EP3 (*S. zeae*) and type strains *S. yabuuchiae* NS355 and *S. sanguinis* NS319. The detection of flavonoid biosynthesis genes in the *P. incarnata*-associated endophytes genome confirms the hypothesis that the flavonoid-producing plant species are candidate host of bacteria with potential for flavonoids production. The search for these bacteria can be performed through bioprospecting of endophytic isolates or functional screening of endophyte metagenomes. Thus, these endophytes represent an alternative biological resource for the sustainable production of plant-associated natural products, such as flavonoids.

8.5 Conclusion

Genetic screening revealed endophyte-derived flavonol synthase gene fragments with a putative role in the biosynthesis of flavonoids. This study offers further proof that endophytes show biosynthetic functions originally assigned to their host plant. It is necessary to complement this investigation with biochemical studies for a characterization of the metabolic pathway of flavonoid synthesis performed exclusively by bacteria or in synergy with plants. The endophytic bacteria producing flavonoids increase the interest for microorganisms with this lifestyle, for large-scale flavonoid production with lower reduce costs environmental impact.

Acknowledgements Financial support and scholarships provided by FAPESP (2015/02395-8), CNPq and PRONABEC are gratefully acknowledged.

References

Britsch, L. (1990). Purification and characterization of flavone synthase I, a 2-oxoglutarate-dependent desaturase. *Archives of Biochemistry and Biophysics, 282*(1), 152–160.

Caruso, M., Colombo, A., Crespi-Perellino, N., Fedeli, L., Malyszko, J., Pavesi, A., Quaroni, S., Saracchi, M., & Ventrella, G. (2000). Studies on a strain of Kitasatospora sp. paclitaxel producer. *Annals of Microbiology, 50*(2), 89–102.

Dhawan, K., Dhawan, S., & Sharma, A. (2004). Passiflora: A review update. *Journal of Ethnopharmacology, 94*(1), 1–23. https://doi.org/10.1016/j.jep.2004.02.023.

Du, H., Huang, Y., & Tang, Y. (2010). Genetic and metabolic engineering of isoflavonoid biosynthesis. *Applied Microbiology and Biotechnology, 86*(5), 1293–1312.

Ebada, S. S., Eze, P., Okoye, F. B. C., Esimone, C. O., & Proksch, P. (2016). The fungal endophyte Nigrospora oryzae produces quercetin Monoglycosides previously known only from plants. *ChemistrySelect, 1*(11), 2767–2771. https://doi.org/10.1002/slct.201600478.

Falcone Ferreyra, M. L., Rius, S. P., & Casati, P. (2012). Flavonoids: Biosynthesis, biological functions, and biotechnological applications. *Frontiers in Plant Science, 3*, 222. https://doi.org/10.3389/fpls.2012.00222.

Flores-Bustamante, Z. R., Rivera-Orduña, F. N., Martínez-Cárdenas, A., & Flores-Cotera, L. B. (2010). Microbial paclitaxel: Advances and perspectives. *The Journal of Antibiotics, 63*(8), 460–467.

Fowler, Z. L., Gikandi, W. W., & Koffas, M. A. (2009). Increased malonyl coenzyme A biosynthesis by tuning the Escherichia coli metabolic network and its application to flavanone production. *Applied and Environmental Microbiology, 75*(18), 5831–5839.

Guindon, S., Gascuel, O., & Rannala, B. (2003). A simple, fast, and accurate algorithm to estimate large phylogenies by maximum likelihood. *Systematic Biology, 52*(5), 696–704.

Hall, T. A. (1999). BioEdit: A user-friendly biological sequence alignment editor and analysis program for windows 95/98/NT. *Nucleic Acids Symposium Series, 41*, 95–98.

Horinouchi, S. (2008). Combinatorial biosynthesis of non-bacterial and unnatural flavonoids, stilbenoids and curcuminoids by microorganisms. *The Journal of Antibiotics, 61*(12), 709–728.

Hwang, E. I., Kaneko, M., Ohnishi, Y., & Horinouchi, S. (2003). Production of plant-specific flavanones by Escherichia coli containing an artificial gene cluster. *Applied and Environmental Microbiology, 69*(5), 2699–2706.

Kaneko, M., Hwang, E. I., Ohnishi, Y., & Horinouchi, S. (2003). Heterologous production of flavanones in Escherichia coli: Potential for combinatorial biosynthesis of flavonoids in bacteria. *Journal of Industrial Microbiology and Biotechnology, 30*(8), 456–461. https://doi.org/10.1007/s10295-003-0061-1.

Katsuyama, Y., Funa, N., Miyahisa, I., & Horinouchi, S. (2007). Synthesis of unnatural flavonoids and stilbenes by exploiting the plant biosynthetic pathway in Escherichia coli. *Chemistry & Biology, 14*(6), 613–621.

Kumar, S., & Pandey, A. K. (2013). Chemistry and biological activities of flavonoids: An overview. *The Scientific World Journal, 2013*, 1–16. https://doi.org/10.1155/2013/162750.

Leonard, E., Lim, K. H., Saw, P. N., & Koffas, M. A. G. (2007). Engineering central metabolic pathways for high-level flavonoid production in Escherichia coli. *Applied and Environmental Microbiology, 73*(12), 3877–3886. https://doi.org/10.1128/AEM.00200-07.

Marchart, E., Krenn, L., & Kopp, B. (2003). Quantification of the flavonoid glycosides in Passiflora incarnata by capillary electrophoresis. *Planta Medica, 69*(05), 452–456.

Marín, L., Gutiérrez-del-Río, I., Yagüe, P., Manteca, Á., Villar, C. J., & Lombó, F. (2017). De novo biosynthesis of Apigenin, Luteolin, and Eriodictyol in the Actinomycete Streptomyces albus and production improvement by feeding and spore conditioning. *Frontiers in Microbiology, 8*(921). https://doi.org/10.3389/fmicb.2017.00921.

Mierziak, J., Kostyn, K., & Kulma, A. (2014). Flavonoids as important molecules of plant interactions with the environment. *Molecules, 19*(10), 16240–16265.

Miroddi, M., Calapai, G., Navarra, M., Minciullo, P. L., & Gangemi, S. (2013). Passiflora incarnata L.: Ethnopharmacology, clinical application, safety and evaluation of clinical trials. *Journal of Ethnopharmacology, 150*(3), 791–804. https://doi.org/10.1016/j.jep.2013.09.047.

Owens, D. K., Alerding, A. B., Crosby, K. C., Bandara, A. B., Westwood, J. H., & Winkel, B. S. (2008). Functional analysis of a predicted flavonol synthase gene family in Arabidopsis. *Plant Physiology, 147*(3), 1046–1061.

Pan, F., Su, T.-J., Cai, S.-M., & Wu, W. (2017). Fungal endophyte-derived Fritillaria unibracteata var. wabuensis: Diversity, antioxidant capacities in vitro and relations to phenolic, flavonoid or saponin compounds. *Scientific Reports, 7*, 42008. https://doi.org/10.1038/srep42008.

Pandey, R. P., Parajuli, P., Koffas, M. A. G., & Sohng, J. K. (2016). Microbial production of natural and non-natural flavonoids: Pathway engineering, directed evolution and systems/synthetic biology. *Biotechnology Advances, 34*(5), 634–662. https://doi.org/10.1016/j.biotechadv.2016.02.012.

Park, S. R., Paik, J. H., Ahn, M. S., Park, J. W., & Yoon, Y. J. (2010). Biosynthesis of plant-specific flavones and flavonols in Streptomyces venezuelae. *Journal of Microbiology and Biotechnology, 20*(9), 1295–1299.

Patel, S., Verma, N., & Gauthaman, K. (2009). *Passiflora incarnata Linn*: A review on morphology, phytochemistry and pharmacological aspects. *Pharmacognosy Reviews, 3*(5), 175–181.

Qiu, M., Xie, R.-s., Shi, Y., Zhang, H., & Chen, H.-m. (2010). Isolation and identification of two flavonoid-producing endophytic fungi from Ginkgo biloba L. *Annals of Microbiology, 60*(1), 143–150.

Samanta, A., Das, G., & Das, S. (2011). Roles of flavonoids in plants. *International Journal of Pharmaceutical Science and Technology, 6*, 12–35.

Seetharaman, P., Gnanasekar, S., Chandrasekaran, R., Chandrakasan, G., Kadarkarai, M., & Sivaperumal, S. (2017). Isolation and characterization of anticancer flavone chrysin (5,7-dihydroxy flavone)-producing endophytic fungi from Passiflora incarnata L. leaves. *Annals of Microbiology, 67*(4), 321–331. https://doi.org/10.1007/s13213-017-1263-5.
Stierle, A., Strobel, G., & Stierle, D. (1993). Taxol and taxane production by Taxomyces andreanae, an endophytic fungus of Pacific yew. *Science, 260*(5105), 214–216.
Sugisawa, T., Miyazaki, T., & Hoshino, T. (2005). Microbial production of L-ascorbic acid from D-sorbitol, L-sorbose, L-gulose, and L-sorbosone by Ketogulonicigenium vulgare DSM 4025. *Bioscience, Biotechnology, and Biochemistry, 69*(3), 659–662. https://doi.org/10.1271/bbb.69.659.
Taghavi, S., Barac, T., Greenberg, B., Borremans, B., Vangronsveld, J., & van der Lelie, D. (2005). Horizontal gene transfer to endogenous endophytic bacteria from poplar improves phytoremediation of toluene. *Applied and Environmental Microbiology, 71*(12), 8500–8505.
Tan, R. X., & Zou, W. X. (2001). Endophytes: A rich source of functional metabolites. *Natural Product Reports, 18*(4), 448–459.
Trantas, E. A., Koffas, M. A. G., Xu, P., & Ververidis, F. (2015). When plants produce not enough or at all: Metabolic engineering of flavonoids in microbial hosts. *Frontiers in Plant Science, 6*(January), 1–16. https://doi.org/10.3389/fpls.2015.00007.
Van Soolingen, D., De Haas, P. E., Hermans, P. W., Groenen, P. M., & Van Embden, J. D. (1993). Comparison of various repetitive DNA elements as genetic markers for strain differentiation and epidemiology of Mycobacterium tuberculosis. *Journal of Clinical Microbiology, 31*(8), 1987–1995.
Ververidis, F., Trantas, E., Douglas, C., Vollmer, G., Kretzschmar, G., & Panopoulos, N. (2007). Biotechnology of flavonoids and other phenylpropanoid-derived natural products. Part I: Chemical diversity, impacts on plant biology and human health. *Biotechnology Journal, 2*(10), 1214–1234. https://doi.org/10.1002/biot.200700084.
Weston, L. A., & Mathesius, U. (2013). Flavonoids: Their structure, biosynthesis and role in the rhizosphere, including allelopathy. *Journal of Chemical Ecology, 39*(2), 283–297. https://doi.org/10.1007/s10886-013-0248-5.
Yan, Y., Kohli, A., & Koffas, M. A. G. (2005). Biosynthesis of natural flavanones in Saccharomyces cerevisiae. *Applied and Environmental Microbiology, 71*(9), 5610–5613. https://doi.org/10.1128/AEM.71.9.5610-5613.2005.
Yang, X., Guo, S., Zhang, L., & Shao, H. (2003). Select of producing podophyllotoxin endophytic fungi from podophyllin plant. *Natural Product Research and Development, 15*(5), 419–422.
Yoon, et al. (2017). Introducing EzBioCloud: a taxonomically united database of 16S rRNA gene sequences and whole-genome assemblies. *International Journal of Systematic and Evolutionary Microbiology, 67*(5), 1613–1617.
Zhao, J., Zhou, L., Wang, J., Shan, T., Zhong, L., Liu, X., & Gao, X. (2010). Endophytic fungi for producing bioactive compounds originally from their host plants. *Current Research, Technology, and Education Topics in Applied Microbiology and Microbial Biotechnology, 1*, 567–576.

Chapter 9
Isolation and Characterization of Antifungal Secondary Metabolites Produced by Rhizobacteria from Common Bean

Miriam Memenza-Zegarra and Doris Zúñiga-Dávila

Abstract Conventional management of crops implies excessive use of fungicides. Rhizobacteria with antagonistic effects against phytopathogens constitute an alternative to chemical control. The aim of this work was to isolate and characterize secondary metabolites with antifungal activity produced by rhizobacteria. Several bacterial strains isolated from the rhizosphere of common bean plants were evaluated for their ability to suppress the growth of fungal phytopathogens due to the production of volatile and non-volatile compounds. Among them, *Bacillus* sp. B02 was selected as an antagonistic microorganism with capacity to produce non-volatile metabolites. Antifungal compounds were produced at 72 h in mineral broth supplemented with starch (20 g l^{-1}) and yeast extract (5 g l^{-1}). The metabolites were isolated from cell free supernatants by precipitation after they were acidified to pH 2 and characterized by their polarity and by thin layer chromatography (TLC) analysis. Lipophilic compounds were extracted using a chloroform:methanol (2:1, v/v) system. In addition, polar antifungal compounds were presented in butanol, acetone and methanol crude extracts. The extracts inhibited the mycelial growth of *R. solani*. TLC revealed the presence of lipid compounds when iodine vapor was used as reagent. This result showed the amphiphilic character of the antifungal compounds produced by *Bacillus* sp. B02 which probably correspond to lipopeptides.

Keywords Rhizobacteria · Antifungal secondary metabolites · TLC · Solvent extraction

---

M. Memenza-Zegarra · D. Zúñiga-Dávila (✉)
Laboratorio de Ecología Microbiana y Biotecnología, Departamento de Biología-Facultad de Ciencias, Universidad Nacional Agraria La Molina, Lima, Peru
e-mail: dzuniga@lamolina.edu.pe

D. Zúñiga-Dávila et al. (eds.), *Microbial Probiotics for Agricultural Systems*, Sustainability in Plant and Crop Protection,
https://doi.org/10.1007/978-3-030-17597-9_9

9.1 Introduction

The common bean (*Phaseolus vulgaris*) represents an important food source due to its high nutritional content with a direct incidence in food security. However, this crop develops within a framework of small-scale family farming with low levels of yields. This is mainly due to the effects of climate changes that favor the development of fungal diseases reducing crop production up to 70% (Velasquez et al. 2018). Chemical control is the most common way to prevent the occurrence of such diseases. However, the use of chemicals is considered harmful because of environmental impacts that include adverse health effects, degradation of the natural environment and development of pesticide/fungicide resistance.

Biological control has been considered a viable alternative to manage plant diseases (Heydari and Pessarakli 2010). Microbial antagonistic strains are able to produce both volatile and non-volatile compounds, which can exhibit strong inhibitory activity against plant pathogens. Volatile compounds (VOCs) produced by bacteria could inhibit growth and spore germination of pathogenic fungi (Chaves-Lopez et al. 2015). Likewise, there is now unequivocal evidence that antibiotics play a key role in the suppression of various soilborne plant pathogens (generally fungi) by antagonistic microorganisms (Compant et al. 2015).

A number of *Bacillus* species can produce various kinds of volatile and non-volatile compounds with strong inhibitory activity against fungal plant pathogens (Arrebola et al. 2010; Hossain et al. 2016; Kim et al. 2015; Nam et al. 2016). Among various diffusible compounds, cyclolipopeptides produced by strains of *Bacillus* can protect cells from attacks by other microorganisms (Shafi et al. 2017). The cyclic lipopeptides (CLPs) belong to the iturin (such as bacillomycin D/F/L/Lc, iturin A/C, and mycosubtilin), fengycin (fengycin A/B, and plipastatin A/B), and surfactin (halobacillin, pumilacidin and surfactin) classes. These all share a common structure consisting of a lipid tail linked to a short cyclic peptide (Zeriouh et al. 2011; Tanaka et al. 2015; Gong et al. 2015).

Lim et al. (2017) reported diffusible antifungal compounds produced by *Bacillus velezensis* G341 that included bacillomycin L with antagonistic activity against rice blast, rice sheath blight and red pepper anthracnose. Similarly, fengycin is able to reduce the development of rice sheath blight and tomato gray mold. Another study reported that *B. amyloliquefaciens* PGPBacCA1 produced different homologues of the lipopeptides surfactin, iturin and fengycin in presence of *Sclerotinia sclerotiorum* and *Fusarium solani*. These compounds were identified as the main responsible for the antagonistic effect against the development of chlamydospores and sclerotia of *Fusarium* and *Sclerotinia*, respectively. The authors concluded that strain PGPBacCA1can be applied to bean seeds as a potential bioprotection agent (Torres et al. 2017). However, it is clear that environmental factors and plant determinants may influence the production of antibiotics in general, and of CLPs in particular, in plant beneficial *Bacillus* spp. (Chowdhury et al. 2015; Debois et al. 2015).

Several abiotic factors such as oxygen, temperature, specific carbon and nitrogen sources, and microelements have been shown to influence antibiotic production by bacterial biocontrol agents. Among biotic factors that may play a determinant role in antibiotic production are the host plant, the pathogen, the indigenous microflora, and the cell density of the producing strain. The aim of this work was to screen rhizobacteria isolated from common bean for antifungal compound production, and to proceed to an initial characterization of the antagonistic metabolites produced by one selected strain.

9.2 Materials and Methods

9.2.1 Isolation of Rhizobacteria

Rhizospheric soil sample was collected from common bean cv. Centenario plants grown at the Ica valley, Peru (14°3′50″S, 75°43′45″W) at 143 m above sea level. The soil presented a sandy-loam texture and was characterized as moderately alkaline and poor because of its high content of calcium carbonate, low percentage of organic matter, low amounts of phosphorous and medium potassium levels. Ten grams of rhizospheric soil were suspended in 90 ml of 0.85% sterile saline solution. The suspension was heated at 80 °C on a water bath for 30 min. Then, samples were serially diluted and 1 ml of each dilution was incorporated into TGE medium (composition per liter: triptone 5 g, glucose 10 g, beef extract 3 g, agar 15 g). The incubation was at 28 °C for 24–48 h (Calvo et al. 2010). Bacterial colonies with different morphological characteristics were selected, Gram stained and subjected to microscopic observations. Pure colonies were preserved in 25% glycerol at −80 °C.

9.2.2 Screening for Antifungal Compound Production

Antagonistic activity was determined by the production of volatile and non-volatile antifungal compounds against *Rhizoctonia solani*, *Sclerotinia sclerotiorum* and *Fusarium oxysporum*. Production of volatile organic compounds was tested by the sealed plate method (Fernando et al. 2005). Bacteria previously grown in Nutrient Broth were spread on the bottom dish of a Petri plate containing Tryptic Soy Agar and then immediately inverted over the fungus plate and sealed with parafilm. The plate with fungi contained a 5 mm agar plug with mycelium placed in the center of a fresh Potato Dextrose Agar (PDA plate. Non-volatile antifungal compounds were tested by using the agar well diffusion method (Venkadesan and Sumathi 2015). Aliquots (150 μl) of cell-free supernatant were placed in 9 mm diam. Wells on PDA plates. A 1-cm agar disc from a fresh culture of the fungus was placed at the center of the plate. Control plates without bacteria inoculation were also prepared. Fungal growth inhibition was quantified using the percentage inhibition (PI) formula: $((R-r) \times R^{-1} \times 100)$, where r is the radius of the fungal colony that grew towards the bacterial colony and *R* is the maximum radius of the fungal colony away from the bacterial colony (the maximum growth achieved by the fungus in the Petri dish) (Calvo et al. 2010). The bacterial strain with highest non-volatile antifungal compound production was chosen for further assays in the present study.

9.2.3 16S rRNA Gene Analysis

Total DNA was extracted from liquid cultures with the Invitrogen kit following the manufacturer's instructions. 16S rRNA gene was PCR amplified using primers fD1 and rD1 (Weisburg et al. 1991) and Sanger sequenced at Macrogen (Seoul, South Korea). Analysis was performed by comparing the sequences produced with those of related type strains retrieved for the GenBank database. Phylogenetic analysis was performed by the neighbor joining (NJ) method with genetic distances computed with the Kimura's two-parameter model using Mega7 (Tamura et al. 2007).

9.2.4 Production of Antifungal Secondary Metabolites

9.2.4.1 Effect of the Incubation Periods

The bacterial strain was grown in a mineral medium (MM) with the following composition (in g l^{-1}): glucose 10, yeast extract 5, KH_2PO_4 1, $MgSO_4$ $7H_2O$ 0.5, NaCl 0.01 and $FeSO_4$ $7H_2O$ 0.01. Incubation was at 28 °C for 192 h on a rotary shaker at 150 rpm. At every 24 h interval, biomass concentration and the presence of non-volatile antifungal compounds were measured. Biomass concentration was estimated by optical density at 600 nm. Production of antifungal compounds was determinated by percentage of growth inhibition of *R. solani*, as previously described (Calvo et al. 2010).

9.2.4.2 Effect of Carbon and Nitrogen Sources

In order to increase the production of antifungal secondary metabolites, different carbon and nitrogen sources were evaluated. Individually, different carbon sources (at 20 g l^{-1}) were added to the MM: glucose, sucrose, mannitol, glycerol and starch. Similarly, different nitrogen sources (at 5 g l^{-1}) were added: yeast extract, meat extract, peptone and glutamic acid. The 50 ml flasks were incubated at 28 °C on a rotary shaker at 150 rpm for 72 h. Biomass concentration was determinated after drying at 60 °C until a constant weight was obtained and expressed as g l^{-1}. Production of metabolites with antifungal activity was determined by the size (mm) of the halo of mycelial growth inhibition of *R. solani* through the agar well diffusion method and by the dry weight of antifungal metabolites, expressed as gl^{-1} of fermentation broth. In the last case, the fermentation broth was centrifuged at 9000 rpm for 20 min at 4 °C. The supernatant was acidified to pH 2 using concentrated HCl. The pellet was collected by centrifugation at 9000 rpm for 10 min at 4 °C and then oven dried at 100 °C for 30 min (Zhang et al. 2016). Production medium was optimized using the carbon and nitrogen sources which allowed the maximum production of antifungal compounds.

9.2.5 Isolation of Antifungal Secondary Compounds

The bacterial isolate was grown in 2 l of mineral medium in a stirred tank bioreactor with a capacity of 3 l. The mineral medium was supplement with starch (20 g l^{-1}) and yeast extract (5 g l^{-1}). Incubation was performed at 300 rpm, 0.5 vvm and 28 °C for 3 days. At the end of the incubation time, the cell-free supernatant was obtained by centrifugation at 9000 rpm for 20 min at 4 °C. The supernatant was acidified to pH 2 using concentrated HCl and subsequently kept at 6 °C overnight, to facilitate proteins or peptides precipitation. The pellet was resuspended in 0.02 mol^{-1} PBS buffer solution and the antagonistic activity was determined through the agar well diffusion method.

9.2.6 Characterization of Antifungal Secondary Metabolites

The bioactive compounds were characterized based on their polarity. The pellet was liquid-liquid partitioned using different non-polar and polar solvents such as hexane, petroleum ether, ethyl acetate, butanol, acetone, methanol and chloroform: methanol (2:1 v/v). The pellet was extracted three times with an equal volume of solvent. The solvent extraction was performed in a separatory funnel (at room temperature). The mixture was vigorously shaken for several min and allowed to rest for 10 min until phase separation. The organic phase was dried in a rotary evaporator at 60 °C and resuspended in 0.02 mol^{-1} PBS buffer solution. Solvent ability to extract the antifungal metabolites was determinated by the agar well diffusion method. In order to preliminary determinate the type of antifungal metabolites, a portion of active crude extract, previously resuspended in methanol, was separated by thin layer chromatography (TLC) on silica gel plates, using the system chloroform: methanol: water (65:15:1, v/v) as running solvent agent (Aparna et al. 2012). The spots were visualized after placing the plates in a closed jar saturated with iodine vapor.

9.2.7 Statistical Analysis

Treatments were organized in a completely randomized design with 2 repetitions. The data were processed using the one-way analysis of variance (ANOVA) and the Duncan's multiple range test. The level of statistical significance was set at 0.05.

9.3 Results

9.3.1 Isolation and Screening for Antagonistic Activity

Thirteen heat resistant strains were isolated from the rhizosphere of common bean. The strains were characterized as G+ bacilli by microscopic observations. In order to evaluate the involvement of volatile and non-volatile compounds in the antifungal activity against to *S. sclerotiorum, R. solani* and *F. oxysporum*, all 13 rhizosphere bacteria isolated were tested by the sealed plate method and diffusion agar technique. Evaluation of the involvement of volatile compounds in the antagonist activity revealed that 8 isolates (42%) inhibited mycelial growth of *R. solani*, 10 (53%) reduced the growth of *S. sclerotiorum* and 10 (53%) inhibited the mycelial growth of *F. oxysporum*. Besides, grown inhibition, mycelia morphological changes were observed such as mycelial translucency and no pigment formation. On the other hand, strains Bac1.4, B02 and Bac2.12 were able to suppress the mycelial growth of all three fungal phytopathogens tested due to the presence of non-volatile antimicrobial compounds in their cell-free supernatants (Table 9.1). The B02 strain showed the highest antifungal activity, compared with all the bacterial strains tested.

Table 9.1 Antagonistic activity by volatile and non-volatile antifungal compounds

	Percentage inhibition (%) by					
	Volatile compounds against			Non-volatile compounds against		
Strains	*R. solani*	*S. sclerotiorum*	*F. oxysporum*	*R. solani*	*S. sclerotiorum*	*F. oxysporum*
Bac1.4	0	0	0	65 ± 5.3	61 ± 3.5	62 ± 3.5
B02	0	0	0	76 ± 1.7	74 ± 1.1	70 ± 2.1
Bac2.9	14 ± 5.5	39 ± 5.7	13 ± 3.5	0	0	0
Bac2.12	0	0	0	70 ± 1.1	68 ± 1.5	66 ± 2.1
Bac2.13	45 ± 1.5	52 ± 2.3	13 ± 0.5	0	0	0
Bac2.14	35 ± 1.0	17 ± 1.1	0	0	0	0
Bac2.15	35 ± 1.1	45 ± 1.1	12 ± 1.1	0	0	0
Bac2.16	48 ± 1.5	0	25 ± 1.1	0	0	0
Bac2.21	0	91 ± 3.5	17 ± 2.1	0	0	0
Bac2.22	0	54 ± 2.1	14 ± 1.4	0	0	0
Bac2.23	16 ± 3.6	59 ± 1.4	12 ± 2.1	0	0	0
Bac3.3	36 ± 3.5	67 ± 2.1	16 ± 3.5	0	0	0
Bac3.4	5 ± 0	51 ± 1.4	69 ± 5.6	0	0	0

Values are the mean ± SE (n = 2)

We decided to continue the study with non-volatile compounds as they are easier to produce and formulate as a commercial product. Strain B02 was selected for further experiments because it showed not only the highest antagonistic activity but also maintained this ability upon successive subculturing (data not shown). Additionally, all the following tests for antagonism where performed with *R. solani* because it is, as common bean phytopathogen, more important than *F. oxysporum* and *S. sclerotiorum*.

9.3.2 Phylogenetic Analysis of Strain B02

The phylogenetic analysis presented in Fig. 9.1 indicated that strain B02 belong to the *Bacillus* genus and grouped within the *subtilis* clade. It was closely related with the type strains of *B. halotolerans*, *B. mojavensis*, *B. axarquiensis* and *B. malacitensis*. However, the 16S rRNA gene sequence did not show enough phylogenetic resolution to discriminate between these species. Further analyses will be needd to determine the species affiliation of strain B02 which is now referred to as *Bacillus* sp.

9.3.3 Production of Antifungal Secondary Metabolites

Culture conditions are critical for the production of antifungal secondary metabolites by antagonistic bacteria. Manipulating these conditions can increase the biosynthesis of secondary metabolites. The effect of incubation period on antifungal

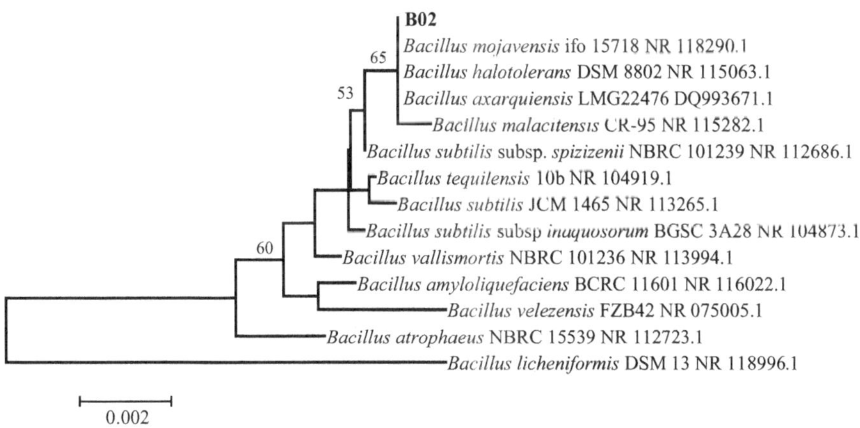

Fig. 9.1 Neighbor Joining phylogenetic tree of 16S RNA gene sequences of the selected antagonist B02 strain (shown in bold). Only bootstrap values higher than 50% are shown (1000 pseudoreplicates)

activity of *Bacillus* sp. B02 was recorded over a period of 144 h. The isolate activity was observed from the second day of incubation and reached maximum inhibition at 72 h corresponding to the stationary growth phase (Fig. 9.2). Afterwards, different sources of carbon and nitrogen were tested to increase the production of antifungal compounds by *Bacillus* sp. B02. Production of biomass was monitored as well as of antifungal compounds by measuring the size of the inhibition zones and the dry weight of antifungal metabolites after 72 h of incubation (Fig. 9.3). Maximum activity (24.8 mm inhibition halo and 0.666 g l^{-1} dry weight of antifungal metabolites) and growth (0.1312 g l^{-1}) were observed when the strain was cultivated in mineral broth supplemented with starch (20 g l^{-1}) (Fig. 9.3a) with significant differences compared with the other carbon sources. Despite the ability to grow well using other carbon sources, isolate B02 did not show the same production of metabolites with antifungal activity against *R. solani* in all of them. For example, B02 showed good growth using sucrose (0.0504 g l^{-1}), however it presented low values in the inhibition halo (21.1 mm) and dry weight of antifungal metabolites (0.117 g l^{-1}).

When different nitrogen sources were tested, yeast extract was the best source for growth (0.0746 g l^{-1}) with maximum production of antifungal metabolites (21.8 mm inhibition zone and 0.471 g l^{-1} dry weight of antifungal metabolites) (Fig. 9.3b) with significant differences compared to other nitrogen sources. The control medium only showed 18 mm inhibition halo and 0.288 g l^{-1} of crude antifungal metabolite yield. Given these results, a production medium was optimized using starch (20 g l^{-1}) and yeast extract (5 g l^{-1}). The optimized medium highly increased the activity (26.1 mm inhibition halo and 1.76 g l^{-1} dry weight of antifungal metabolites) and growth (1.44 g l^{-1}) with significant differences compared to the other treatments.

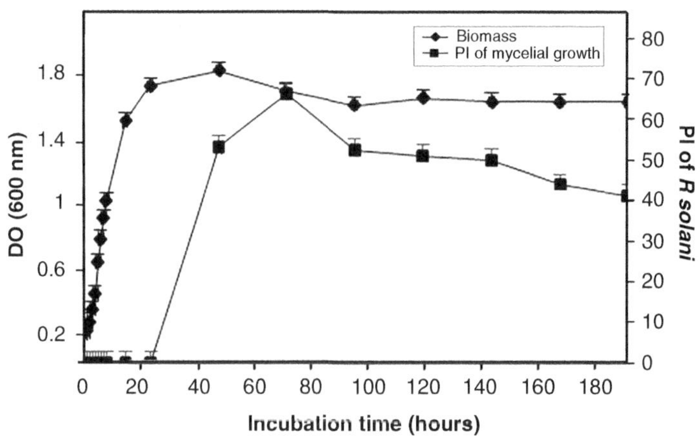

Fig. 9.2 Effect of incubation period on percentage of growth inhibition against *R. solani* by *Bacillus* sp. B02. PI: Percentage inhibition. Each value represents the mean ± SE (n = 2)

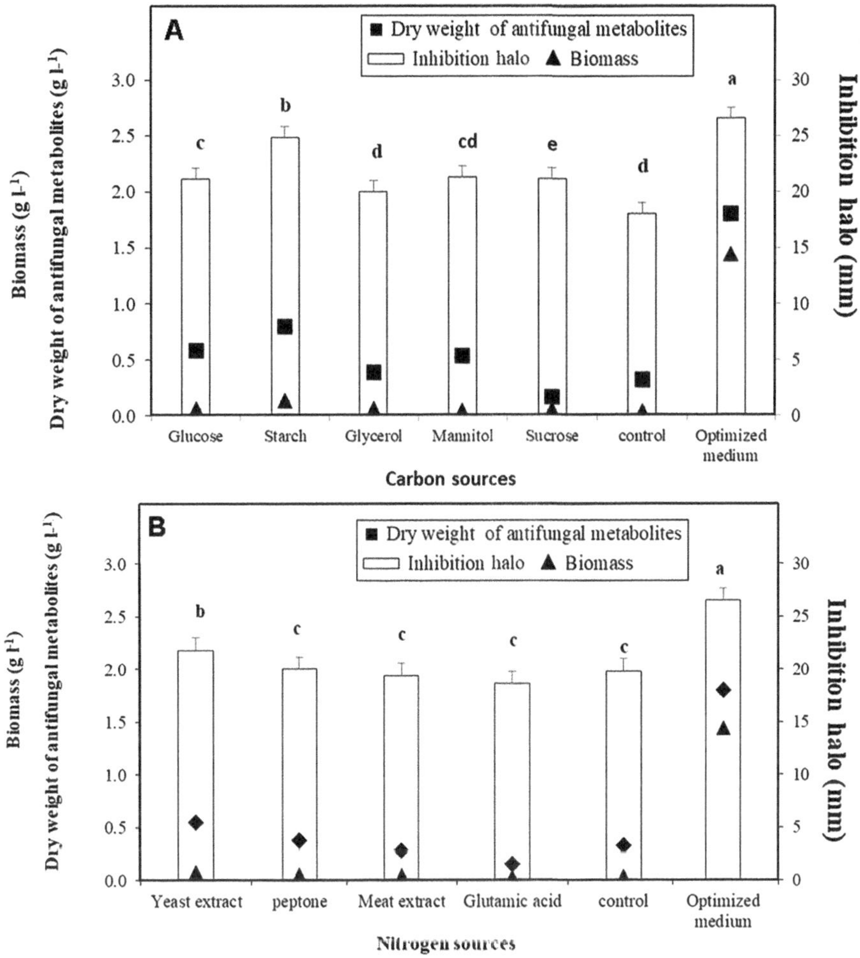

Fig. 9.3 Effect of carbon (**a**) and nitrogen sources (**b**) on growth and antifungal compounds production by *Bacillus* sp. B02 after 72 h of incubation. Bars represent the mean ± SE values. Different letters indicate statistically significant differences. (Duncan's multiple range test, $p \leq 0.05$)

9.3.4 Characterization of Antifungal Secondary Metabolites

The pellet isolated from cell free supernatants showed antifungal activity against *R. solani*. Liquid-liquid partitioning using different solvents determined that chloroform, butanol, acetone, methanol and chloroform: methanol (2:1, v/v) allowed extraction of antifungal metabolites present in the fermentation broth. Butanol and chloroform: methanol (2:1, v/v) extracts showed the major inhibition of mycelial growth of *R. solani*. On the other hand, no antagonistic activity was observed when hexane and petroleum ether extracts were tested (Fig. 9.4a, b). The results revealed

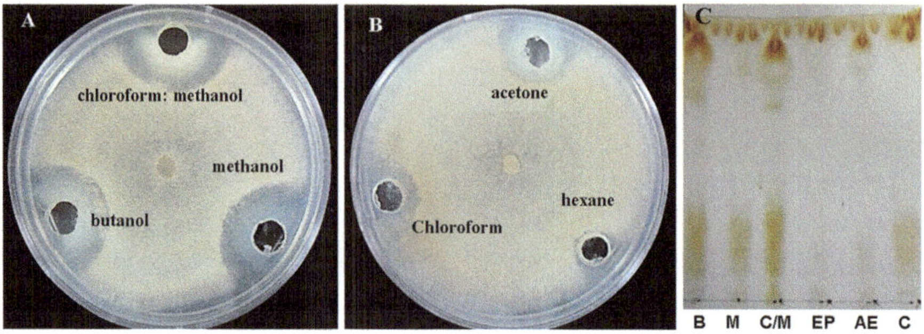

Fig. 9.4 Effect of antifungal metabolites extracted using non-polar and polar solvents (**a** and **b**). (**c**) Thin layer chromatography analysis of the solvent extraction (*B* butanol, *M* methanol, *C/M* chloroform: methanol (2:1, v/v), *EP* petroleum acetate, *AE* ethyl acetate, *C* chloroform)

the presence of lipophilic and polar antifungal compounds. For butanol and chloroform: methanol (2:1, v/v) extracts, TLC revealed 4 dark brown spots when iodine vapor was used as visualization reagent, while with methanol and chloroform extracts only 3 spots were observed corresponding to polar compounds (Fig. 9.4c). TLC revealed the presence of lipid components in the chemical structure of the extracted antifungal compounds.

9.4 Discussion

Bacteria isolated from the rhizosphere of common bean plants were screened, based on volatile and non-volatile compound production, for antagonistic activity against the fungal pathogens *S. sclerotiorum*, *R. solani* and *F. oxysporum*. Some of the tested bacteria were previously shown to reduce disease incidence cause by *S. sclerotiorum* in field experiments (Memenza et al. 2016). Based on its highest antimicrobial activity by non-volatile compounds, strain B02 was chosen for further studies. Phylogenetic analysis showed that B02 is a *Bacillus* sp. strain closely related to *B. halotolerans, B. mojavensis, B. axarquiensis* and *B. malacitensis.* Characterization of antifungal secondary metabolites produced by *Bacillus* sp. B02 was determined using conventional screening methods including acid precipitation, solvent extraction and TLC. Maximum production was obtained at 72 h of incubation time using individually 20 g l^{-1} of starch (0.666 g l^{-1} of dry weight of antifungal metabolites) and 5 g l^{-1} of yeast extract (0.0746 g l^{-1} of dry weight of antifungal metabolites) as carbon and nitrogen sources, respectively. The mineral medium supplemented with starch and yeast extract supported the highest production of antifungal metabolites (1.76 g l^{-1}) and growth (1.44 g l^{-1}). Pellets obtained from cell-free supernatants, after sample acidification to pH 2, also inhibited mycelial growth of *R. solani*, indicating that the antifungal compounds likely corresponded to a protein or peptide.

In this study, solvent extraction revealed the presence of polar compounds soluble in solvents such as butanol, acetone and methanol, as well as lipophilic components extracted with a chloroform: methanol (2:1, v/v) system. Characterization of the crude extract using TLC detected the presence of lipid components. The protocol used for extraction showed the amphiphilic character of the antifungal compounds produced by strain B02, a property also displayed by lipopeptides.

Lipopeptides are often extracted from culture broth by classical methods including acid precipitation and extraction with organic solvents as used here. Lipopeptides are peptide antibiotics studied for their potent antagonistic activities against various phytopathogens (Ongena and Jacques 2008). They exhibit highly rigid, hydrophobic and/or cyclic structures with unusual constituents like D-amino acids and are generally resistant to hydrolysis by peptidases and proteases (Cawoy et al. 2015). Generally, bactericidal activity of lipopeptides increases with the addition of a lipid moiety of appropriate length (typically C10–C12). Moreover, those containing higher number of carbon atoms, such as 14 or 16, in their lipid tails exhibit enhanced antifungal activity in addition to antibacterial activity (Mandal et al. 2013).

Lipopeptide antibiotics identified so far have been divided into three main groups according to their structure: surfactin, iturin and fengycin groups (Tsuge et al. 2005). All these agents occur as families of closely related isoforms which differ in length and branching of the fatty acid side chains, and in amino-acid substitutions in the peptide rings (Vater et al. 2002). These are non-ribosomally generated peptides whose synthesis is directed by large multienzyme complexes. These have a modular structural organization and are thought to orderly link the amino-acid residues of the final peptide (Challis et al. 2000).

The experiments carried out showed an antimicrobial effect of cell free supernatant and indicated the presence of antimicrobial putative lipopeptides in liquid cultures. The assays carried out showed an antimicrobial effect of the cell free supernatants supporting the contention of that antibiosis is likely the main mechanism by which *Bacillus* sp. B02 strain may affect the development and insurgence of fungal agents responsible of plant diseases.

9.5 Conclusion

Bacillus sp. B02 isolated from common bean plants produced non-volatile antifungal compounds able to suppress the mycelial growth of *R. solani*. The antifungal secondary metabolites, produced at 72 h in mineral broth supplemented with starch (20 g l^{-1}) and yeast extract (5 g l^{-1}), were isolated from the cell-free supernatant after acidification to pH 2 indicating the presence of protein or peptides. In the characterization, solvent extraction revealed the presences of polar antifungal compounds. In addition, TLC revealed the presence of lipid compounds. This result shows the amphiphilic character of the antifungal compounds produced by *Bacillus* sp. B02, suggesting a likely correspondence to lipopeptides.

Acknowledgements This research was supported by Programa Nacional de Innovación para la Competitividad y Productividad (Innovate Perú) via grant contract 158-PNCIP-PIAP-2015.

References

Aparna, A., Srinikethan, G., & Hegde, S. (2012). Isolation, screening and production of biosurfactant by *Bacillus clausii* 5B. *Research in Biotechnology, 3*, 49–56.

Arrebola, E., Sivakumar, D., & Korsten, L. (2010). Effect of volatile compounds produced by *Bacillus* strains on postharvest decay in citrus. *Biological Control, 53*, 122–128.

Calvo, P., Ormeño, E., Martínez, E., & Zúñiga, D. (2010). Characterization of *Bacillus* isolates of potato rhizosphere from andean soils of Peru and their potential PGPR characteristics. *Brazilian Journal of Microbiology, 41*, 899–906.

Cawoy, H., Debois, D., Franzil, L., De Pauw, E., Thonart, P., & Ongena, M. (2015). Lipopeptides as main ingredients for inhibition of fungal phytopathogens by *Bacillus subtilis/amyloliquefaciens*. *Microbial Biotechnology, 8*, 281–295.

Challis, G., Ravel, J., & Townsend, C. (2000). Predictive structure based model of amino acid recognition by nonribosomal peptide synthetase adenylation domains. *Chemistry & Biology, 7*, 211–224.

Chaves-Lopez, C., Serio, A., Gianotti, A., Sacchetti, G., Ndagijimana, M., Ciccarone, C., Stellarini, A., Corsetti, A., & Paparella, A. (2015). Diversity of food-borne Bacillus volatile compounds and influence on fungal growth. *Journal of Applied Microbiology, 119*, 487–499.

Chowdhury, S. P., Hartmann, A., Gao, X., & Borriss, R. (2015). Biocontrol mechanism by root-associated *Bacillus amyloliquefaciens* FZB42. A review. *Frontiers in Microbiology, 6*, 780.

Compant, S., Duffy, B., Nowak, J., Clément, C., & Ait Barka, E. (2015). Use of plant growth-promoting bacteria for biocontrol of plant diseases: Principles, mechanisms of action, and future prospects. *Applied and Environmental Microbiology, 71*, 4951–4959.

Debois, D., Fernandez, O., Franzil, L., Jourdan, E., De Brogniez, A., Willems, L., Clément, C., Dorey, S., & De Pauw, E. (2015). Plant polysaccharides initiate underground crosstalk with bacilli by inducing synthesis of the immunogenic lipopeptide surfactin. *Environmental Microbiology Reports, 7*, 570–582.

Fernando, D. W. G., Ramarathnama, R., Krishnamoorthy, A. S., & Savchuk, S. C. (2005). Identification and use of potential bacterial organic antifungal volatiles in biocontrol. *Soil Biology and Biochemistry, 37*, 955–964.

Gong, A. D., Li, H. P., Yuan, Q. S., Song, X. S., Yao, W., He, W. J., Zhang, J. B., & Liao, Y. C. (2015). Antagonistic mechanism of Iturin A and Plipastatin A from *Bacillus amyloliquefaciens* S76-3 from wheat spikes against *Fusarium graminearum*. *PLoS One, 10*, 1–18.

Heydari, A., & Pessarakli, M. (2010). A review on biological control of fungal plant pathogens using microbial antagonists. *Journal of Biological Sciences, 10*, 272–290.

Hossain, M. T., Khan, A., Chung, E. J., Rashid, M. H. O., & Chung, Y. R. (2016). Biological control of rice bakanae by an endophytic *Bacillus oryzicola* YC7007. *Plant Pathology Journal, 32*, 228–241.

Kim, J. S., Lee, J., Lee, C. H., Woo, S. Y., Kang, H., Seo, S. G., & Kim, S. H. (2015). Activation of pathogenesis-related genes by the rhizobacterium, *Bacillus* sp. JS, which induces systemic resistance in tobacco plants. *Plant Pathology Journal, 31*, 195–201.

Lim, S. M., Yoon, M. Y., Choi, G. J., Choi, Y. H., Jang, K. S., Shin, T. S., Park, H. W., Yu, N. H., Kim, Y. H., & Kim, J. C. (2017). Diffusible and volatile antifungal compounds produced by an antagonistic *Bacillus velezensis* G341 against various Phytopathogenic Fungi. *Plant Pathology Journal, 33*, 488–498.

Mandal, S. M., Sharma, S., Pinnaka, A., Kumari, A. K., & Korpole, S. (2013). Isolation and characterization of diverse antimicrobial lipopeptides produced by *Citrobacter* and *Enterobacter*. *BMC Microbiology, 13*, 152–160.

Memenza, M., Mostacero, E., Camarena, F., & Zuñiga, D. (2016). Disease control and plant growth promotion (PGP) of selected bacterial strains in *Phaseolus vulgaris*. In F. Gonzales-Andres & E. James (Eds.), *Biological nitrogen fixation and beneficial plant – Microbe interactions* (pp. 237–245). Cham: Springer.

Nam, H. S., Yang, H.-J., Oh, B. J., Anderson, A. J., & Kim, Y. C. (2016). Biological control potential of *Bacillus amyloliquefaciens* KB3 isolated from the feces of *Allomyrina dichotoma* larvae. *Plant Pathology Journal, 32*, 273–280.

Ongena, M., & Jacques, P. (2008). *Bacillus* lipopeptides: Versatile weapons for plant disease biocontrol. *Trends in Microbiology, 16*, 115–125.

Shafi, J., Tian, H., & Ji, M. (2017). *Bacillus* species as versatile weapons for plant pathogens: A review. *Biotechnology & Biotechnological Equipment, 31*, 446–459.

Tamura, K., Dudley, J., Nei, M., & Kumar, S. (2007). MEGA4: Molecular evolutionary genetics analysis (MEGA) software version 4.0. *Molecular Biology and Evolution, 24*, 1596–1599.

Tanaka, K., Amaki, Y., Ishihara, A., & Nakajima, H. (2015). Synergistic effects of [Ile7] surfactin homologues with bacillomycin D in suppression of gray mold disease by *Bacillus amyloliquefaciens* biocontrol strain SD-32. *Journal of Agricultural and Food Chemistry, 63*, 5344–5353.

Torres, M. J., Pérez, C., Daniela, B., Sabaté, D. C., Petroselli, G., Erra-Balsells, R., & Audisio, M. C. (2017). Biological activity of the lipopeptide-producing *Bacillus amyloliquefaciens* PGPBacCA1 on common bean *Phaseolus vulgaris* L. pathogens. *Biological Control, 105*, 93–99.

Tsuge, K., Inoue, S., Ano, T., Itaya, M., & Shoda, M. (2005). Horizontal transfer of iturin A operon, *itu*, to *Bacillus subtilis* 168 and conversion into an iturin A producer. *Antimicrobial Agents and Chemotherapy, 49*, 4641–4648.

Vater, J., Kablitz, B., Wilde, F. P., Mehta, N., & Cameotra, S. S. (2002). Matrix-assisted laser desorption ionization—Time of flight mass spectrometry of lipopeptide biosurfactants in whole cells and culture filtrates of *Bacillus subtilis* C-1 isolated from petroleum sludge. *Applied and Environmental Microbiology, 68*, 6210–6219.

Velasquez, A., Castroverde, C., & Yamg He, S. (2018). Plant–pathogen warfare under changing climate conditions. *Current Biology, 10*, 19–34.

Venkadesan, D., & Sumathi, V. (2015). Screening of lactic acid bacteria for their antibacterial activity against milk borne pathogens. *IJAR, 1*, 970–973.

Weisburg, W., Barns, S., Pelletier, D., & Lane, D. (1991). 16S ribosomal DNA amplification for phylogenetic study. *Journal of Bacteriology, 173*, 697–703.

Zeriouh, H., Romero, D., Garcia-Gutierrez, L., Cazorla, F. M., De Vicente, A., & Perez-García. (2011). The iturin-like lipopeptides are essential components in the biological control arsenal of *Bacillus subtilis* against bacterial diseases of cucurbits. *Molecular Plant-Microbe Interactions, 24*, 1540–1552.

Zhang, J., Xue, Q., Gao, H., Lai, H., & Wang, P. (2016). Production of lipopeptide biosurfactants by *Bacillus atrophaeus* 5-2a and their potential use in microbial enhanced oil recovery. *Microbial Cell Factories, 15*, 1–11.

Chapter 10
Rhizobia Promote Rice (*Oryza sativa* L.) Growth: First Evidence in Cuba

Ionel Hernández Forte and María C. Nápoles García

Abstract There are few studies on rhizobia from rice plants. Thus, the objectives of this study were to characterize rhizobia from the rhizosphere of rice plants cv. INCA LP-5 and to assess the effect of their inoculation on rice growth under controlled conditions. Eleven rhizobia isolates were studied to determine if they were capable of producing polyhydroxybutyrate, indolic compounds, ammonium and phosphate solubilization. The effect of rhizobial inoculation on height, shoot dry weight, root dry weight root and shoot/root dry weight of rice plants was assessed too. Isolate Rf7 produced 9.6 mg L^{-1} of polyhydroxybutyrate. When tryptophan concentrations were increased in the culture medium, all isolates, except Rf1, showed increased production of indolic compounds up to 125 µg ml^{-1} with Rpd38 isolate. All isolates produced ammonium from amino acids and Rpd8 produced 8.4 µg ml^{-1} of ammonium. Rf7 and Rpr2 solubilized tricalcium phosphate. Isolates Rf7, Rpr1, Rpr2 and Rpr11 increased significantly the shoot plant growth at 50 days after inoculation. The results obtained provide the first evidence on the plant-growth promotion effect of rhizobia on rice in Cuba.

Keywords *Rhizobium* · Indolic compounds · Ammonia production · Phosphorus solubilization · Rice growth

10.1 Introduction

Rice (*Oryza sativa* L.) is the cereal with the highest supply of calories per hectare. After wheat it occupies the largest cultivated extension in the world (Acevedo et al. 2006). In 2017/18, rice global production was 503 million tonnes. However, rice demands surpass rice production (FAO 2017). The difference between the slow

I. Hernández Forte (✉) · M. C. Nápoles García
Group of Bioactive Products, Department of Plant Physiology and Biochemistry, National Institute of Agricultural Science, Mayabeque, Cuba
e-mail: ionel@inca.edu.cu; tere@inca.edu.cu

D. Zúñiga-Dávila et al. (eds.), *Microbial Probiotics for Agricultural Systems*, Sustainability in Plant and Crop Protection,
https://doi.org/10.1007/978-3-030-17597-9_10

increase in production and the fast population growth in consuming countries is one of the main nutritional problems to solve. It is necessary hence to obtain better rice varieties and, at the same time, to increase production (Pérez et al. 2002).

Nitrogen fertilization in rice is subject to losses of up to 67% due to denitrification and volatilization processes, as well as for the inability of plants to absorb more than 50 kg N ha^{-1} (Rivero and Suárez 2014). Additionally, these plants only absorb less than 20% of available soil phosphorus (Khan et al. 2007). The fertilizer not used by the plant pollutes soil, atmosphere and water, causing serious diseases problems (Arumi et al. 2006).

Rhizobia are soil bacteria which are studied because of their ability to induce root nodules on legume plants of agricultural importance, providing these plant hosts with fixed nitrogen, enabling them to grow productively in nitrogen-limited agricultural soils (Long 2001). These microorganisms belong to a bacterial group named Plant Growth Promoting Rhizobacteria (PGPR). They promote the growth of grasses such as corn and rice when present in their rhizosphere and/or as endophytes, without displaying morphological changes such as the nodules induced in the root system of legumes (Sessitsch et al. 2002).

There is evidence of the positive effects of rhizobia inoculation in the development, growth and yield of rice. Rhizobia strains increased stomatal conductance, net photosynthetic rate, root growth and seeds germination (Chi et al. 2005; Hasan et al. 2014). In Cuba, rhizobia were recovered from the rice rhizosphere (Hernández and Nápoles 2017) but their effects on rice growth was not yet evaluated.

Rhizobial characterization is a requisite for further development of inoculants. Besides the advent of molecular biology analyses for bacteria characterization, some traditional phenotypic analysis are still applied because they evaluate bacterial physiology and potential ecological functions. The objectives of this study were to characterize rhizobia isolated from the rice rhizosphere (Hernández and Nápoles 2017) and to evaluate the effect of their inoculation on growth of rice plants, under controlled conditions.

10.2 Materials and Methods

10.2.1 Characterization as Plant Growth Promoting Rhizobacteria

Rhizobia isolated from rhizospheric soil and rhizoplane of rice plants (*Oryza sativa* L.) cv. INCA LP-5 (Hernández and Nápoles 2017) were used. The taxonomic assignation of the isolates was determined by morphocultural and physiological characteristics. *In vitro* nodulation assays were also performed in *Macroptilium atropurpureum* (DC.) Urb. (siratro).

Inoculants were prepared in 20 mL Erlenmeyer flasks containing 5 mL of yeast mannitol (YM) liquid medium (Vincent 1970), pH 6.8. Flasks were kept during

18–20 h under stirring conditions at 150 rpm, at 29 ± 1 °C. Bacterial culture absorbance was adjusted to 0.1 ($\lambda = 600$ nm) in a spectrophotometer, corresponding to $10^7 - 10^8$ CFU ml^{-1}.

10.2.1.1 Polyhydroxybutyrate (PHB) Production

PHB quantitative determination was performed by spectrophotometric methods (Finkelstein et al. 1997). Three repetitions were performed for each isolate.

10.2.1.2 Ammonium Production

Fresh cultures were inoculated into peptone water medium and maintained under stirring conditions in orbital shaker at 150 rpm for 16 h at 29 ± 1 °C. Culture optical density was adjusted to 0.5 ($\lambda = 600$ nm) and then centrifuged at 10000 rpm for 15 min. Broth was collected and the amount of ammonium in the supernatant was estimated by means of the Nesslerization reaction. The development of brown to yellow color was positive for ammonium production.

Neesler reagent is a chemical compound used to detect small amounts of ammonia (NH_3) or ammonium cation (NH_4^+), and is used for the analysis, research and specific chemistry of nitrogen. Parameters such as specificity, linearity, system range, limit of quantification, accuracy and precision of the established Nesslerization reaction allow to determine, with precision, the presence of these ions in culture media, without interference with other substances such as amides (Miyares et al. 2015). Optical density was measured at 450 nm (Demutskaya and Kalinichenko 2010). The total ammonium concentration was estimated based on a standard curve of ammonium sulphate ranging from 0.8 to 10.0 μgmL^{-1}. Five replicates were performed for each rhizobium isolate.

10.2.1.3 Indolic Compounds Production

Ten ml of YM medium (pH 6.8), supplemented with two tryptophan concentrations (200 and 500 μg mL^{-1}) were inoculated with each isolate. A third treatment without tryptophan was used as control. The cultures were maintained under stirring conditions in orbital shaker at 150 rpm for 24 h at 29 ± 1 °C. Culture optical density was adjusted to 0.5 ($\lambda = 600$ nm). Indolic compounds concentration was determined by colorimetric measurement at 530 nm, using Salkowski reagent (Patten and Glick 2002) and estimated based on a standard curve of synthetic IAA (indol acetic acid, range: 10–400 μg ml^{-1}). Three replicates were performed for each rhizobium isolate and tryptophan concentration.

10.2.1.4 Calcium Phosphate Solubilisation

Five aliquots of 10 μl of bacterial cultures were placed on Petri dishes with National Botanical Research Institute's Phosphate medium (NBRIP) (Nautiyal 1999). Aliquots of the same volume of sterile YM medium were used as control. Inoculated plates were incubated for 15 days at 29 ± 1 °C. Colony diameter (mm) and clearly visible haloes (mm) around the colonies were measured, visualized as a translucent area around the colony when the microorganism had the ability to solubilize the calcium phosphate.

10.2.2 Effect of Rhizobia Isolates Inoculation on Rice Plants

Rice seeds cv. INCA LP-5 were used. The seeds were disinfected with Tween 80 solution for 5 min, ethanol at 70%, for 1 min and Tween 80 solution, with calcium hypochlorite, for 20 min. Seeds were washed six times with sterile distilled water under aseptic conditions and placed in Petri dishes containing sterile filter paper with 2 ml of sterile distilled water. The plates were incubated at 29 ± 1 °C for 2 days in the dark.

Two disinfected and pre-germinated seedlings were placed in pots with 0.27 kg of sterile silica sand and inoculated with 0.3 ml of each inoculum at a concentration of 10^8 CFU ml^{-1}. Seedling inoculated with YM sterile medium were used as negative control.

Pots were placed in trays containing Hoagland nutrient solution (Awais et al. 2018), modified removing solution A (containing nitrogen salts). Plants were maintained under controlled conditions of light/darkness (16/8 h) at 25 °C and 70% RH. Fifty days after, plant height (cm), shoot dry mass (mg), root dry mass (mg) and the relationship between shoot and root dry mass (SDM/RDM), were determined. Ten plants were established for each treatment.

10.2.3 Experimental Design and Statistical Analysis

Data from all trials were subjected to normality test (Bartlett test) and variance homogeneity (Kormogorov-Smirnov test). Simple classification variance analysis was applied, using Tukey comparison test ($p \leq 0.05$) to determinate differences among means (Sigarroa 1985).

The variables evaluated in the bioassay with rice plants were further processed by principal components analysis (PCA) using the package SPSS version 19.0. The average value of each variable in each treatment was considered. The location of the variables in the components was determined considering that their values were above 0.7 (Gray and Kinnear 2012). Data from isolates characterization tests related to positive attributes were plotted with SigmaPlot 2001.

10.3 Results

10.3.1 Characterization of Rhizobia as Plant Growth Promoting Rhizobacteria

10.3.1.1 PHB Production

All rhizobia isolates produced PHB (Table 10.1). Rf7 and Rpr3 produced the highest concentrations. Rpr11 produced the lowest concentration of PHB detected in this trial, without significant differences with Rpd3, Rpd16 and Rpr1.

10.3.1.2 Total Ammonium Production

The orange colour in the supernatant of all bacterial isolates showed their ability to produce ammonium, with significant differences in concentrations (Table 10.1). Rpd8 produced the highest concentration of ammonium, followed by Rpd3. While Rf7, Rpd7 and Rpr3 produced low amounts, below 2.5 μg ml^{-1}.

10.3.1.3 Indolic Compound Production

All isolates obtained produced indolic compounds, although in different concentrations. When the tryptophan concentrations in the culture medium were increased all isolates, except Rf1, showed increased production of indolic compounds (Fig. 10.1).

Table 10.1 PHB and ammonium production by rhizobia isolates

Isolates	PHB concentration (mg l^{-1})	Ammonium concentration (μg ml^{-1})
Rf1	8.4521 b	3.8996 c
Rf7	9.6144 a	1.8980 d
Rpd3	7.2787 cde	6.2631 b
Rpd7	7.4965 cd	2.4400 d
Rpd8	8.8784 b	8.3662 a
Rpd 16	7.3001 cde	4.2510 c
Rpd38	7.7878 c	4.1948 c
Rpr1	6.8963 de	4.6546 c
Rpr2	8.6740 b	4.3032 c
Rpr3	9.6650 a	2.4398 d
Rpr11	6.7011 e	3.8494 c
ESx	0.3026***	0.2280*

Different letters among treatments indicate significant differences in the Tukey multiple rank test, $p \leq 0.05$ (PHB concentration, n = 3; ammoniun concentration, n = 5)

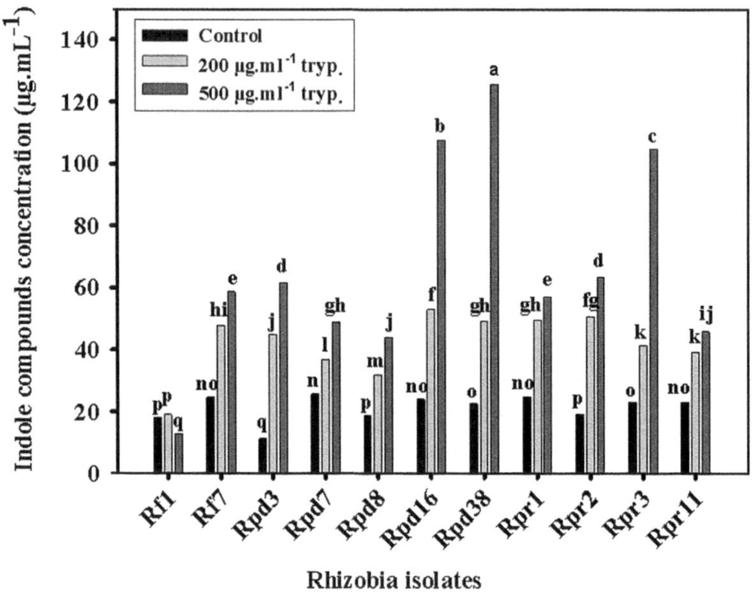

Fig. 10.1 Indolic compound production (µg ml^{-1}) by rhizobia isolates cultured in YM medium with 0, 200 and 500 µg mL^{-1} tryptophan. Different letters among treatments indicate significant differences in the Tukey multiple rank test ($p \leq 0.05$, ES***= 4.36; n = 3)

Rf7, Rpd7, Rpd16, Rpd38, Rpr1 and Rpr11 produced indolic compounds even in culture medium without tryptophan and Rpd16 and Rpr2 produced the highest amounts when they were grown with 200 µg ml^{-1} of tryptophan, reaching values of 52.90 and 50.68 µg ml^{-1}, respectively. In medium supplemented with a larger amount of tryptophan (500 µg ml^{-1}), Rpd38 produced the highest concentrations of indolic compounds (125.55 µg ml^{-1}), followed by Rpd16 and Rpr3, reaching 107.62 and 104.62 µg ml^{-1}, respectively. Rf1 was the only isolate that showed a decrease in production of these metabolites in culture medium with tryptophan 500 µg ml^{-1}.

10.3.1.4 Calcium Phosphate Solubilization

All isolates grew in NBRIP medium with colony diameter from 5.6 mm (Rpd8) to 21.5 mm (Rpr3). The largest colonies corresponded to Rpr3, Rpr1, Rpr2 and Rf1. After 15 days of incubation only Rf7 and Rpr2 (18.2% of all isolates) produced halo around their colonies (Fig. 10.2). The colony and halo diameters of Rpr2 were higher than Rf7. However, the relationship of diameter colony/diameter halo of Rf7 was higher than Rpr2. The solubilisation rate of Rf7 (1.4) was also higher than Rpr2 (Fig. 10.2).

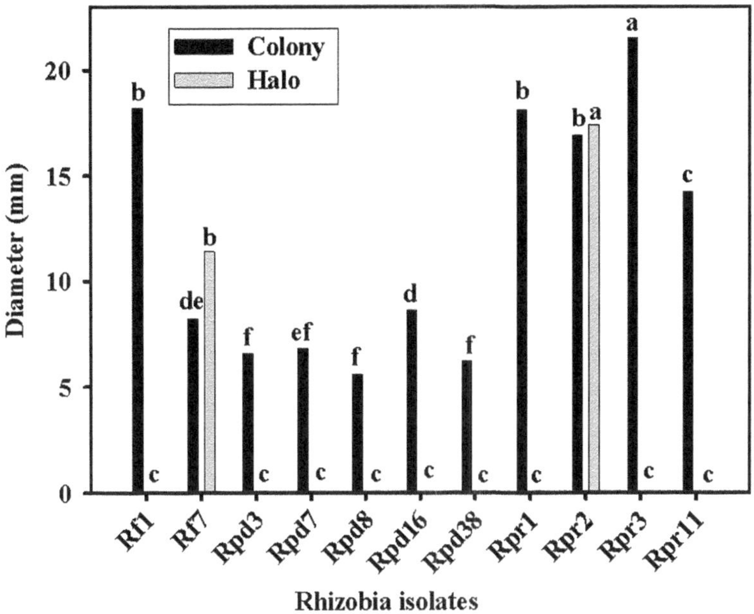

Fig. 10.2 Colony and halo diameters of rhizobia isolates cultured in NBRIP medium. Different letters among treatments indicate significant differences in the Tukey multiple rank test ($p \leq 0.05$, $n = 5$)

10.3.2 Effect of Rhizobia Inoculation on Rice Plants

Inoculation with the different isolates increased rice plant height and their shoot dry mass, and decreased root dry mass, 50 days after inoculation (Table 10.2).

Isolate Rf7 produced the largest increase in rice plants height. Rpd16, Rpr11 and Rpr2 also outperformed the control height, without significant differences between them. Rf7 produced the largest increase in shoot dry mass. Rpr2, Rpr11, Rpr1, Rpd16 and Rpd38 also increased shoot dry mass. In contrast, inoculation with most isolates, except Rf1, caused a reduction in root dry weight.

The SDM/RDM ratio showed that plant inoculation caused changes in their dry matter pattern distribution at 50 days of culture. Rf7 followed by Rpd7, Rpd16, Rpr1 and Rpr2 had a major contribution in the analysed relationship.

The PCA of growth indicators showed that component 1 mostly accounted for height, shoot dry mass and SDM/RDM relationship (73.46%) than in component 2 (26.54%), the latter mostly reflecting root dry mass (23.08%) (Fig. 10.3). The two components plot showed that Rf7, Rpr2, Rpr11, Rpr1 and Rpd16 were related to the height and shoot dry mass of plants. Control plants and those inoculated with Rf1, Rf7, Rpd38 and Rpr11 had a stronger relation to root development, than the rest of treatments (component 2).

Table 10.2 Rhizobia isolates inoculation effect on growth of rice plants cv. INCA LP-5

Treatments[a]	Height (cm)	Shoot dry mass (SDM) (mg)	Root dry mass (RDM) (mg)	SDM/RDM[b]
Control	24.83 d	91.401 f	164.362 b	0.551 h
Rf1	25.06 d	88.801 f	201.147 a	0.415 h
Rf7	31.76 a	432.010 a	126.987 cd	3.249 a
Rpd3	26.42 bcd	101.500 f	81.694 ghi	1.126 fg
Rpd7	26.52 bcd	114.518 ef	64.368 i	2.035 b
Rpd8	24.75 d	98.256 f	960.980 ef	1.085 fg
Rpd16	27.74 b	124.363 cde	780.589 hi	1.651 c
Rpd38	26.19 bcd	125.582 cde	110.587 de	1.202 efg
Rpr1	27.09 bcd	136.698 bcd	90.685 efg	1.590 cd
Rpr2	27.58 bc	148.239 b	97.697ef	1.415 cde
Rpr3	25.31 cd	115.895 def	89.456 fgh	1.330 def
Rpr11	27.85 b	145.691 bc	144.584 c	1.025 g
Esx[c]	0.87***	8.361***	7.369**	0.108*

[a]Different letters among treatments indicate significant differences by Tukey's multiple rank test ($p \leq 0.05$, $n = 10$)
[b]SDM/RDM = ratio between shoot dry mass and root dry mass
[c]Esx = ANOVA standard error

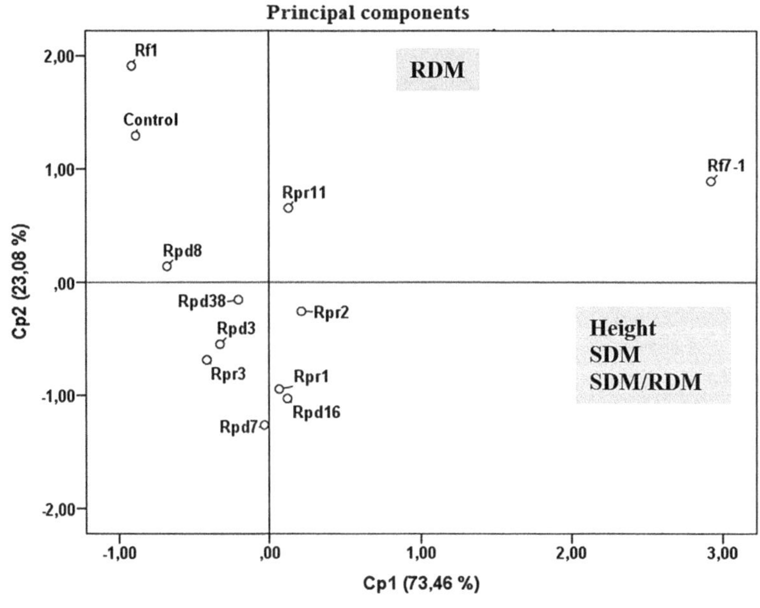

Fig. 10.3 Graphic representation of first two PCA axes, based on the average value of each variable in each treatment. SDM: shoot dry mass, RDM: root dry mass, SDM/RDM: relationship between shoot dry mass and root dry mass

10.4 Discussion

10.4.1 Characterization as Plant Growth Promoting Rhizobacteria

All rhizobia isolates produced PHB (Table 10.1). Similar results confirm results obtained with rhizobia isolated from other legumes (Dhingra and Priya 2013).

The ability to synthesize and degrade PHB by rhizobia confers a survival advantage in the rhizosphere. Rhizobia use PHB as carbon and energy source (Hamieh et al. 2013) and to stabilize the redox potential and tricarboxylic acid cycle (Trainer and Charles 2006). Isolates Rf7, Rpr3, Rf1, Rpd8 and Rpr2 which produced the highest concentrations of PHB, likely rely on higher intracellular carbon and energy reserves, mobilized to survive saprophytically in nutrient deficient soils. Besides, these isolates could use this polymer to protect against high temperatures, UV radiation, osmotic stress, drainage and presence of oxidizing agents in soil (Kadouri et al. 2003).

PHB play key role in bacterial ecology, as indicated by a reported decrease in competitiveness of a *Sinorhizobium meliloti* isolae, due to its inability to synthesize PHB and colonize the legume rhizosphere (Aneja et al. 2006). This bacterium uses PHB in cell division and to remain viable (Ratcliff and Denison 2010). The relatively high levels of PHB produced by Rf7 and Rpr3, might increase their rhizosphere competitiveness during rice root colonization.

In rhizobia, the synthesis of ammonium from amino acids is performed through deamination/deamidation reactions (Patriarca et al. 2002). The main amino acids used by these microorganisms are glutamine and histidine, from which glutamate is formed (Lu and Abdelal 2001).

Peptone water medium was used in this work to determine the ability of the rhizobia isolates to produce ammonium. Peptone contains high concentrations of polypeptides and free amino acids (Ozdal and Kurbanoglu 2018). High concentrations of amino acids in peptone could have been used by rhizobia as a nitrogen source for the ammonium formation. It could explain the relatively high concentrations of the ion in the bacterial supernatants. The ability of rhizobia to use amino acids such as glutamine, asparagine and histidine as unique sources of carbon and energy (Lu and Abdelal 2001), supports this hypothesis.

High nitrogen concentration inhibits the *glnk-amtB* operon expression, which codes for an external ammonium transporter and activates the glutamate dehydrogenase, with subsequent production of ammonium from glutamate (Patriarca et al. 2002). The cell concentration of ammonium is regulated and that produced by deamination is released outside, reducing its concentration and toxicity (Duran et al. 1995). This mechanism explains the presence of ammonium in the cultured supernatant.

All rhizobia isolates (except Rf1) produced higher indolic compound amounts with increasing tryptophan concentrations. Similar results were obtained by Raut et al. (2017). In culture media with high concentrations of tryptophan, *Azospirillum* simultaneously produced IAA through two tryptophan-dependent pathways: indolic-3-pyruvic acid (IPA) and indolic-3-acetamide (IAM) (Nonhebel et al. 1993).

Additionally, results showed that rhizobia isolates produced indolic compounds in absence of tryptophan. Besides tryptophan-dependent pathways, there is a further way that does not rely on this amino acid as aprecursor for IAA synthesis (Nonhebel et al. 1993). Ammonium is a precursor of IAA synthesis in *Streptomyces* sp. and *Flavobacterium* sp. (Merzaeva and Shirokikh 2010). The indolic compounds production by rhizobia isolates, even in absence of tryptophan, suggests the presence of alternative synthesis pathways, similar to what occurs in *Azospirillum*, *Streptomyces* and *Flavobacterium*.

Rf1 did not increase IAA production when the tryptophan concentrations increased in culture medium, as also observed for other rhizobia (Dubler et al. 2011). Another feature of Rf1 was that in the highest tryptophan concentration it produced less indolic compounds, suggesting that tryptophan had an inhibitory feedback effect on their production (Mohite 2013).

The fast-growing rhizobia studied in this research use glucose as alternative carbon sources (Graham et al. 1982). This explains their growth in NBRIP medium, where glucose is the major component.

Rhizobium meliloti and *Rhizobium leguminosarum* strains solubilize phosphate salts by excreting 2-ketogluconic acid (Halder et al. 1990). It seems possible that Rpr2 and Rf7 solubilized tricalcium phosphate by the same mechanism. Similar results were obtained with *Bacillus* strains isolated from rice-plant rhizosphere (Sarkar et al. 2018). The haloes produced by Rpr2 and Rf7 around their colonies had different size. The intensity of solubilization of calcium phosphate in NBRIP medium is proportional to the diameter of the halo in relation to the colony diameter (Nautiyal 1999). This suggests that Rf7 had a greater capacity for solubilizing calcium phosphate than Rpr2. However, the presence (or absence) of the halo are not enough to classify microorganisms as phosphorus solubilizing (Bashan et al. 2013). It is possible that some rhizobia herein studied did not produce haloes but may solubilize calcium phosphate, as shown by reported bacteria solubilizing phosphate in absence of any visible halo (López et al. 2011).

A positive effect on rice was observed 50 days after inoculation. Rice plants inoculated with four rhizobial isolates significantly increased their height over the non-inoculated control. Other reports showed increases in sugarcane-plant height when they were inoculated with rhizobia (Mirza et al. 2001). Similar results were also obtained in wheat, corn and rice (Flores et al. 2013).

There is a direct action of IAA produced by rhizobacteria on stem growth. Furthermore the indirect action of IAA also produces similar effects by increasing endogenous gibberellins (Ford et al. 2002). Isolates Rf7, Rpd16 and Rpr11which produced high concentrations of indolic compounds (Fig. 10.1) increased rice plants height. This may indicate a positive effect of indolic compounds stimulating cell division and stem elongation.

The deficit of nitrogen salts in Hoagland solution promotes bio-fertilization by rhizobia. It is known that low concentrations of nitrogen are not inhibitory for nitrogenase enzyme activity (Xuan et al. 2017). Nitrogen deficiency inhibits growth of the shoot and leaves, with premature aging of plant tissues resulting in a decrease in rice leaf development (Azcón- Vieto and Talón 2013). Non-inoculated plants presented less dry shoot mass than those inoculated with Rpr1, Rpd38, Rf7, Rpr2, Rpr11 and Rpd16. This result suggest a contribution through bio-fertilization by the tested rhizobial isolates.

The analysis of SDM/RDM relationship showed that inoculation with the isolates produced an inhibition of root growth (Table 10.2). Similar results were reported in *Sinorhizobium meliloti* as increases in IAA synthesis reduced root sdevelopment (Perrine et al. 2005). All rhizobia studied here produced indolic compounds, which may be inhibiting rice root development. The highest values of dry root mass was in fact detected in the non-inoculated plants or those inoculated with Rf1, an isolate with lowest compound indolic concentrations.

Some researchers have showed an inhibitory effect on the length, density and emergence of rice plant roots treated with ammonium. The presence of ammonium in roots increases the acropetal transport (stem to root) of endogenous IAA, increasing IAA concentration in roots (Song et al. 2013). On the other hand, the root exudates have a great diversity of amino acids (Bacilio et al. 2003) that could be used by the rhizobia isolates, associated to plants producing ammonium, through a mechanism similar to that observed in peptone water medium.

All isolates produced IAA-like metabolites (Fig. 10.1) and ammonium from amino acids (Table 10.1) may affect the mobilization of endogenous IAA to the root system. This would produce relatively high concentrations of IAA thus inhibiting root growth (Azcón- Vieto and Talón 2013).

PHB may help rhizobia promote plant growth by reducing oxygen tension that enhances nitrogenase activity, with higher amounts of nitrogen in the host plant (Bergersen et al. 1991). Rf7 was able to produce relatively high levels of PHB (Table 10.1) and produced the largest increases in shoot plant weight. Rpr11, Rpr2, Rpd16 and Rpr1 also significantly increased rice growth. These bacteria present several attributes that allow plant growth promotion, which may explain the positive effects observed on rice growth.

10.5 Conclusion

A number of rhizobia isolated from rice showed potentialities for plant growth promotion. PHB, indolic compounds and ammonium production, as well as calcium phosphate solubilisation, could act in a cooperative way to stimulate the growth of rice plants. Isolates Rf7, Rpr11, Rpr2, Rpd16 and Rpr1 are promising rhizobia to bio-fertilize rice crop. These results constitute the first evidence in Cuba on the natural association of rhizobia with a non-legume plant as rice.

Acknowledgements The authors would like to express their gratitude to PhD Inés M. Reynaldo Escobar of the National Institute of Agricultural Science for her valuable support in the plant physiology study.

References

Acevedo, M. A., Castrillo, W. A., & Belmonte, U. C. (2006). Origen, evolución y diversidad del arroz. *Agronomía Trop, 56*(2), 151–170.

Aneja, P., Zachertowska, A., & Charles, T. C. (2006). Comparison of the symbiotic and competition phenotypes of *Sinorhizobium meliloti* PHB synthesis and degradation pathway mutants. *Canadian Journal of Microbiology, 51*(7), 599–604.

Arumi, J. L., Núñez, J., Salgado, L., et al. (2006). Evaluación del riego de contaminación con nitrato de pozos de suministro de agua potable rural en Chile. *Revista Panamericana Salud Pública, 20*(6), 385–392.

Awais, R., Ashraf, M., & Mansoor. (2018). Response of foliar-applied nutrient solution with and without soil-applied fertilizers on growth and yield of mung bean. *Journal of Plant Nutrition, 41*(9), 1083–1093.

Azcón- Vieto, J., & Talón, M. (2013). Fundamentos de Fisiología Vegetal. Valencia.

Bacilio, M., Aguilar, S., Ventura, E., et al. (2003). Chemical characterization of root exudates from rice (Oryza sativa) and their effects on the chemotactic response of endophytic bacteria. *Plant and Soil, 249*(2), 271–277.

Bashan, Y., Kamnev, A. A., & de-Bashan, L. E. (2013). Tricalcium phosphate is inappropriate as a universal selection factor for isolating and testing phosphate-solubilizing bacteria that enhance plant growth: a proposal for an alternative procedure. *Biology and Fertility of Soils, 49*, 465–479.

Bergersen, F. J., Peoples, M. B., & Turner, G. L. (1991). A role for polybhydroxybutyrate in bacteroids of soybean nodules. *Proceedings of the Royal Society of London B, 245*, 59–64.

Chi, F., Shen, S., Cheng, H., et al. (2005). Ascending migration of endophytic rhizobia, from roots to leaves, inside rice plants and assessment of benefits to rice growth physiology. *Applied and Environmental Microbiology, 71*(11), 7271–7278.

Demutskaya, L. N., & Kalinichenko, I. E. (2010). Photometric determination of ammonium nitrogen with the nessler reagent in drinking water after its chlorination. *Journal of Water Chemistry and Technology, 32*(2), 90–94.

Dhingra, H. K., & Priya, K. (2013). Physiological and molecular identification of polyhydroxybutyrates (PHB) producing microorganisms isolated from root nodules of leguminous plants. *African Journal of Microbiology Research, 7*(30), 3961–3967.

Dubler, M. R., Sacool da Sá, E. L., & Goulart, R. (2011). Caracterización e influencia de rizobios aislados de alfalfa en la geminación y crecimiento de plántulas de arroz. *Ciencia Rural, 41*(10), 1738–1743.

Duran, S., Du Pont, G., Huerta-Zepeda, A., et al. (1995). The role of glutaminase in *Rhizobium etli*: Studies with a new mutant. *Microbiology, 141*, 2883–2889.

Finkelstein, R. A., Boesman-Finkelstein, M., Sengupta, D. K., et al. (1997). Colonial opacity variations among the *Cholera genic vibrios*. *Microbiology, 143*, 23–34.

Flores, J. D., Menéndez, E., Rivera, L. P., et al. (2013). Use of *Rhizobium leguminosarum* as a potential biofertilizer for *Lactucasativa* and *Daucuscarota* crops. *Journal of Plant Nutrition and Soil Science, 176*, 876–882.

Food and Agriculture Organization of the United Nations. (2017). *FAO rice market monitor*. http://www.fao.org. Accessed 11 Apr 2018.

Ford, Y., Taylor, J., Blake, P., et al. (2002). Gibberellin A3 stimulates adventitious rooting of cuttings from cherry (Prunus avium). *Plant Growth Regulation, 3*(2), 127–133.

Graham, P. H., Viteri, S. E., Mackie, F., et al. (1982). Variation in acid soil tolerance among strains of Rhizobium phaseoli. *Field Crops Research, 5*, 121–128.

Gray, C. D., & Kinnear, P. R. (2012). *IBM SPSS statistics 19 made simple*. Hove/New York. Psychology Press.

Halder, A. K., Mishra, A. K., & Bhattacharyya, P. (1990). Solubilization of rock phosphate by *Rhizobium* and *Bradyrhizobium*. *The Journal of General and Applied Microbiology, 36*(2), 81–92.

Hamieh, A., Olama, Z., & Holail, H. (2013). Microbial production of polyhydroxybutyrate, a biodegradable plastic using agro-industrial waste products. *Global Advanced Research Journal of Microbiology, 2*(3), 054–064.

Hasan, M., Bano, A., Hassan, S. G., et al. (2014). Enhancement of rice growth and production of growth-promoting phytohormones by inoculation with *rhizobium* and other rhizobacteria. *World Applied Sciences Journal, 31*(10), 1734–1743.

Hernández, I., & Nápoles, M. C. (2017). Rizobios residentes en la rizosfera de plantas de arroz (Oryza sativa L.) cultivar INCA LP-5. *Cultivos Tropicales, 38*(1), 39–49.

Kadouri, D., Jurkevitch, E., & Okon, Y. (2003). Involvement of their serve material poly-b-hydroxybutyrate in *Azospirillum brasilense* stress endurance and root colonization. *Applied and Environmental Microbiology, 69*, 3244–3250.

Khan, M. S., Zaidi, A., & Wani, P. A. (2007). Role of phosphate-solubilizing microorganisms in sustainable agricultura. A review. *Agronomy for Sustainable Development, 27*, 29–43.

Long, S. R. (2001). Genes and signals in the *Rhizobium*-legume symbiosis. *Plant Physiology, 125*, 69–72.

López, B. R., Bashan, Y., & Bacilio, M. (2011). Endophytic bacteria of *Mammillaria fraileana*, an endemic rock-colonizing cactus of the Southern Sonoran Desert. *Archives of Microbiology, 193*, 527–541.

Lu, C. D., & Abdelal, A. T. (2001). The gdhB gene of *Pseudomona aeruginosa* encodes an arginine-inducible NAD-dependent glutamate dehydrogenase which is subject to allosteric regulation. *Journal of Bacteriology, 183*, 490–499.

Merzaeva, O. V., & Shirokikh, I. G. (2010). The production of auxins by the endophytic bacteria of winter rye. *Applied Biochemistry and Microbiology, 46*(1), 44–50.

Mirza, M. S., Ahmad, W., Latif, F., et al. (2001). Isolation, partial characterization and the effect of plant growth-promoting bacteria (PGPB) on micro- propagated sugarcane in vitro. *Plant and Soil, 237*, 47–54.

Miyares, M., Torres, D., Padrón, S., et al. (2015). Aplicación del reactivo de Neesler en la cuantificación de amonio para las fermentaciones de productos biotecnológicos. *VacciMonitor, 24*(1), 33–44.

Mohite, B. (2013). Isolation and characterization of indolic acetic acid (IAA) producing bacteria from rhizospheric soil and its effect on plant growth. *Journal of Soil Science and Plant Nutrition, 13*(3), 638–649.

Nautiyal, C. S. (1999). An efficient microbiological growth medium for screening phosphate solubilizing microorganisms. *FEMS Microbiology Letters, 170*, 265–270.

Nonhebel, H., Cooney, T., & Simpson, R. (1993). The route, control and compartmentation of auxin synthesis. *Australian Journal of Plant Physiology, 20*, 527–539.

Ozdal, M., & Kurbanoglu, E. B. (2018). Citric acid production by *Aspergillus niger* from agro-industrial by products: Molasses and chicken feather peptone. *Waste and Biomass Valorization, 9*(47), 1–10.

Patriarca, E. J., Tatè, R., & Laccarino, M. (2002). Key role of bacterial NH_4^+ metabolism in *rhizobium*-plant symbiosis. *Microbiology and Molecular Biology Reviews, 66*(2), 203–222.

Patten, C., & Glick, B. (2002). Role of *Pseudomonas putida* indolic acetic acid in development of the host plant root system. *Applied and Environmental Microbiology, 68*, 3795–3801.

Pérez, N., González, M. C., & Castro, R. I. (2002). Validación de nuevas variedades cubanas de arroz (*Oryza sativa* L.) para la provincia de Pinar del Río. *Cultivos Tropicales, 23*(2), 51–54.

Perrine, F. M., Hocart, C. H., Hynes, M. F., et al. (2005). Plasmid-associated genes in the model micro-symbiont *Sinorhizobium meliloti*1021 affect the growth and development of young rice seedlings. *Environmental Microbiology, 7*(11), 1826–1838.

Ratcliff, W. C., & Denison, R. F. (2010). Individual-level bet hedging in the bacterium *Sinorhizobium meliloti*. *Current Biology, 20*, 1740–1744.

Raut, V., Shaikh, I., Naphade, B., et al. (2017). Plant growth promotion using microbial IAA producers in conjunction with azolla: a novel approach. *Chemical and Biological Technologies in Agriculture, 4*(1), 1–11.

Rivero, L. E., & Suárez, E. (2014). *Instructivo Técnico. Cultivo de Arroz*. La Habana: Instituto de Investigaciones de Granos.

Sarkar, A., Kumar, P., Pramanik, K., et al. (2018). Ahalotolerant *Enterobacter* sp. displaying ACC deaminase activity promotes rice seedling growth under salt stress. *Research in Microbiology, 169*, 20–32.

Sessitsch, A., Howieson, J. G., Perret, X., et al. (2002). Advances in *Rhizobium* research. *Critical Reviews in Plant Sciences, 21*, 323–378.

Sigarroa, A. (1985). Biometría y Diseño Experimental. La Habana.

Song, W., Sun, H., Li, J., et al. (2013). Auxin distribution is differentially affected by nitrate in roots of two rice cultivars differing in responsiveness to nitrogen. *Annals of Botany-London, 112*, 1383–1393.

Trainer, M., & Charles, T. (2006). The role of PHB metabolism in the simbiosis of rhizobia with legumes. *Applied Microbiology and Biotechnology, 1*, 377–386.

Vincent, J. M. (Ed.). (1970). *A manual for the practical study of root-nodule bacteria*. London: Blackwell.

Xuan, X., Chunmei, M., Shoukun, D., et al. (2017). Effects of nitrogen concentrations on nodulation and nitrogenase activity in dual root systems of soybean plants. *Plant Nutrition, 63*(5), 470–482.

Chapter 11
Arbuscular Mycorrhizae and PGPR Applications in Tropical Savannas

Edith Mora, Danilo Lopez-Hernández, and Marcia Toro

Abstract In Venezuela, most of agricultural soils are acid with low fertility. In these soils available forms of P are scarce and not readily available. Most of P is inorganic associated with iron and aluminum phosphates (45%), and is available by solubilization. To improve production and soil fertility it is recommended to fertilize with rock phosphate, organic amendments as cover crops and use of biofertilizers. Microorganisms related to P nutrition were studied in an acid Ultisol, located in central-northern Venezuela. 25% of native plants were mycorrhized and had phosphate solubilizing bacteria (PSB) in their rhizospheres, highlighting *Burkholderia* sp. solubilizing iron, aluminum and calcium phosphates and rock phosphate. Native Glomeromycota fungi (NGF) were reproduced in trap pots to obtain a consortia inoculum. Greenhouse tests were carried out with *Zea mays* L. co-inoculated with NGF in consortium and/or *Burkholderia* sp. Rock phosphate was added in recommended doses. Cover crops were also applied into the soil as organic matter similar to suggested agroecological management. Inoculation of NGF and PSB improved maize biomass when fertilized with rock phosphate. In all cases of BSP-AM interaction, the percent of active mycorrhizal root colonization with succinate dehydrogenase activity was higher, suggesting cooperation of this bacteria in Glomeromycota colonization and functionality, and a possible role as *mycorrhiza-helper bacteria*. Fertilization indicated that rock phosphate was adequate. Its use together with beneficial microorganisms appears as an alternative to improve crop production in low fertility soils.

Keywords Arbuscular mycorrhiza · Acid soils · Savanna · P solubilizing bacteria · Maize · *Mycorrhiza-helper bacteria*

E. Mora · D. Lopez-Hernández · M. Toro (✉)
Laboratorio de Estudios Ambientales, Instituto de Zoología y Ecología Tropical, Facultad de Ciencias, Universidad Central de Venezuela, Caracas, Venezuela
e-mail: marcia.toro@ciens.ucv.ve

D. Zúñiga-Dávila et al. (eds.), *Microbial Probiotics for Agricultural Systems*, Sustainability in Plant and Crop Protection,
https://doi.org/10.1007/978-3-030-17597-9_11

11.1 Introduction

In Venezuela, 29% of agricultural soils are acidic with strong nutrient deficiencies (Rodríguez et al. 2003; Weil and Brady 2016). As acidity increases, concentration of aluminum (Al^{+3}) gets toxic and nutrients such as P, Ca and Mg are scarce (Weil and Brady 2016). Phosphorus is fixed, precipitating as aluminum and iron phosphates, with scanty total and plant-available P (Novais et al. 2007). Total P range is 250–270 ppm, of which 75% is organic, with P-Fe around 28 ppm and P-Alaround 12 ppm. Only 1 ppm is given by soluble P, thus restricting plant growth. This is one of the most frequent problems encountered in tropical soils (Sarmiento et al. 2006).

The use of rock phosphate has been recommended to fertilize acid soils due to its low cost and availability in tropics, its slow dissolution in acid pH and low cost with respect to soluble P sources (Casanova et al. 2002). Cover crops such as *Centrosema* sp. and *Brachiaria dictyoneura* (known as *Urochloa dictyoneura* Fig. & De Not.) are used as organic fertilizers. Combined with inorganic sources they provide alternative fertilization modalities recommended to improve soil fertility (Powlson et al. 2014). Likewhise, the use of beneficial microorganisms has been recommended for tropical soils (Cardoso and Kuyper 2006).

Several authors have outlined the presence of Arbuscular Mycorrhizae (AM) and shown the importance of their application in savannas for agricultural and restoring practices (Lovera and Cuenca 2007; Tchabi et al. 2008; Mora et al. 2013). AM and plant growth promoting bacteria (PGPR) can be an alternative for the management of tropical agricultural systems (Barea 2001; Nath and Meena 2018). Moreover, the interaction of AM and PSB is beneficial (Barea et al. 2002; Jlabbé et al. 2014). Bacteria stimulating mycorrhizal colonization constitute the group of Mycorrhiza Helper Bacteria (Kurth et al. 2013; Lies et al. 2018).

In order to apply agroecological managements and improve soil fertility a greenhouse experiment inoculating beneficial microorganisms was designed with an acid savanna soil. The use of rock phosphate combined with organic amendments, arbuscular mycorrhizae and phosphate solubilizing bacteria were evaluated on maize productivity, to assess the feasibility of using native biofertilizers for a sustainable agriculture.

11.2 Materials and Methods

Glomeromycota fungi and PSB were evaluated individually and in combination with PSB, inorganic fertilizers and cover crops in a greenhouse experiment in which P was added to soil as diamonic phosphate and/or rock phosphate, in 2 kg microcosms. N was applied as urea and K as potassium chloride. NPK doses were calculated based on soil characterization and crop requirements. The following fertilization treatments were applied: **T0**: without fertilization; **T1**: 100 kg N, 80 kg P_2O_5 as rock phosphate, 70 kg K_2O; **T2**: 100 kg N, 80 kg P_2O_5 (40 kg as rock phosphate +40 kg as F diamonic phosphate), 70 kg K_2O; **T3**: 100 kg N, 20 kg P_2O_5

as phosphate rock (1/4 P dose used in T1), 70 kg K_2O. Soil was collected from field (0–20 cm) and sieved to retire coarse material, using 2 mm sieved samples for chemical and physical analyses. For microcosms experiment, soil was handled similarly to the field: *C. macrocarpum* and *U. dictyoneura* cover crops were sowed in microcosms. After flowering, they were cut and incorporated into the soil with roots. A treatment without cover crops was also considered as control (no cover crop). Two IMECA 3005 maize seeds/pot were sowed in 2 kg pots amended with T0, T2, T3 and T4 treatments. Microbial inoculation treatments were: Control (no microorganisms); BSP (plants inoculated with 10^9 cfu/ml of *Burkholderia* sp. isolated from rhizospheric soil of dominant plant species in the plot, with ability to solubilize phosphate); Native Glomeromycota Fungi (NGF) consortia (30 g of NGF pot^{-1}, containing 150 spores g^{-1} dry soil) and BSP + NGF, as indicated before. Three repetitions per treatment were made for a total of 144 microcosms evaluated. Plants were grown in a greenhouse for 6 weeks. They were harvested by separating the stem from the root and dried in the oven at 60 °C to determine dry weight. Subsequently, samples were milled and N foliar content was analyzed by the micro-Kjeldhal method (Bremner and Mulvaney 1982). P foliar content was analyzed by colorimetry (Murphy and Riley 1962). Percentage of mycorrhizal root length and vital staining of succinate dehydrogenase were measured in a subsample of harvested roots. The activity of succinate dehydrogenase in roots (SDH) was measured as described by Smith and Gianinazzi-Pearson (1990) and Schaffer and Peterson (1993). This method allows detecting structures in colonized roots in which the fungus is alive. Quantification of SDH AM colonization percentage (% active mycorrhizae) was performed with the intersection method (Giovanetti and Mosse 1980), quantifying each point that had colored fungal structure with enzyme activity. The data obtained were statistically analyzed using MANOVA (three-way Analysis of Multivariate Variance) transformed to the square root to meet the assumptions of homogeneity of variances. Cochran test showed the variances were homogeneous. A Tukey test with 95% confidence was applied as *a posteriori* analyses.

11.3 Results and Discussion

Table 11.1 shows chemical and physical characteristics of soil, standing out its low fertility (scarce P, 11.30 – 8.67 mg kg $^{-1}$), low organic matter (1.33 – 1.04) and acidic pH (5.01–4.75), typical of distrophic Ultisol soils from savanna ecosystems (Sarmiento et al. 2006).

11.3.1 Effect of Cover Crops and Microbial Inoculations

The use of cover crops such as legumes and grasses, single or in combination, favored soil fertility and apeared as an excellent strategy to obtain a better crop yield (Fageria et al. 2005). In our results, the general trend was to obtain higher maize

Table 11.1 Soil chemical and physical characteristics

Depth (cm)	Texture	pH	Inorganic N (mg Kg^{-1})	P	K	Ca	Mg	Na	Exchangeable Al (cmol.Kg^{-1})	CEC	Organic matter (%)
0–5	SL	5.01	20.36	11.30	41.95	57.73	35.13	7.2	0.34	3.06	1.33
5–15	SL	4.81	16.28	10.01	22.26	45.76	32.58	8.1	0.36	2.30	1.23
15–30	LS	4.75	14.30	8.67	19.12	29.74	39.94	7.4	0.56	1.81	1.04

SL Sandy loam, *LS* Loamy sand

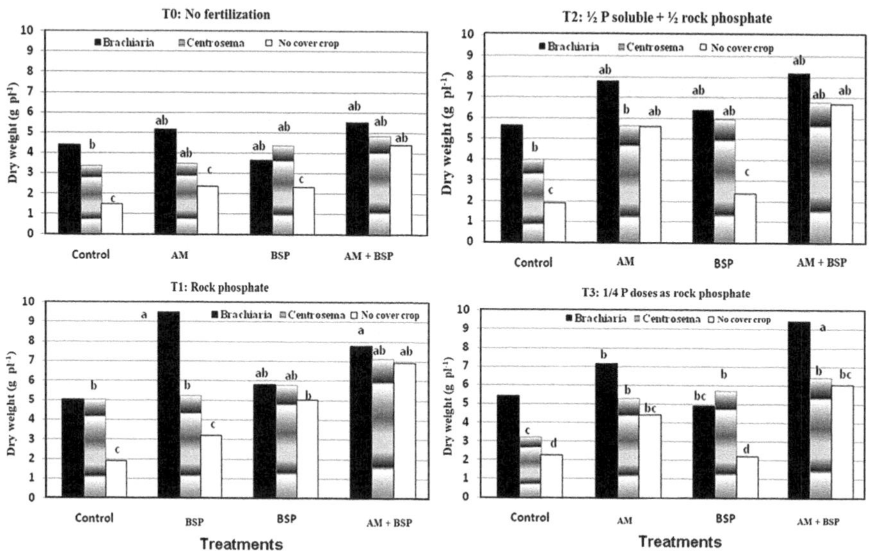

Fig. 11.1 Dry weight of maize (g/plant) grown with cover crops combined with fertilization doses. Values with the same letters do not show statistically significant differences, according to the Tukey test with 95% confidence

biomass production when *Urochloa* was applied as a cover crop, followed by biomass values with *Centrosema* and control without cover crop (Fig. 11.1). Dry weight of maize inoculated with AM had significant differences grown with *Brachiaria* and rock phosphate (T1 and T3) and in combination with soluble P (T2). Inoculation with PSB had significant response with T1 and the two cover crops, and T2 combined with *Brachiaria*. Significant differences were observed in the AM-BSP interaction with T3 and *Brachiaria*. Treatments without cover crops had significant differences with values lower than the other treatments (Fig. 11.1).

These results stand out the properties of *Brachiaria* as a cover crop, which may be related with changes in soil characteristics, as reported by Lozano et al. (2010), Bassegio et al. (2015) and Janegitz et al. (2017), who observed increase of pH, reduction of exchangeable and total acidity, increase of cation exchange capacity, inorganic nitrogen, ammonia ($N–NH_4$ contents) and available S. Lozano et al.

(2012) also highlighted the greater volume of fine roots of *Brachiaria* which tends to increase the rate of decomposition and nutrient release of its residues, stimulating microorganisms development. All these characteristics may be affecting favorably soil fertility stimulating maize dry weight, P and N foliar contents (data not shown).

Simoneti Foloni et al. (2016) observed that in the presence of *B. dyctioneura* the agronomic efficiency of rock phosphate improved. Likewise, Janegitz et al. (2016) observed that combining soluble P and rock phosphate favored soybean yield. In our case, maize biomass was always higher in treatments with AM (T1) and/or in interaction with BSP and AM (T1 and T2). However, best biomass values were obtained with T3 with the interaction of BSP and AM (Fig. 11.1), suggesting that the microbial interaction was effective in taking maximum advantage of P with a moderate dose of rock phosphate. Similar results have been obtained using isotope techniques (^{32}P), corroborating the favorable interaction between PSB and AM for the use of rock phosphate (Toro et al. 1997; Casanova et al. 2002; Barea et al. 2002).

11.3.2 Active Mycorrhiza

In all cases, the interaction of BSP with AM favored mycorrhizal colonization and/or was equal to single AM inoculation (Fig. 11.2). Highest AM colonization values (80%) were obtained with the interaction of PSB in T1 and *Brachiaria*, suggesting the effectiveness of this interaction on the use of rock phosphate. Positive BSP + AM interaction has been described by Toro et al. (1997) and Barea et al. (2002) as highly effective on the use of rock phosphate. In this case, the interaction also favored the physiological activity of AM, as evidenced by the activity of the enzyme succinate dehydrogenase. According to Smith and Gianinazzi-Pearson (1990) the presence of this mitochondrial enzyme is observed histochemically to detect the part of the infected root in which the fungus is alive. Our results show that PSB favoured AM colonization and suggest the role of *Burkholderia* sp. as mycorrhiza helper bacteria (MHB), as reported (Compant et al. 2008; Lies et al. 2018).

The concept of MHB was introduced by Garbaye (1994) referring to bacteria that promote the establishment of the fungus-root symbiosis of mycorrhiza or the increase of the efficiency of mycorrhization. In relation to AM, many examples of the interaction with this kind of bacteria have been described (Barea et al. 2005; Artursson et al. 2006; Pivato et al. 2009, Jlabbé et al. 2014), constituting an interesting interaction to apply for agricultural purposes. *Burkholderia* sp. has been considered as a MHB (Lies et al. 2018), as also confirmed by our results.

AM and PSB are important biological resources to take advantage and improve savanna soils fertility (Tchabi et al. 2008; Mora et al. 2013; Nath and Meena 2018). According to Johansson et al. (2004), Barea et al. (2005) and Toljander et al. (2007), microbial cooperation in the rhizosphere is common and may be used to favour plant growth and nutrition, as well as controlling plant pathogens. According to our results native mycorrhizal fungi and PSB cooperate and can be used for the formulation of biofertilizers, suitable for savanna soils.

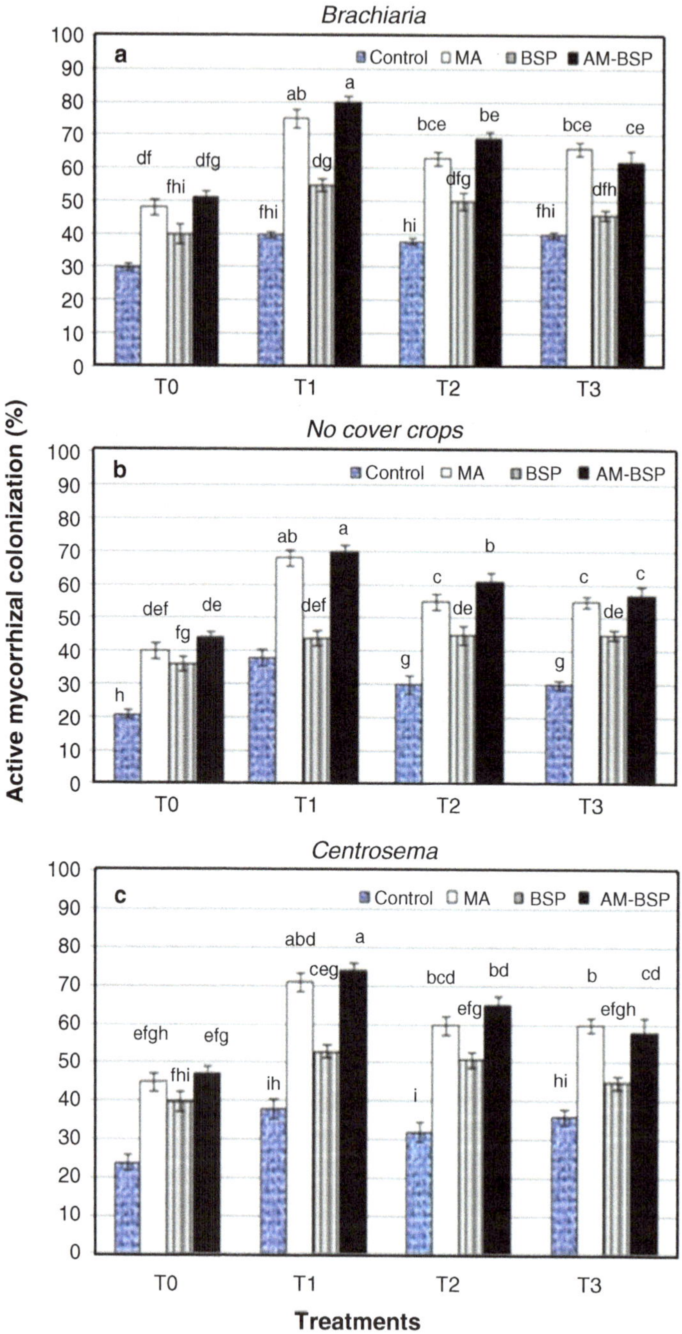

Fig. 11.2 Active mycorrhizal colonization (% of root length) of maize grown with *Brachiaria* (**a**), no cover crops (**b**) and *Centrosema* (**c**). Values with the same letter do not show significant statistical differences, according to the Tukey test with 95% confidence

11.4 Conclusions

Fertilization showed that rock phosphate and cover crops are appropriate chemical amendments for acid soils and maize. The best cover crop to improve maize production was *B. dictyoneura*, resulting in a considerable improvement of maize biomass and nutritional values. Its use together with microorganisms appears as an alternative to improve the production of crops in low fertility savanna soils. Fertilization with rock phosphate, in combination with mycorrhizae and PSB indicated that microbial inoculation favored the use of phosphate and development of the plant. This approach can be suggested as an agricultural practice to improve AM inoculum efficiency and as an alternative for a sustainable management of these agroecosystems, decreasing expenses for farmers. The inoculation of PSB with Glomeromycota fungi favored plant mycorrhization indicating the positive effect of this bacteria on the colonization and establishment of the fungi in the root. *Burkholderia* sp. stimulated AM colonization and behaved as "mycorrhiza helper bacteria". Before using and recommend *Burkholderia* sp. as a growth promoting bacteria, it is important to carry out more detailed studies and a definitive identification. Despite its potential as a biofertilizer, safety of use must be checked before recommending application, including discarding in case of positive pathogenicity tests.

Acknowledgements The authors wish to express their gratitude to Mrs. Maria Auxi Toro Garcia for editing and manuscript corrections.

References

Artursson, V., Finlay, R. D., & Jansson, J. K. (2006). Interactions between arbuscular mycorrhizal fungi and bacteria and their potential for stimulating plant growth. *Environmental Microbiology, 8*, 1–10. https://doi.org/10.1111/j.1462-2920.2005.00942.x.

Barea, J. M. (2001). Interacciones ecológicas de los microorganismos en el suelo y sus implicaciones en agricultura. In J. L. Moreno & M. A. Altieri (Eds.), *Agroecología y Desarrollo* (pp. 165–184). Cáceres: Ediciones Mundi-Prensa.

Barea, J. M., Toro, M., Orozco, M. O., Campos, E., & Azcón, R. (2002). The application of isotopic (^{32}P and ^{15}N) dilution techniques to evaluate the interactive effect of phosphate-solubilizing rhizobacteria, mycorrhizal fungi and Rhizobium to improve the agronomic efficiency of rock phosphate for legume crops. *Nutrient Cycling in Agroecosystems, 63*(1), 35–42.

Barea, J. M., Pozo, M. J., Azcón, R., & Azcón-Aguilar, C. (2005). Microbial co-operation in the rhizosphere. *Journal of Experimental Botany, 56*(417), 1761–1778.

Bassegio, D., Ferreira Santos, R., Secco, D., Zanão Junior, L. A., Werncke, I., & Mansano Sarto, M. V. (2015). Short-term effects of crop rotations on soil chemical properties under no-tillage condition. *Australian Journal of Crop Science, 9*(1), 49–54.

Bremner, J., & Mulvaney, C. (1982). Total nitrogen. In A. L. Page, R. H. Miller, & D. R. Keeney (Eds.), *Agronomy monograph number 9. Methods of soil analysis, Part 2: Chemical and biological properties* (2nd ed., pp. 595–642). Madison: ASA-SSSA.

Cardoso, I., & Kuyper, T. (2006). Mycorrhizas and tropical soil fertility. *Agriculture, Ecosystems and Environment, 116*, 72–84.

Casanova, E., Salas, A. M., & Toro, M. (2002). The use of nuclear and related techniques for evaluating agronomic effectiveness of phosphate fertilizers, in particular rock phosphate, in Venezuela. In F. Sikora (Ed.), *Assessment of soil phosphorus status and management of phosphate fertilizers to optimize crop production* (pp. 93–100). International Atomic Energy Agency- TECDOC 1272.

Compant, S., Nowak, J., Coenye, T., Clément, C., & Barka, E. (2008). Diversity and occurrence of *Burkholderia* spp. in the natural environment. *FEMS Microbiology Reviews, 32*, 607–626.

Fageria, N. K., Baligar, V. C., & Bailey, B. A. (2005). Role of cover crops in improving soil and row crop productivity. *Communications in Soil Science and Plant Analysis, 36*, 2733–2757.

Garbaye, J. (1994). Helper bacteria—A new dimension to the mycorrhizal symbiosis. *The New Phytologist, 128*, 197–210.

Giovanetti, M., & Mosse, B. (1980). An evaluation of techniques for measuring vesicular-arbuscular infection in roots. *New Phytologist, 84*, 489–500.

Janegitz, M. C., Souza, E. A., & Rosolem, C. A. (2016). Brachiaria as a cover crop to improve phosphorus use efficiency in a no-till oxisol. *Revista Brasileira de Ciência do Solo, 40*, e0150128.

Janegitz, M., Martins, A. R. H., & Rosolem, C. A. (2017). Cover crops and soil phosphorus availability. *Communications in Soil Science and Plant Analysis, 48*, 1240–1246. https://doi.org/10.1080/00103624.2017.1341918.

Jlabbé, J., Weston, D. J., Dunkirk, N., Pelletier, D. A., & Tuskan, G. A. (2014). Newly identified helper bacteria stimulate ectomycorrhizal formation in Populus. *Frontiers in Plant Science, 5*, 1–10.

Johansson, J. F., Paul, L. R., & Finlay, R. D. (2004). Microbial interactions in the mycorrhizosphere and their significance for sustainable agriculture. *FEMS Microbiology Ecology, 48*, 1–13.

Kurth, F., Zeitler, K., Feldhahn, L., Neu, T., Weber, T., Krištůfek, V., Wubet, T., Herrmann, S., Buscot, F., & Tarkka, M. (2013). Detection and quantification of a mycorrhization helper bacterium and a mycorrhizal fungus in plant-soil microcosms at different levels of complexity. *BMC Microbiology, 13*, 205–215.

Lies, A., Delteil, A., Prin, Y., & Duponnois, R. (2018). Using mycorrhiza helper microorganisms (MHM) to improve the mycorrhizal efficiency on plant growth. In V. S. Meena (Ed.), *Role of rhizospheric microbes in soil volume 1: Stress management and agricultural sustainability* (pp. 277–298). Singapore: Springer.

Lovera, M., & Cuenca, G. (2007). Diversity of arbuscular mycorrhizal fungi (AMF) and mycorrhizal potential of the soil from a natural and a disturbed savannah from La Gran Sabana. *Venezuela Interciencia, 32*, 108–114.

Lozano, Z. P., Mogollón, A., Hernández, R., Bravo, C., Ojeda, A., Torres, A., Rivero, C., & Toro, M. (2010). Cambios en las propiedades químicas de un suelo de sabana luego de la introducción de pasturas mejoradas. *Bioagro, 22*(2), 135–144.

Lozano, Z., Hernández-Hernández, R. M., Bravo, C., Rivero, C., Toro, M., & Delgado, M. (2012). Availability of phosphorus in Venezuelan savannas with well drained soil, under different cover crops and of fertilization. *Interciencia, 37*(11), 820–827.

Mora, E., Toro, M., & López-Hernández, D. (2013). Exploring arbuscular mycorrhizae, rhizobium and phosphate solubilizing bacteria in low fertility savanna soils in central Venezuela: Characterization and isolation of potential biofertilizers. In M. Miransari (Ed.), *Soil microbiology and biotechnology* (pp. 97–114). Studium Press LLC.

Murphy, J., & Riley, J. P. (1962). A modified single solution method for determination of phosphate in natural waters. *Analytica Chimica Acta, 27*, 31–36.

Nath, D., & Meena, V. S. (2018). Mycorrhizae: A potential microorganism and its implication in agriculture. In V. S. Meena (Ed.), *Role of rhizospheric microbes in soil volume 1: Stress management and agricultural sustainability* (pp. 251–276). Singapore: Springer.

Novais, R. F., Smyth, T. J, & Nunes, F. N. (2007). Fosforo. In R. F. Novais, V. H. Alvarez, N. F. Barros, R. L. F. Fontes, R. B. Cantarutti, & J. C. L. Neves (Eds.), Fertilidade so solo. Chapter 8. Sociedade brasilera do ciencias do solo (pp. 472–550). Vicosa MG.

Pivato, B., Offre, P., Marchelli, S., Barbonaglia, B., Mougel, C., Lemanceau, P., et al. (2009). Bacterial effects on arbuscular mycorrhizal fungi and mycorrhiza development as influenced by the bacteria, fungi,and host plant. *Mycorrhiza, 19*, 81–90. https://doi.org/10.1007/s00572-008-0205-2.

Powlson, D. S., Stirling, C. M., Jat, M. L., Gerard, B. G., Palm, C., Sanchez, P. A., & Cassman, K. G. (2014). Limited potential of no-till agriculture for climate change mitigation. *Nature Climate Change, 4*, 678–683.

Rodríguez, T., Sanabria, D., & Navarro, L. (2003). *Nuevos enfoques en el Manejo de Sabanas en los llanos Orientales Venezolanos*. http://www.ceniap.gov.ve/bdigital/fdivul/fd52/sabanas.htm

Sarmiento, G., da Silva, M., Naranjo, M. E., & Pinillos, M. (2006). Nitrogen and phosphorus as limiting factors for growth and primary production in a flooded savanna in the Venezuelan Llanos. *Journal of Tropical Ecology, 22*, 203–212.

Schaffer, G. F., & Peterson, R. L. (1993). Modifications to clearing methods used in combination with vital staining of roots colonized with vesicular-arbuscular mycorrhizal fungi. *Mycorrhiza, 4*, 29–35.

Simoneti Foloni, J. S., Calonego, J. C., Barbosa, A., Catuchi, T. A., & Tiritan, C. S. (2016). Availability of phosphorus in soil and straw in successive tropical grasses crop fertilized with different phosphates. *Journal of Agronomy, 15*(3), 104–113.

Smith, S. E., & Gianinazzi-Pearson, V. (1990). Phosphate uptake and vesicular-arbuscular activity in mycorrhiza *Allium cepa* L.: Effect of photon irradiance and phosphate nutrition. *Australian Journal of Plant Physiology, 17*, 177–188.

Tchabi, A., Coyne, D., Hountondji, F., Lawouin, L., Wiemken, A., & Oehl, F. (2008). Arbuscular mycorrhizal fungal communities in sub-Saharan Savannas of Benin, West Africa, as affected by agricultural land use intensity and ecological zone. *Mycorrhiza, 18*, 181–195.

Toljander, J. F., Lindahl, B., Paul, L., Elfstrand, M., & Finlay, R. (2007). Influence of arbuscular mycorrhizal mycelial exudates on soil bacterial growth and community structure. *FEMS Microbiology Ecology, 61*, 295–304.

Toro, M., Azcón, R., & Barea, J. M. (1997). Improvement of arbusular mycorrhizal development by inoculation with phosphate solubilizing rhizobacteria to benefit rock phosphate bioavailability (^{32}P) and nutrient cycling. *Applied and Environmental Microbiology, 63*(11), 4408–4412.

Weil, R. R., & Brady, N. C. (2016). *The nature and properties of soils* (15th ed.). New York: Pearson.

Chapter 12
Early Effects of Land Use Intensity on Arbuscular Mycorrhizal Fungi in Rice-Soybean Rotations

Andrea Rodríguez-Blanco and Matías Giménez

Abstract Rice crop in Uruguay is known for soil usage in rotations with pasture. During the last years soybean was introduced looking for rotation alternatives in rice systems. Arbuscular mycorrhizal fungi (AMF) contribute to plant growth and productivity, and provide other ecosystem services that sustain the environment. The aim of this study was to assess the early impact of soil use intensity in AMF in rice-soybean contrasting rotations. The study was conducted in a long-term field experiment, which started in 2012. Two rotations were selected: rice-crops (RC, $rice_1$ + sorghum + $rice_2$ + soybean), and the less intensive rice-soybean-pasture (RSP, $rice_1$ + $soybean_1$ + $soybean_2$ + $rice_2$ + 2 years pasture. All rotation phases were present at the same time. Soil and soybean roots were sampled in RC, RS_1P (first soybean in RSP) and RS_2P (second soybean in RSP). Mycorrhizal colonization, soil glomalin content and AMF diversity were determined using T-RFLP. AMF diversity and colonization in soybean roots with rice as previous crop were lower than plants sown after soybean crop (RS_2P). In this initial stage of the long-term experiment, soil use intensity did not affect AMF colonization and diversity in soybean roots. However, it did affect the AMF community structure, differentiating the intensive rotation (RC) from the less intensive ones (RS_1P and RS_2P).

Keywords Diversity · Colonization · T-RFLP · Sustainable intensification · Crop rotations · Glomalin

12.1 Introduction

Rice (*Oryza sativa* L.) is one of the most economically important plants worldwide and the main food crop cultivated in wetland environments. Rice crop productivity in Uruguay is one of the highest worldwide. Yields have evolved from 5 t/ha, at the

A. Rodríguez-Blanco (✉) · M. Giménez
Microbiología, Facultad de Agronomía, Universidad de la República, Montevideo, Uruguay
e-mail: andrearb@fagro.edu.uy

D. Zúñiga-Dávila et al. (eds.), *Microbial Probiotics for Agricultural Systems*, Sustainability in Plant and Crop Protection,
https://doi.org/10.1007/978-3-030-17597-9_12

end of the 1980s, to more than 8 t/ha in the last harvests. Historically, rice crop has alternated soil usage with pastures, which confers the system with some environmental and productive advantages (Pittelkow et al. 2016). However, during the last years there has been a trend towards a system intensification through the increase in rice crop frequency, inclusion of other crops, and/or the addition of short productive pastures. Soybean inclusion in rice-pasture rotations is an important alternative as it allows to boost and diversify incomes, intensify soil usage, reduce soil laboring costs and add alternatives for weed control.

Arbuscular mycorrhizal fungi (AMF) (phylum *Glomeromycota*) are mutualistic symbionts living in association with the roots of the majority of land plants (Smith and Read 2008), including many crop plants. AMF play a key role in sustainable production systems. In exchange for the plant-assimilated carbon, the fungi benefit the host by facilitating plant access to mineral nutrients and water. Apart from that, it has been observed that AMF improve the host plant resistance to pathogens (Smith and Read 2008). The external AMF mycelium can also secrete a protein called glomalin, which has a positive effect on soil aggregate formation, improving soil structure. In addition, the AMF community composition and diversity may influence plant productivity and community structure (van der Heijden et al. 2008).

Intensification processes are sustainable as long as they maintain or increase productivity, increase the use efficiency of agricultural inputs and preserve natural resources such as water and soil. To know the course of a system in sustainability terms it is important to identify the impact of intensification, not only in terms of productivity, but also as concerns soil quality, pests dynamics, weeds, diseases and nutrients. Taking this into account, in the year 2012 a long-term experiment with six contrasting rice rotations was set at the *Paso de la Laguna* station of the National Agricultural Research Institute (INIA), Treinta y Tres, Uruguay. The 12 ha experiment is expected to run for at least 20 years and will constitute an important platform for research in the following years. The aim of this study was to assess the early impact of soil use intensity in AMF communities, in contrasting rotations including soybean and rice crops.

12.2 Materials and Methods

12.2.1 Study Site Description

The study was carried out in a rice rotation long-term experiment established in 2012 at *Paso de la Laguna* station of the National Agricultural Research Institute (INIA), located in Treinta y Tres, Uruguay (33° 14′ 58″ S, 54° 29′ 24″ W). Before starting the experiment the field has been under rice-pasture rotations since 1979. The experiment has six contrasting rice rotations, with 3 replications in space and with all phases of each rotation present simultaneously. For this study, two rice-soybean rotations were selected, the first one, named rice-row crop (RC), consists of a 4-year rotation of rice-soybean-rice-sorghum, with a *Trifolium alexandrinum* pasture between crops. The second one, named rice-soybean-pasture (RSP), consists of a 6-year rotation with rice-soybean1-soybean2-rice-2 years of pasture (mixture of *Festulolium* and *Lotus corniculatus)*. For the present study three soybean phases were evaluated: RC (soybean in rice-crop rotation), RS_1P (first soybean in RSP), RS_2P (second soybean in RSP) with three replicas distributed in blocks of 18 × 60 m each. Soybean (*Glycine max* L.) variety NS 6909 IPRO seeds were treated with anti-fungi and inoculated with soybean recommended rhizobia. Crop management was the same for the three rotations (same herbicide, insecticide and irrigation dosages). Phosphate fertilization was based on sufficiency levels. P Bray levels before sowing were 8 ppm in RC and RS_2P and 11 ppm in RS_1P. Therefore, 85 kg/ha of P_2O_5were added for RC and R_2SP and 20 kg/ha for RS_1P. Average yield for soybean crops in the season was 1630 kg/ha; yields were severely affected by weather conditions.

12.2.2 Soil and Root Sampling and Processing

Soil and root sampling was conducted in March 2016, just before soybean harvest. Three different plants of *Glicine max* randomly selected were taken per plot, at a minimum distance between plants no shorter than 5 m. Roots were washed, cut in fragments and stored in vinegar until evaluation of AMF colonization. The remaining portions of roots of three plants from each block were mixed thoroughly to obtain a single representative sample. Surface-disinfected roots were pulverized in sterilized mortars with liquid nitrogen and stored at −20 °C for DNA extraction.

Ten soil core samples were taken per plot to a depth of 10 cm by using a soil core with a 3 cm diameter. Each sample was carefully ground by hand, mixed, and air dried. To form composite samples per block, ten soil core samples were mixed. Sub-samples were kept at 4 °C until glomalin extraction. The remaining portions of soil samples were stored at −20 °C for DNA extraction.

12.2.3 Soybean Root Colonization by AMF

Soybean roots were immersed in 10% KOH and heated at 90 °C for 1 h. After washing with water, they were immersed in trypan blue 0.05% at 90 °C for 15 min and washed. Finally, roots were immersed overnight in lactoglycerol solution (lactic acid: glycerol: water 1:1:1). Thirty segments of about 1 cm length of each plant were placed on a slide, and then observed under a light microscope at 400× for estimation of AMF colonization. Roots were examined for the presence of AMF characteristic structures, including hyphae, arbuscules and vesicles, to determine colonization prevalence (percentage). All determinations were repeated with three separate samples.

12.2.4 Soil Glomalin Content

The extraction of total glomalin (TG) and easily extractable glomalin (EEG) were done following the method of Wright et al. (1996). Protein content was determined by the Bradford dye binding assay using bovine serum albumin as a standard.

12.2.5 DNA Extraction

Total DNA was extracted from 250 mg of roots following Gollotte et al. (2004) with modifications (García et al. 2017). Soil DNA was extracted from 250 mg with UltraClean Soil DNA Isolation Kit (Mo Bio Laboratories, Carlsbad, CA). The DNA was resuspended in 25 μL of TE (10 mMTris [pH 8.0] 1 mM EDTA). The quality and amount of the extracted DNA were checked on a 1.0% (w/v) agarose gel. The DNA was then stored at −20 °C for subsequent analysis.

12.2.6 Terminal Restriction Fragment Length Polymorphism (T-RFLP)

The community structure of soil and root-colonizing AMF was assessed by T-RFLP amplifying a region of the 18S rDNA of Glomeromycota fungi. A nested PCR strategy was performed. Primers LR1 (5′-GCA TAT CAA TAA GCG GAG GA-3′) and FLR2 (5′-GTC GTT TAA AGC CAT TAC GTC-3′), which are specific for the fungal division (Trouvelot et al. 1999), were employed in the first amplification reaction. Primers FLR3 (5′-TTG AAA GGG AAA CGA TTG AAG T-3′) and FLR4 (5′-TAC GTC AAC ATC CTT AAC GAA-3′), which are specific for Glomeromycota (Gollotte et al. 2004), were used in a second PCR reaction. PCR reactions were performed following García et al. (2017). In order to decrease variations in the PCR process, samples were amplified in triplicate, pooled and then purified with a Gene JET Gel Extraction kit (Thermo Scientific, USA). The obtained DNA was quantified in 1% agarose gel using the molecular weight marker Gene Ruler 100 bp DNA (Fermentas, USA) and in Nanodrop (Thermo Fisher Scientific, USA). Approximately, 100 ng of purified amplicons were digested with 5 U of MboI (Fermentas, USA) at 37 °C for 5 h, followed by 20 min at 65 °C to denature the enzyme. The samples were analyzed by Macrogen (S. Korea) using an ABI3730X (Applied Biosystems) and GS500LIZ as a standard.

12.2.7 Analysis of the T-RFLP Data

GeneScan 3.7 software (Applied Biosystems, USA) was used to analyze fragment sizes and peak fluorescence intensities. T-RFs between 30 and 360 bp were included in the analysis. The relative frequency of each peak in a given sample was calculated using the T-Rex program (Culman et al. 2009). Peaks that represented at least 1% of the total peak area for a given sample were included in our analysis. Peaks below this threshold were assigned a value of 0%. The richness (S) of AMF in each sample was estimated based on a matrix considering presence (1) or absence (0) of each T-RF. The Shannon diversity index (H′) and the Pielou evenness index (J′) were estimated based on the relative abundance of each T-RF in the community, using the software PAST 2.07 (Hammer et al. 2001).

12.2.8 Statistical Analyses

Analysis of variance (ANOVA) was performed on the content of TG and EEG. The variables: total colonization percentage, vesicles percentage, arbuscules percentage and T-RF's S were analyzed using a Mixed Generalized Linear Model (MGLM), Poisson family. The diversity indexes H′ and J′ for T-RFs were analyzed using a Generalized Linear and Mixed Model. Differences between the compared averages were analyzed through the Least Significant Difference (LSD) test (significance p-value = 0.05). All analyses were carried out using the statistics program INFOSTAT (Di Rienzo et al. 2014).

A Bray-Curtis dissimilarity matrix among samples was calculated using the matrix of T-RF's relative abundance. Differences in community structure were analyzed with Non Metric Multidimensional Scaling (NMDS) using the software PAST 2.07. The same matrix and software were used to perform analysis of similarity (ANOSIM) to test significant differences among AMF community composition. The similarity percentage (SIMPER) was also run on this data to determine which T-RFs contribute more to differences between groups. The Pearson's correlation coefficients among AMF variables and soybean yield were performed using INFOSTAT. Data from soybean yield were kindly provided by Ignacio Macedo, INIA, Treinta y Tres (personal communication).

12.3 Results

12.3.1 AMF Colonization and Soil Glomalin Content

All treatments showed a high prevalence of roots colonized by AMF, ranging from 52% to 67% of root length. AMF hyphae and vesicles colonization in soybean roots were significantly lower in RC rotation than in RS_2P rotation (Table 12.1).

Table 12.1 The content of soil total glomalin (TG), easily extractable glomalin (EEG) and soybean root colonization by arbuscular mycorrhizal fungi (AMF) in rice-soybean rotations

	Glomalin content (mg/g dry soil)		% of AMF colonization by		
Rotation	TG	EEG	Arbuscules	Vesicles	Hyphae
RC	57,8a	14,6a	21a	19a	52a
RS_1P	59,3a	13,2a	27a	21a	60ab
RS_2P	53,9a	13,5a	24a	37b	67b

RC, soybean in rice-row crops rotation; RS_1P, 1st soybean in rice-soybean-pasture rotation; RS_2P, 2nd soybean in rice-soybean-pasture rotation. The data represent the mean of three independent replicas with three repetitions each. Means in each column followed by the same letters are not significantly different ($p < 0.05$, Fisher's LSD)

Table 12.2 Arbuscular mycorrhiza fungal richness (S) and Shannon diversity index (H) calculated from T-RFLP data of soybean root and soil from rice-soybean rotations

	Root		Soil	
Rotation	S	H	S	H
RC	8.0A	1.62A	13.0A	2.30A
RS_1P	7.7A	1.65A	13.7A	2.26A
RS_2P	11.3B	2.10B	12.0A	2.08A
Mean	9 a	1.80 a	13 b	2.21 b

RC, soybean in rice row crops rotation; RS_1P, 1st soybean in rice-soybean-pasture rotation; RS_2P, 2nd soybean in rice-soybean-pasture rotation. The data are the mean of three independent replicates. Means in each column followed by the same capital letters are not significantly different ($P > 0.05$, Fisher's LSD). Means followed by the same lower-case letters are not significantly different ($P > 0.05$, Fisher's LSD) between root and soil, for each parameter evaluated

The content of total and easy extractable glomalin did not vary significantly among soybean rotations (Table 12.1).

12.3.2 *AMF Community Diversity and Structure (T-RFLP)*

After normalization of T-RFLP profiles, 25 different biotypes or T-RFs were identified varying from a minimum of 6 to a maximum of 16 T-RFs per sample. Two T-RFs (63 bp and 272 bp) were found to be common to all T-RFLP profiles. Soil AMF communities were more diverse than root communities (Table 12.2), four T-RFs were present in soil AMF communities but not in roots communities (146, 228, 271 and 356 bp).

Richness and Shannon diversity indexes in soybean roots differed among rotations, while soil AMF communities did not (Table 12.2). Soybean roots of RS_2P showed higher T-RF richness and Shannon index values than RS_1P and RC (Table 12.2). Communities did not show differences in evenness ($p = 0.3967$) (data not showed).

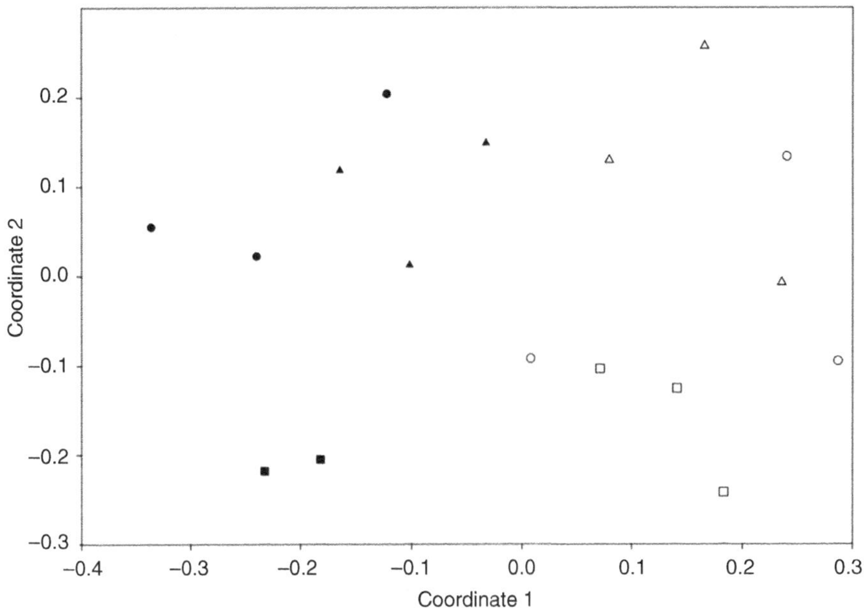

Fig. 12.1 Multidimensional scaling (Bray-Curtis distances) plot of the AMF community composition found in soil (empty symbols) and roots (filled symbols) of soybean plants grown under the different rotations (square: RC, soybean in rice-row crops rotation; circles: RS_1P, 1st soybean in rice-soybean-pasture rotation; triangles: RS_2P, 2nd soybean in rice-soybean-pasture rotation)

Multivariate analysis (NMDS) allowed to distinguish AMF soil communities from AMF root communities. The AMF community structure found in roots from RC plants was significantly different from the community structure of RS_1P and RS_2P plant roots (Fig. 12.1). However, the R values comparison (ANOSIM test) showed that distance between RC and RS_2P ($R = 1$) was greater than that between RC and RS_1P ($R = 0.5$), both of these with rice as previous crop. AMF communities from soybean plants of RSP rotations were not different when the previous crop was soybean (RS_2P) or rice (RS_1P) ($R = 0.2593$). Soil communities were not different in the three evaluated treatments ($R < 0.5$).

Simper test showed that 5 T-RFs contributed in more than 50% to differences between AMF communities. The 63 bp T-RF was highly abundant in roots samples while 153 bp T-RF was dominant in AMF soil communities. Abundance of this T-RFs in soybean roots significantly differed between the more intensive rotation (RC) and the less intensive one (RSP). Soybean roots in RC rotation presented lower abundance of 273 and 272 bpT-RFs than RS_1P and RS_2P. However, the relative abundance of 63 bp T-RF was higher in RC roots than in RSP roots (Fig. 12.2).

Soybean yield, root fungal variables (AMF colonization, Shannon index and richness) and soil variables (TG, EEG) were analyzed for their correlation (Table 12.3). EEG was negatively correlated with percentage of AMF colonization

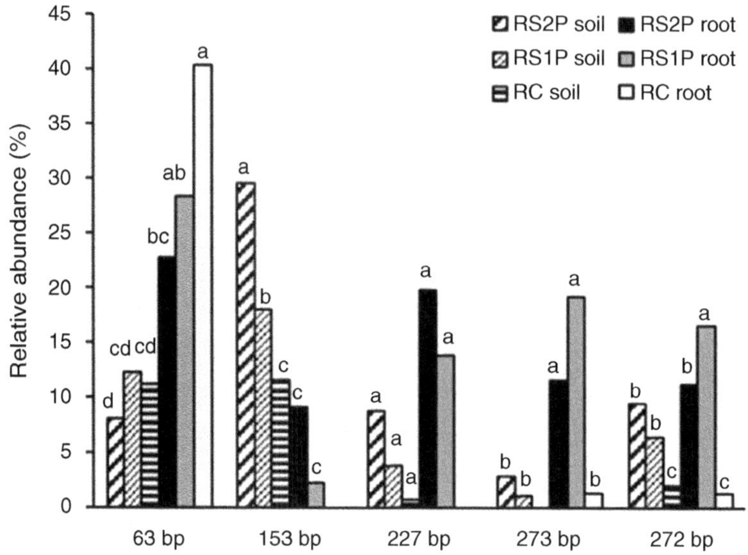

Fig. 12.2 Relative abundance of 5 T-RFs of AMF in soil and roots of rice-soybean rotations. RC, soybean in rice row crops rotation; RS1P, 1st soybean in rice-soybean-pasture rotation; RS2P, 2nd soybean in rice-soybean-pasture rotation. Means follow by different letters in each T-RF are significantly differences (Fisher p<0.05)

Table 12.3 Pearson's correlation coefficients among soil glomalin content (total glomalin: TG, and easy extractable glomalin: EEG), % of root colonized by hyphae, vesicles, arbuscules, AMF richness, Shannon diversity index and soybean yield

Pearson's coefficients	TG	EEG	% hyphae	% vesicles	% arbuscules	AMF richness	Shannon index
EEG	−0.22						
% hyphae	−0.22	−0.72*					
% vesicles	−0.35	−0.32	0.82*				
% arbuscules	−0.11	−0.74*	0.56	0.22			
AMF richness	−0.12	0.29	0.27	0.68*	−0.42		
Shannon index	−0.29	0.36	0.31	0.70*	−0.45	0.96*	
Soybean yield	−0.63	0.07	0.61	0.79*	0.01	0.74*	0.81*

*Significant correlation (p ≤ 0.05)

by hyphae and arbuscules. Percentage of AMF colonization by hyphae was positively correlated with colonization by vesicles but not with arbuscules. AMF richness and Shannon diversity index were highly correlated between them and along with percentage of root colonized by vesicles. The soybean yield was positively correlated with colonization by vesicles, AMF richness and Shannon index (Table 12.3).

12.4 Discussion

Soybean is the main grain crop in Uruguay, with a mean yield of 2 t/ha and a planting area extended to non-traditional agricultural soils. During the last years soybean has been included in rice rotations as an important alternative for rice system intensification. Crop yield depends not only on the crop plant and its physical and chemical environment, but also on its biological interactions. Soybean is an arbuscular mycorrhizal biological nitrogen fixer. This work studied the colonization and diversity of arbuscular mycorrhizal fungi (AMF) in soybean roots and soil in two rice-soybean contrasting rotations: an intensive one, which includes 4 years of crops (RC), and a less intensive one, including rice-soybean-2 years pasture (RSP).

Our results showed that roots of soybean plants were highly colonized by diverse communities of AMF, even under intensive soil use. These observations are in accordance with Pontes et al. (2017) and García de León et al. (2018) who evaluated AMF diversity in soybean in Brazil and Argentina. This result is not surprising considering that rotations evaluated in our study include diverse crop sequences and no-tillage. These factors had been reported to favor spore germination, root colonization and AMF diversity (Alguacil et al. 2008; Bainard et al. 2012; Dai et al. 2015). This is an important result if we consider that AMF are not only important for crop productivity, but also for promoting more resilient and sustainable rice systems.

AMF colonization, richness and diversity in soybean roots were affected by the previous crop in the rotation. We found that AMF colonization and diversity in soybean roots were lower when rice was the previous crop than when soybean was the previous one. Vallino et al. (2013) found that rice roots were poorly colonized by AMF in waterlogged paddy fields. The results obtained by Oka et al. (2010) also showed that AMF colonization and diversity were lower when the previous crop was a non-mycorrhizal crop. The decrease of colonization and diversity observed in our study could be explained by the fact that in Uruguay, as well as in most other places, rice is grown under waterlogged conditions. Soybean in rotation with rice still retain high colonization and fungal diversity, confirming the results obtained in other studies which found that AMF are able to survive in waterlogged paddy fields (Watanarojanaporn et al. 2013).

Our results agree with previous studies which demonstrated that AMF community composition and root colonization depend on previous crops (Hill 2006; Wang et al. 2012). In concordance with results reported by Jie et al. (2013), we observed that when the years of continuous soybean cropping were increased, the diversity indexes of colonizing AMF also raised. This is consistent with an increasing trend in colonization rates when the crop is repeated. Continuous cropping of soybean can lead to changes in soil nutrients, microorganisms, enzyme activities and other factors. In general, these changes are not beneficial for the growth of crops such as soybean, but could be beneficial for AMF (Wang et al. 2012). AMF stimulate symbiotic nitrogen fixation in legumes (Chalk et al. 2006), therefore a positive effect on soybean performance may be expected, as AMF has been reported to increase soybean biomass (Bainard et al. 2013). We found a positive correlation between soybean yield, AMF colonization, richness and diversity. As soybean yield was significantly higher when soybean was the previous crop (RS_2P) than when rice was the previous one (RS_1P and RC), this correlation could indicate the contribution of AMF to soybean growth and productivity, even in conditions of full nutrient supply.

Several studies demonstrated that agricultural management practices, such as cultivation intensity, fertilizer application and water management, can affect AMF communities (Lin et al. 2012; Lumini et al. 2011; Moora et al. 2014; Gottshall et al. 2017). In this work, soil use intensity did not alter the colonization nor the richness of AMF in soybean roots. Our results contrast with previous studies where high soil use intensity was linked with low AM fungal diversity (Alguacil et al. 2008; Lumini et al. 2011; Verbruggen et al. 2012). There are also other studies where no changes in diversity were observed (Mathimaran et al. 2007, Moora et al. 2014). The lack of response of AMF diversity to soil use intensity may be related to the age of the field experiment, which is still in its initial phase. On the other hand, soil use intensity produced changes in AMF community structure, which differed between the more intensive rotation (rice-crop) and the less intensive one (rice-soybean-pasture).

Results also showed that the abundance of five T-RFs in AMF communities differed between RC and RSP rotations. Consistently with Moora et al. (2014), Alguacil et al. (2014) and Gottshall et al. (2017), we found that management alters the composition, rather than the diversity of AMF communities. The inclusion of 2 years of pasture in RSP rotation could explain the changes in AMF communities composition, as highly diverse AMF communities have been reported in permanent grasslands (Horn et al. 2014; Moora et al. 2014).

The effects of soil use intensity on soil physical-chemical properties are widely recognized. The interaction between soil use intensity and soil chemistry cannot be discarded as a possible source of variation in AMF community composition (García de León et al. 2018). The performance of soybean depends in fact not only on the presence, but also on the taxon composition of the symbiotic AM fungal community (Cely et al. 2016). More efforts are needed to understand the importance of individual AMF taxon responses to agricultural management and how it affects soil aggregation, plants health and growth, and other agroecosystem functions and services.

Glomalin content in soil was not affected by soil use intensity nor by previous crop. This result may be related to the recalcitrant nature of glomalin. Production of glomalin is not always correlated with AM hyphal density in soils or with hyphal root colonization (Auge et al. 2003). We found higher concentrations of glomalin in our system compared to other agroecosystems studied. However, there are also reports of much higher glomalin concentrations, especially in natural ecosystems (Rillig et al. 2001; Borie et al. 2010). Glomalin is reported to be highly stable in soils and there are certain factors that increase its stability (Wright and Upadhyaya 1998). In our system there are important periods of flooding and little is known about the recalcitrance of glomalin under this condition.

12.5 Conclusion

Results showed that soybean plants with rice as previous crop had lower AMF diversity and colonization, compared with plants that had soybean as previous crop. In this initial stage of the long term experiment, soil use intensity did not affect AMF colonization and diversity in soybean plants. However, it did affect the AMF community structure allowing to differentiate between the intensive rotation, which includes 4 years of crops, and the less intensive one which includes rice-soybean-2 years pasture. The present study represents a first insight into AMF communities in the long-term rice rotations experiment. It is important to continue assessing mid- and long-term variations in AMF communities in this experiment, as they play a fundamental role in soil biology. These results can contribute to design sustainable and intensive rice production systems which combine high productivity rates with resources conservation practices.

Acknowledgement The authors thank the Instituto Nacional de InvestigaciónAgropecuaria (INIA), for allowing use of the station infrastructure and for the logistic support provided to carry out this study. Ignacio Macedo and the staff of Paso de la Laguna Station for field work collaboration. This work was financied by Comisión Sectorial de Investigación Científica (CSIC), Universidad de la República, Uruguay.

References

Alguacil, M. M., Lumini, E., Roldan, A., Salinas-Garcia, J. R., Bonfante, P., & Bianciotto, V. (2008). The impact of tillage practices on arbuscular mycorrhizal fungal diversity in subtropical crops. *Ecological Applications, 18*, 527–536.

Alguacil, M. M., Torrecillas, E., García-Orenes, F., & Roldán, A. (2014). Changes in the composition and diversity of AMF communities mediated by management practices in a Mediterranean soil are related with increases in soil biological activity. *Soil Biology and Biochemistry, 76*, 34–44.

Auge, R. M., Moore, J. L., Cho, K. H., Stutz, J. C., Sylvia, D. M., AlAgely, A. K., & Saxon, A. M. (2003). Relating foliar dehydration tolerance of mycorrhizal *Phaseolus vulgaris* to soil and root colonization by hyphae. *Journal of Plant Physiology, 160*, 1147–1156.

Bainard, L. D., Koch, A. M., Gordon, A. M., & Klironomos, J. N. (2012). Temporal and compositional differences of arbuscular mycorrhizal fungal communities in conventional monocropping and tree-based intercropping systems. *Soil Biology and Biochemistry, 45*, 172–180.

Bainard, L. D., Koch, A. M., Gordon, A. M., & Klironomos, J. N. (2013). Growth response of crops to soil microbial communities from conventional monocropping and tree-based intercropping systems. *Plant and Soil, 363*, 345–356.

Borie, F., Rubio, R., Morales, A., Curaqueo, G., & Cornejo, P. (2010). Arbuscular mycorrhizae in agricultural and forest ecosystems in Chile. *Journal of Soil Science and Plant Nutrition, 10*, 185–206.

Cely, M. V. T., de Oliveira, A. G., de Freitas, V. F., de Luca, M. B., Barazetti, A. R., dos Santos, I. M. O., Gionco, B., Garcia, G. V., Prete, C. E. C., & Andrade, G. (2016). Inoculant of arbuscular mycorrhizal fungi (Rhizophagusclarus) increase yield of soybean and cotton under field conditions. *Frontiers in Microbiology, 7*, 720.

Chalk, P. M., Souza, R., Urquiaga, S., Alves, B. J. R., & Boddey, R. M. (2006). The role of arbuscular mycorrhiza in legume symbiotic performance. *Soil Biology and Biochemistry, 38*, 2944–2951.

Culman, S., Bukowski, R., Gauch, H., Cadillo-Quiroz, H., & Buckley, D. (2009). T-REX:Software for the processing and analysis of T-RFLP data. *BMC Bioinformatics, 10*, 171.

Dai, J., Hu, J., Zhu, A., Bai, J., Wang, J., & Lin, X. (2015). No tillage enhances arbuscular mycorrhizal fungal population, glomalin-related soil protein content, and organic carbon accumulation in soil macroaggregates. *Journal of Soils and Sediments, 15*, 1055–1062.

Di Rienzo, J., Casanoves, F., Balzarini, M., Gonzalez, L., Tablada, M., & Robledo, C. (2014). InfoStat versión 2014. Grupo InfoStat, Universidad Nacional de Córdoba, Argentina. http://www.infostat.com.ar

García, S., Pezzani, F., & Rodríguez-Blanco, A. (2017). Long-term phosphorus fertilization effects on arbuscular mycorrhizal fungal diversity in Uruguayan grasses. *Journal of Soil Science and Plant Nutrition, 17*, 1013–1024.

García de León, D., Cantero, J. J., Moora, M., Öpik, M., Davison, J., Vasar, M., Jairus, T., & Zobel, M. (2018). Soybean cultivation supports a diverse arbuscular mycorrhizal fungal community in Central Argentina. *Applied Soil Ecology, 124*, 289–297.

Gollotte, A., van Tuinen, D., & Atkinson, D. (2004). Diversity of arbuscular mycorrhizal fungi colonizing roots of the grass species *Agrostis capillaris* and *Lolium perenne* in a field experiment. *Mycorrhiza, 14*, 111–117.

Gottshall, C. B., Cooper, M., & Emery, S. M. (2017). Activity, diversity and function of arbuscular mycorrhizae vary with changes in agricultural management intensity. *Agriculture Ecosystems & Environment, 241*, 142–149.

Hammer, O., Harper, D., & Ryan, P. (2001). PAST: Paleontological statistics software package for education and data analysis. *Palaeontologia Electronica, 4*, 9.

Hill, J. (2006). Inhibition of vesicular-arbuscular-mycorrhizae on soybean roots following Brassica cover crop. *Journal of Natural Resources and Life Sciences Education, 35*, 158–160.

Horn, S., Caruso, T., Verbruggen, E., Rillig, M. C., & Hempel, S. (2014). Arbuscular mycorrhizal fungal communities are phylogenetically clustered at small scales. *The ISME Journal, 11*, 2231–2242.

Jie, W., Liu, X., & Cai, B. (2013). Diversity of rhizosphere soil arbuscular mycorrhizal fungi in various soybean cultivars under different continuous cropping regimes. *PLoS ONE, 8*(8), e72898. https://doi.org/10.1371/journal.pone.0072898.

Lin, X. G., Feng, Y. Z., Zhang, H. Y., Chen, R. R., Wang, J. H., Zhang, J. B., & Chu, H. Y. (2012). Long-term balanced fertilization decreases arbuscular mycorrhizal fungal diversity in an arable soil in north China revealed by 454 pyrosequencing. *Environmental Science & Technology, 46*, 5764–5771.

Lumini, E., Vallino, M., Alguacil, M. M., Romani, M., & Bianciotto, V. (2011). Different farming and water regimes in Italian rice fields affect arbuscular mycorrhizal fungal soil communities. *Ecological Applications, 21*, 1696–1707.

Mathimaran, N., Ruh, R., Jama, B., Verchot, L., Frossard, E., & Jansa, J. (2007). Impact of agricultural management on arbuscular mycorrhizal fungal communities in Kenyan ferralsol. *Agriculture, Ecosystems and Environment, 119*, 22–32.

Moora, M., Davison, J., Öpik, M., Metsis, M., Saks, U., Jairus, T., Vasar, M., & Zobel, M. (2014). Anthropogenic land use shapes the composition and phylogenetic structure of soil arbuscular mycorrhizal fungal communities. *FEMS Microbiology Ecology, 90*, 609–621.

Oka, N., Karasawa, T., Okazaki, K., & Takeke, M. (2010). Maintenance of soybean yield with reduced phosphorus application by previous cropping with mycorrhizal plants. *Soil Science & Plant Nutrition, 56*, 824–830.

Pittelkow, C. M., Zorrilla, G., Terra, J., Riccetto, S., Macedo, I., Bonilla, C., & Roel, A. (2016). Sustainability of rice intensification in Uruguay from 1993 to 2013. *Global Food Security, 9*, 10–18.

Pontes, J. S., Oehl, F., Pereira, C. D., Cynthia Torres, C. M., Coyne, D., Alves da Silva, D. K., & Costa, L. M. (2017). Diversity of arbuscular mycorrhizal fungi in the Brazilian's Cerrado and in soybean under conservation and conventional tillage. *Applied Soil Ecology, 117–118*, 178–189.

Rillig, M. C., Wright, S. F., Nichols, K. A., Schmidt, W. F., & Torn, M. S. (2001). Large contribution of arbuscular mycorrhizal fungi to soil carbon pools in tropical forest soils. *Plant and Soil, 233*, 167–177.

Smith, S. E., & Read, D. J. (2008). *Mycorrhizal symbiosis*. Cambridge: Academic.

Trouvelot, S., Tuinen, D., & Gianinazzi-pearson, M. (1999). Visualization of ribosomal DNA loci in spore interphasic nuclei of glomalean fungi by fluoresence in situ hybridization. *Mycorrhiza, 8*, 203–206.

Vallino, M., Fiorilli, V., & Bonfante, P. (2013). Rice flooding negatively impacts root branching and arbuscular mycorrhizal colonization, but not fungal viability. *Plant, Cell & Environment, 37*, 57–572.

van der Heijden, M. G. A., Bardgett, R. D., & van Straalen, N. M. (2008). Theunseen majority: Soil microbes as drivers of plant diversity and productivity in terrestrial ecosystems. *Ecology Letters, 11*, 296–310.

Verbruggen, E., van der Heijden, M. G. A., Weedon, J. T., Kowalchuk, G. A., & Roling, W. F. M. (2012). Community assembly, species richness and nestedness of arbuscular mycorrhizal fungi inagricultural soils. *Molecular Ecology, 21*, 2341–2353.

Wang, J., Li, X., Zhang, J., Yao, T., Wei, D., Wang, Y., & Wang, J. (2012). Effect of root exudates on beneficial microorganisms: evidence from a continuous soybean monoculture. *Plant Ecology, 213*, 1883–1892.

Watanarojanaporn, N., Boonkerd, N., Tittabutr, P., Longtonglang, A., Young, J. P., & Teaumroong, N. (2013). Effect of rice cultivation systems on indigenous arbuscular mycorrhizal fungal community structure. *Microbes and Environments, 28*, 316–324.

Wright, S. F., & Upadhyaya, A. (1998). A survey of soils for aggregate stability and glomalin, a glycoprotein produced by hyphae of arbuscular mycorrhizal fungi. *Plant and Soil, 198*, 97–107.

Wright, S. F., Franke-Snyder, M., Morton, J. B., & Upadhyaya, A. (1996). Time-course study and partial characterization of a protein on hyphae of arbuscular mycorrhizal fungi during active colonization of roots. *Plant and Soil, 181*, 193–203.

Chapter 13
Evaluation of Bioproducts and Mycorrhizal Inoculation in Asian Soybean Rust Control, Nutrient Leaf Contents and Yield Under Field Conditions

André Riedi Barazetti, Ane Stéfano Simionato, Miguel Octavio Pérez Navarro, Mickely Liuti Dealis, Jean Marcos Soares Matos, Flúvio Modolon, Matheus Felipe de Lima Andreata, Gabriel Liuti, and Galdino Andrade

Abstract Asiatic Soybean Rust (ASR), caused by the fungus *Phakopsora pachyrhizi*, is the main disease affecting soybean crop, generating high losses each year, with insurgence of fungicide resistance. One strategy to decrease the fungus resistance is to use natural antimicrobial compounds to control ASR. Therefore, the objective of this work was to evaluate the potential for ASR control of a fraction produced during secondary metabolism by a *Pseudomonas aeruginosa* LV strain, inoculating plants with *Rhizophagus clarus*, an arbuscular mycorrhizal fungus (AMF) produced *in vitro*, evaluating the treatments effect on grain yields. Plants were treated with two doses of F4A semi- purified fraction (1 μg mL^{-1}, 10 μg mL^{-1}), mixed or not with commercial fungicides (Unizeb Gold® or Sphere Max®). Scanning electron microscopy (SEM) and light microscopy images showed that the association between F4A in both doses plus Sphere Max® provided protection, similarly to Unizeb Gold® or Sphere Max®, against ASR, causing ultrastructural changes in *P. pachyrhizi* hyphae and spores. Mycorrhizal colonization, plant growth and height increased in plants treated with F4A10 + Sphere Max. Plants treated with F4A1 compound produced 4.64 ton ha^{-1}, followed by F4A1 + Sphere (4.51 ton ha^{-1}), F4A10 + Sphere (4.44 ton ha^{-1}) and Unizeb+Sphere (4.36 ton ha^{-1}).

Keywords *Rhizophagus clarus* · Natural compounds · Soybean rust · Inoculant

A. R. Barazetti · A. S. Simionato · M. O. P. Navarro · M. L. Dealis · J. M. S. Matos · F. Modolon · M. F. de Lima Andreata · G. Liuti · G. Andrade (✉)
Department of Microbiology, Microbial Ecology Laboratory, Londrina State University, Londrina, PR, Brazil
e-mail: andradeg@uel.br

D. Zúñiga-Dávila et al. (eds.), *Microbial Probiotics for Agricultural Systems*, Sustainability in Plant and Crop Protection,
https://doi.org/10.1007/978-3-030-17597-9_13

13.1 Introduction

Brazil is one of the main grain producers in the world, being soybean the most important crop with a total area planted around 31,151 hectares and 118,885 thousand tonnes produced in the 2017/18 harvest (Conab 2018). To gain high productivity, a series of managements practices are essential, such as those needed for disease control and the use of chemical fertilizers. Management may increase production costs and cause serious environmental problems such as soil and water contamination, salinization, and also the selection resistant pathogens.

Among the diseases affecting soybean, the Asian Soybean Rust (ASR), caused by the fungus *Phakopsora pachyrhizi*, is responsible for main yield losses. The most common symptoms of ASR are small lesions with burnt yellow color, formed in the abaxial part of the leaf, associated to foliar chlorosis, leading to a premature defoliation and early maturation (Hartman et al. 2015). According to the Fungicide Resistance Action Committee (FRAC), *P. pachyrhizi* is becoming resistant to the main fungicide applied for control. The report (http://www.frac-br.org/) showed that, in areas with intensive use of carboxamide, the efficiency of *P. pachyrhizi* control is rapidly decreasing, in spite of a natural mutation in the subunit C (in position 186F). Also, fungicides specific to Asian Rust act through three action modes: demethylation inhibitors (IDM, triazoles), external quinone inhibitor (IQe, strobirulin) and succinate dehydrogenase inhibitor (ISDH, carboxamide). Other fungicides too, such as triazole and strobirulin, are losing efficacy vs *P. pachyrhizi*. The main risk related to the loss of efficacy of the fungicides available today is that there is no new mode of action to control Asian Rust, in the coming years.

COnsidering other management issues, the scarcity of natural reserves of phosphorus, an essential nutrient for plant development, promoted the search for tools which favor plant P assumption by roots, such as arbuscular mycorrhizae (Tapia-Torres et al. 2016).

The problems described above highlight the challenge to discover new fungicides for ASR management. New practices and technologies are essential to achieve a more sustainable agriculture, eg. the use of microbial inoculants and/or of biological antimicrobial compounds to control plant diseases. The interaction among microorganisms and plants occurs in the rhizosphere, a space under the direct influence of roots growth (Barea et al. 2005). Arbuscular mycorrhizal (AMF) fungi are obligatory symbionts that colonize most terrestrial plants, helping in plant nutrition, growth and disease tolerance, being one of the most important components of the rhizosphere (Smith and Read 2008). AMF act providing a root extension of plant, expanding the nutrient adsorption area beyond the rhizosphere, especially useful for phosphorus, which they release into root cells (Smith and Read 2008). AMF also form stable soil aggregates (Lehmann et al. 2017), increase resistance to water stress and erosion (Garg and Chandel 2009; Gonçalves et al. 2015; Querejeta 2017), protect roots from soil pathogens (Jung et al. 2012), thus globally sustaining agricultural productions (Andrade 2008; Cely et al. 2016).

Members of the *Pseudomonas* genus live in soils of diverse habitats. They have a high adaptive capacity and produce several metabolites with antimicrobial activity, including growth promotion and eliciting of plant defense or resistance pathways.

Among these metabolites, the phenazines are very important. They form a is a group of aromatic heterocyclic compounds, formed during the secondary metabolism, that could help farmers in the competition with pathogens. Phenazines diffuse or enter though the microbial membrane, resulting in oxidative phosphorylation decoupling and generating superoxide radicals and hydrogen peroxide, harmful to the cell. They also affect the electrons flow and the enzymes related to cellular respiration (Navarro et al. 2017). Different types of phenazines are known: phenazine-1-carboxylic acid (PCA), due to its oxy-reduction activity and superoxide accumulation, has a wide antimicrobial activity, mainly against phytopathogenic fungi such as *Botrytis cinerea* (Simionato et al. 2017) and *Phellinus noxius*, the casual agent of brown root rot disease (Huang et al. 2016). PCA is a precursor of other types of phenazine, such as phenazine-1-carboxamide (PCN) that also has antifungal activity against phytopathogenic microorganisms (Zhang et al. 2015). In addition, a natural organo-copper antibiotic compound (OAC), is also produced by *P. aeruginosa* LV strain, with high antimicrobial activity against *Xanthomonas citri* subsp. *citri*, the causal agent of citrus canker (de Oliveira et al. 2016) and others plant diseases.

The objective of this work was to evaluate the effect of a semi-purified fraction produced by *P. aeruginosa* strain LV, called F4A (PCA, PCN, OAC and Indolin-3-one) to control ASR in plants inoculated with AM fungi *R. clarus*.

13.2 Material and Methods

13.2.1 In vitro *Experiment*

An *in vitro* assay was carrie out to evaluate the effect of different doses of F4A (1 μg ml^{-1}, 10 μg ml^{-1} and 100 μg ml^{-1}) on *P. pachyrizis* spore germination. Spores collected from uredia were inoculated on water agar plus F4A, in different concentrations in Petri dishes, in three replicates. 300 μL of a spore suspension with $2.15 \cdot 10^4$ ml^{-1} were inoculated and incubated. After 12 h the spore germination was evaluated with a photomicroscope (40 ×), measuring the percentage of spore germination.

13.2.2 Field Experiment

13.2.2.1 Experimental Area

The experiment was carried out during the 2016/17 harvest at the experimental farm of Londrina State University, PR, Brazil. The soil is classified as Dystroferric Red Ferralsol (FAO 2006).Soil chemical analysis is: pH ($CaCl_2$) 6.44; Al^{3+} 0.00 cmolc Kg^{-1}; Ca^{2+} 8.45 cmolc Kg^{-1}; Mg^{2+} 2.46 cmolc Kg^{-1}; K^+ 0.15 cmolc Kg^{-1}; P 15.60 cmolc Kg^{-1}; C 12.14 cmolc Kg^{-1}. Mean temperature and precipitation from November 2016 to March 2017 were 24.2 °C and 698.8 mm respectively (Iapar 2017).

13.2.2.2 Soybean Experiment

Pre-emerging weed control was carried out with N (phosphonomethyl) glycine (Roundup® WG, Monsanto), using the recommended dose. Pests and insects control was performed with imidacloprid (Imida Gold 700® WG, UPL), tiametoxam + lambda-cialotrina (Engeo TMPleno®, Syngenta) and espiromesifene (Oberon®, Bayer) using the recommended doses.

Soybean seeds NS 5959 IPRO (Nidera Sementes, Brazil) were treated with metalaxyl-M and fludioxonil (Maxim® XL, Syngenta) and Imidacloprid + Tiodicarbe (Cropstar®, Bayer) at the recommended doses, before sowing.

A randomized block design with seven treatments (Uni + Sphe; F4A1; F4A1 + Uni; F4A1 + Sphe; F4A10; F4A10 + Uni and F4A10 + Sphe) was performed, with two doses of F4A (1 and 10 μg mL^{-1}) combined with Unizeb Gold® or Sphere Max®, in five replicates per treatment. The agronomic treatment consisted on applying fungicides Unizeb Gold plus Sphere Max.

The fertilizer dose used was 250 Kg ha^{-1} of $N\text{-}P_2O_5\text{-}K_2O$ 10-15-15 (Fertilize® Fertilizers, Fertilize Agrícola Ltda). All treatments were also inoculated with *Bradyrhizobium japonicum* (SEMIA 5079 and SEMIA 5080, Rizo Plus®, Rhizobacter), according to the recommended doses.

The inoculum of *R. clarus* was obtained from the collection of the Microbial Ecology Laboratory, Londrina, Brazil, produced under *in vitro* conditions in association with transformed carrot (*Daucus carota* L.) roots (Ri T-DNA), in a continuous subculture of young colonized roots fragments in minimal mineral medium M (Bécard and Fortin 1988). The inoculation of *R. clarus* in soybean seeds was mixed with peat according to patent No BR 1020140173897 (07/15/2014). The inoculum concentration used had approximately 47.5 fungal spores, colonized roots and mycelia per g of seed.

Seeds were coated with *B. japonicum, R. clarus* and graphite before sowing to avoid friction during the procedure, providing a uniform stand, and sowed with hand planter. Plot dimension was of 6.0 m × 3.0 m. The area used for data collection was 9 m^2, equivalent to three central lines, excluding 1.0 m from the edges, besides the two external lines. Soybeans were grown from November 2016 to March 2017.

13.2.2.3 Plant Measurements

Five soybean plants from three central adjacent lines were used to measure AM root colonization and root length before flowering at the R1 stage, according to Fehr and Caviness classification for soybean development stages (1977). Plants were brought to the laboratory and the secondary roots selected, cleaned and stained according to Phillips and Hayman (1970). Colonization was determined by the grid line method (Giovanetti and Mosse 1980).

Shoot dry weight was determined from five plants collected as described above, dried at 50 °C until a constant mass was obtained and dry weight was determined. The total height was determined by field measurement of 5 plants per plot of the central lines, in stage R5.1, and the average was calculated. Grain yield was evaluated by harvesting 3 m^2 of the central lines of each plot, and expressed as tonnes ha^{-1}.

Phosphorus and nitrogen foliar contents were evaluated on the third trefoil, starting from the apex without petiole at the R1 stage. Samples were dried at 50 °C and ground. Phosphorus was extracted by nitroperchloric digestion and quantified (Murphy and Riley 1962). Nitrogen was extracted by sulfur digestion and quantified (Sarruge and Haag 1974).

The ASR severity (leaf area covered with symptoms) was estimated weekly after the first application of products, using diagrammatic scale (Godoy et al. 2006) to calculate the area under the disease progress curve (AUDPC).

In addition, soybean leaf samples were dried at CO_2 critical point (BALTEC CPD 030 Critical Point Dryer). After drying, the slides were gold coated (BALTEC SDC 050 Sputter Coater) and visualized by scanning electron microscopy (SEM) (FEI Quanta 200), to verify ultrastructural changes in the spores and mycelium of *P. pachyrhizi*.

13.2.2.4 Statistical Analysis

The Shapiro-Wilk test was applied ($\alpha = 5\%$). Homogeneity of variances was verified by the Bartlett test ($\alpha = 5\%$). After all results met the assumptions, variance analysis was applied. The means were compared by the Duncan test, at $p < 0.1$ (Nelson 1989) using R. Statistical program.

13.3 Results

AM colonization was greater in plants treated with F4A10 + Sphere with 75% of colonization, followed by F4A10, F4A1 and F4A1 + Sphere. AM colonization decreased in plants treated with Unizeb Gold in both doses.when compared to Uni + Sphere Max agronomic treatment (Table 13.1).

The nitrogen foliar contents increased in plants treated with Uni + Sphe, when compared with F4A1 + Sphe. The phosphorus foliar content was higher in plants treated with F4A10 + Sphe, followed by F4A1 + Sphe, Uni + Sphe and F4A10 + Uni. The lowest content was observed in plants treated with F4A1 (Table 13.1).

F4A10 + Sphe showed highest total protein contents, differing significantly from F4A1 + Uni and F4A10. The other treatments did not show any differences (Table 13.1).

F4A10 + Sphe enhanced plant height when compared to the agronomic treatment suggesting a plant growth promotion effect, and was followed by F4A1, F4A10 + Uni and F4A1 + Uni, all differing significantly from the treatment with Uni + Sphe. Shoot dry weight F4A10 + Uni and F4A1 + Uni were the highest. In contrast, F4A1 and F4A10 + Sphe were the lowest (Table 13.1).

Table 13.1 Parameters evaluated under field conditions per treatment: Arbuscular mycorrhizal colonization (AM); nitrogen and phosphorus foliar contents; seed proteins; plant height and shoot dry weight; yield

Treatment	AM (%)	N (mg plant^{-1})	P (mg plant^{-1})	Seed proteins (mg seed^{-1})	Height (cm)	Shoot dry weight (g)	Yield (tonnes ha^{-1})
Uni + Sphe	68.58 b	46.98 a	3.12 a	313.68 ab	52.52 cd	46.44 cd	4.37 a
F4A1	75.81 a	45.47 ab	2.98bc	311.13 ab	62.12 ab	40.48 e	4.64 a
F4A1 + Uni	70.69 b	44.86 ab	3.08 ab	307.93 b	60.96 ab	52.47 ab	4.32 ab
F4A1 + Sph	74.52 a	43.57 b	3.11 a	326.16 ab	58.20bc	44.77 d	4.51 a
F4A10	76.18 a	43.85 ab	3.07 ab	302.81 b	58.28bc	49.78bc	3.68 bc
F4A10 + Uni	69.58 b	43.57 ab	3.06 a	314.64 ab	61.33 ab	55.29 a	4.35 ab
F4A10 + Sph	77.84 a	45.30 ab	3.12 a	335.75 a	66.56 a	38.48 e	4.45 a

Values followed by the same letters had no significant difference ($p < 0.10$) Duncan test

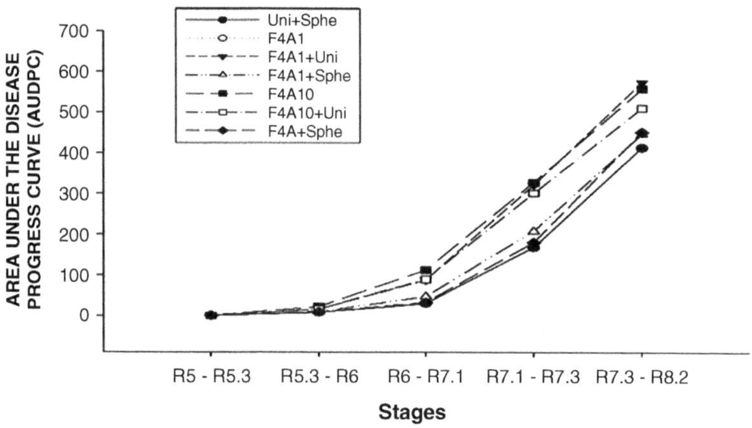

Fig. 13.1 Area under the disease progress curve of treatments for Asian soybean rust, field conditions

Plants treated with F4A1 compound produced highest yield, with F4A1 + Sphe, F4A10 + Sphe and Uni + Sphe, all significantly different from F4A10 (Table 13.1).

Finally, the AUDPC decreased with Uni + Sphe, being the most effective on ASR control, followed by F4A10 + Sphe and F4A1 + Sphe, while F4A1 + Uni, F4A1 and F4A10 (Fig. 13.1).

In the preliminary *in vitro* test, it was observed that the spores of *P. pachyrhizi* had 85% of germination, with intense hyphal growth after 12 h. When applying F4A 1 μg mL^{-1}, a drastic reduction of hyphae development, with 75% of spores germination was observed. F4A 10 μg mL^{-1} was effective to control hyphal growth, showing only 40% of spores germinated. Finally, the 100 μg mL^{-1} dose was 100% effective, completely inhibiting spore germination, and growth of hyphae (Fig. 13.2).

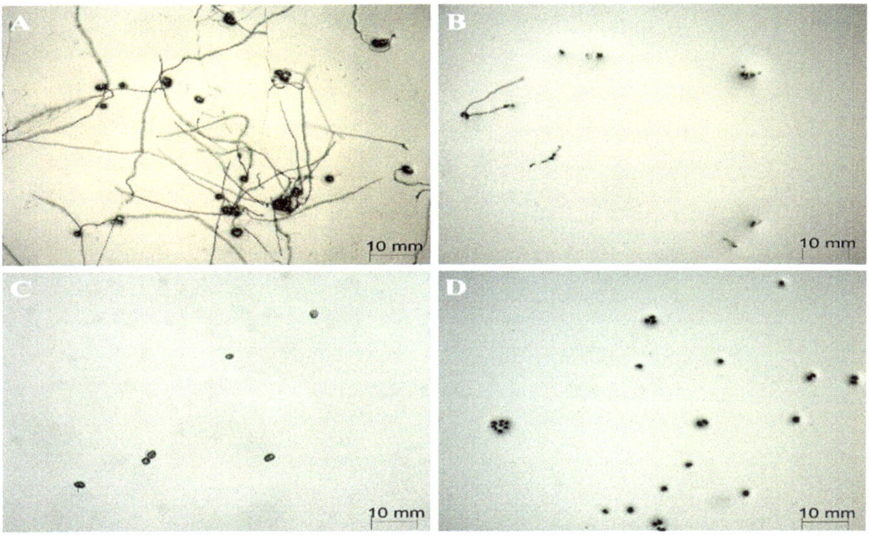

Fig. 13.2 Light microscopy images of *Phakopsora pachyrhizi* spore development under different F4A doses. Treatments are: control (**a**), F4A at 1 μg ml^{-1} (**b**), F4A at 10 μg ml^{-1} (**c**) and F4A at 100 μg ml^{-1}

In addition, SEM images of *P. pachyrhizi* infecting soybean leaves showed spores and uredia in non-treated plants, with high hyphal growth, uredia formation and large number of spores. Dead spores and a lower number of pustules were observed in agronomic treatment (Uni + Sphe) (Fig. 13.3).

F4A1 apparently did not affect *P. pachyrhizi*. F4A1 + Uni presented a large number of unviable spores, while F4A1 + Sphere Max reduced the number of fungal pustules, consequently reducing spore numbers.

The application of F4A10 reduced the development of fungal mycelium and large number of unviable spores were observed. In F4A10 + Uni a great presence of spores, hyphae and pustules were observed, but the spores and mycelia were unviable; F4A10 + Sphe showed greater fungicide activity against *P. pachyrhizi*, with larger number of dead spores and hyphae.

13.4 Discussion

The application of Unizeb Gold (Mancozeb) reduced AM colonization. This product is a fungicide that acts by contact.The same results were observed by other authors, where Mancozeb was applied in the aerial part of plants with a negative effect on root colonization and spore density in the rhizosphere (Channabasava et al. 2015; Mallmann et al. 2018).

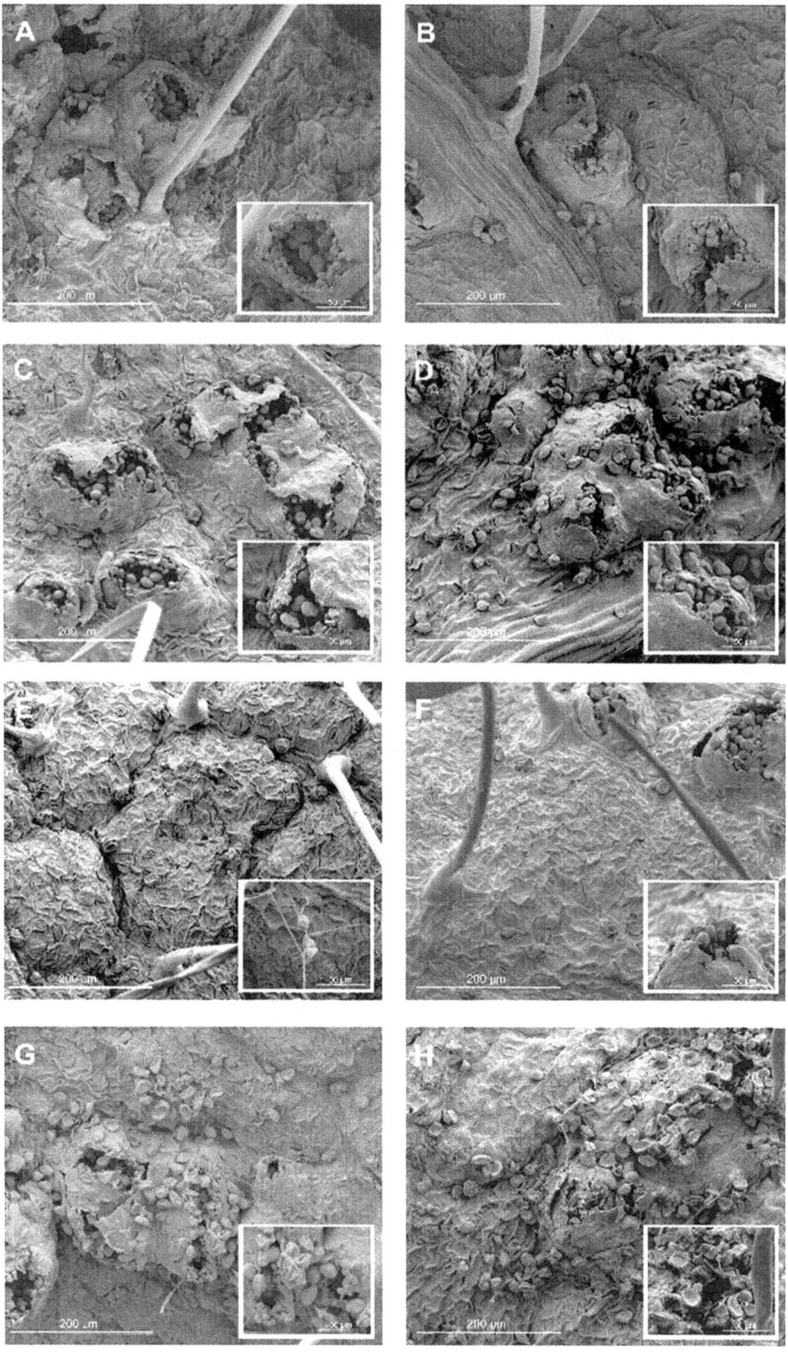

Fig. 13.3 Scanning electron microscopy images of soybean leaf's infected with *Phakopsora pachyrhizi* under different treatments. Treatments are: control (**a**), Uni + Sphe (**b**), F4A1 (**c**), F4A1 + Uni (**d**), F4A1 + Sphe (**e**), F4A10 (**f**), F4A10 + Uni (**g**) and F4A10 + Sphe (**h**). Scale bars: 200 µm (insights magnification = 4×)

During the first weeks of the association between a mycorrhizal fungus and a host plant, the fungus tends to suppress plant growth, since it competes for nutrients, especially phosphorus. In later stages of association the reverse occurs, where symbiosis is already established and transfer nutrients from the fungus to the plant is greater (Bethlenfalvay et al. 1983).

Uni + Sphe application slightly decreased AM colonization after 60 days, but P content increased, suggesting that plants did not suffer growth depletion in the initial stages of the mycorrhizal establishment and consequently had more nutrients, especially P.

Plants treated with F4A10 + Sphe showed highest growth. Sphere Max is a systemic fungicide, composed of trifloxystrobin and ciproconazole.The first one belongs to the strobilurin group, inducers of physiological and metabolic responses, inhibiting ethylene biosynthesis and increasing cytokinin production, extending the time of green pigmentation (Han et al. 2012). Also, F4A10 has an indolin-3-one that may promote plant growth, as observed in others studies in greenhouse experiments (unpublished data).

It was observed that F4A10 alone reduced yield and had low ASR control, which reduces the leaf area for photosynthesis and consequently the production of pods and grain filling. In contrast, F4A1 alone showed the highest values for yield. PCN produced by *Pseudomonas* sp. CMR12a, triggered the induced systemic resistance (ISR) in rice and bean (Ma et al. 2016). However it was highly dependent on the phenazine concentration, being optimum at concentrations of 0.1 mM and 1 mM, and losing effect at lower and higher concentrations. In addition, F4A1 plus Sphere max and F4A10 plus Sphere max were the 2nd and 3rd in yield respectively.

As mentioned, F4A has molecules with oxy-reduction activities, superoxide accumulation and antimicrobial activity against phytopathogens, plus to possible growth promotion in plants and ISR.This, added to the Sphere Max effect, indicates a synergism between the natural compound and the commercial product.

The AUDPC analysis showed that Uni + Sphe, F4A1 + Sphe and F4A10 + Sphe were the most effective for ASR control, which is evident aat the R6 stage. As observed in the *in vitro* assay, 1, 10 and 100 μg mL^{-1} doses of F4A were able to control the development of *P. pachyrhizi*. The same effect was observed in SEM images where F4A that, in association with Sphere Max, caused several changes in the fungal structures, as spores and hyphae.

Natural compounds from different origins have been related to an antimicrobial action against different pathogens in the recent years, including *P. pachyrhizi*. Dorighello et al. (2015) demonstrated that coffee oil combined with half dose of fungicide, inhibited 100% of uredospore germination in greenhouse conditions. In field conditions the severity of disease was reduced by 23%.

In another SEM study, da Silva et al. (2014) observed that the essential oils of *Hyptis marrubioides*, *Aloysia gratissima* and *Cordia verbenacea* completely inhibited spore germination under *in vitro* conditions. However, there are few studies about the antibiotic effect of secondary metabolites from *P. aeruginosa* against phytopathogenic bacteria or fungi, in the field.

F4A semi-purified fraction, specially OAC, showed an effect against many bacteria and fungi (Oliveira et al. 2011; Lopes et al. 2012; Munhoz et al. 2017). PCN did not show activity against *Pectobacterium carotovorum*. However, it increased the productivity and number of fruits in uninfected plants, and also increased peroxidase, polyphenol oxidase and phenylalanine ammonia lyase activities, suggesting that it may induce systemic acquired resistance (SAR) (Munhoz et al. 2017). PCA, another compound present among the F4A metabolites, demonstrated antifungal activity against *Botrytis cinerea* (Simionato et al. 2017).

The results showed that F4A compounds have antimicrobial activity against different microorganisms. In this work, we showed that *P. pachyrhizi* spores and hyphae suffer serious damage when treated with F4A. However, new studies are needed to determine action mechanisms of F4A compounds.

13.5 Conclusion

Unizeb Gold showed a reduction in AM colonization. Nutrient foliar contents and shoot dry weight were higher in agronomic treatment. Plants treated with F4A1, F4A1 + Sphere max and F4A10 + Sphere max showed higher yields. Uni + Sphe, F4A1 + Sphe and F4A10 + Sphe were the most effective treatments to control ASR, causing structural changes in hyphae and spores. However, the application of 10 μg ml^{-1} F4A alone did not show ASR control, and consequently decreased yield.

Acknowledgements The authors are grateful to Laboratory for Electron Microscopy and Microanalysis – LMEM UEL/FINEP for carrying out SEM analysis and to the National Council for Scientific and Technological Development (CNPq) for supporting this project with grants PIBID, MSc, PhD and Research Productivity.

References

Andrade G (2008) Role of functional groups of microorganisms on the rhizosphere microcosm dynamics. In: Varma A, Abbott L, Werner D HR (ed) Plant surface microbiology. Springer, Berlin/Heidelberg, pp 51–69.

Barea, J. M., Pozo, M. J., Azcón, R., & Azcón-Aguilar, C. (2005). Microbial co-operation in the rhizosphere. *Journal of Experimental Botany, 56*, 1761–1778. https://doi.org/10.1093/jxb/eri197.

Bécard, G., & Fortin, J. A. (1988). Early events of vesicular-arbuscular mycorrhiza formation on Ri T-DNA transformed roots. *New Phytologist, 108*(2), 211–218.

Bethlenfalvay, G. J., Bayne, H. G., & Pacovsky, R. S. (1983). Parasitic and mutualistic associations between a mycorrhizal fungus and soybean: the effect of phosphorus on host plant–endophyte interactions. *Physiologia Plantarum, 57*, 543–548. https://doi.org/10.1111/j.1399-3054.1983.tb02783.x.

Cely, M. V. T., de Oliveira, A. G., de Freitas, V. F., et al. (2016). Inoculant of arbuscular mycorrhizal fungi (*Rhizophagus clarus*) increase yield of soybean and cotton under field conditions. *Frontiers in Microbiology, 7*, 1–9. https://doi.org/10.3389/fmicb.2016.00720.

Channabasava, A., Lakshman, H., & Jorquera, M. (2015). Effect of fungicides on association of arbuscular mycorrhiza fungus *Rhizophagus fasciculatus* and growth of Proso millet (*Panicum miliaceum* L.). *Journal of Soil Science and Plant Nutrition, 15*, 35–45. https://doi.org/10.4067/S0718-95162015005000004.

Conab. (2018). *Acompanhamento da safra brasileira*. http://conab.gov.br

da Silva, A. C., de Souza, P. E., Amaral, D. C., et al. (2014). Essential oils from *Hyptis marrubioides*, *Aloysia gratissima* and *Cordia verbenacea* reduce the progress of Asian soybean rust. *Acta Scientiarum Agronomy, 36*, 159–166. https://doi.org/10.4025/actasciagron.v36i2.17441.

de Oliveira, A. G., Murate, L. S., Spago, F. R., Lopes, L., Beranger, J. P. d. O., Martin, J. A. B. S., Nogueira, M. A., de Mello, J. C. P., Andrade, C. G. T. d. J., & Andrade, G. (2011). Evaluation of the antibiotic activity of extracellular compounds produced by the Pseudomonas strain against the Xanthomonas citri pv. Citri 306 strain. *Biological Control, 56*(2), 125–131.

de Oliveira, A. G., Spago, F. R., Simionato, A. S., et al. (2016). Bioactive organocopper compound from *Pseudomonas aeruginosa* inhibits the growth of *Xanthomonas citri* subsp. citri. *Frontiers in Microbiology, 7*, 1–12. https://doi.org/10.3389/fmicb.2016.00113.

Dorighello, D. V., Bettiol, W., Maia, N. B., & de Campos Leite, R. M. V. B. (2015). Controlling Asian soybean rust (*Phakopsora pachyrhizi*) with *Bacillus* spp. and coffee oil. *Crop Protection, 67*, 59–65. https://doi.org/10.1016/j.cropro.2014.09.017.

FAO. (2006). *World Reference Base for soil resources 2006: A framework for international classification, correlation and communication*. Rome: World Soil Resources Reports.

Fehr, W. R., & Caviness, C. E. (1977). *Stages of soybeandevelopment*. Iowa State University Cooperative Extension Service, Special Report 80.

Garg, N., & Chandel, S. (2009). Arbuscular mycorrhizal networks: Process and functions. *Sustain Agric, 2*, 907–930. https://doi.org/10.1007/978-94-007-0394-0_40.

Giovanetti, M., & Mosse, B. (1980). An evaluation of techniques for measuring vesicular arbuscular mycorrhizal infection in roots. *The New Phytologist, 84*, 489–500. https://doi.org/10.1111/j.1469-8137.1980.tb04556.x.

Godoy, C. V., Koga, L. J., & Canteri, M. G. (2006). Diagrammatic scale for assessment of soybean rust severity. *Fitopatologia Brasileira, 31*, 63–68. https://doi.org/10.1590/S0100-41582006000100011.

Gonçalves, P. J. R. O., Oliveira, A. G., Freitas, V. F., et al. (2015). Plant growth-promoting microbial inoculant for Schizolobium parahyba pv. parahyba. *Rev Árvore, 39*, 663–670. https://doi.org/10.1590/0100-67622015000400008.

Han, S. H., Kang, B. R., Lee, J. H., et al. (2012). A trifloxystrobin fungicide induces systemic tolerance to abiotic stresses. *Plant Pathology Journal, 28*, 101–106. https://doi.org/10.5423/PPJ.NT.11.2011.0207.

Hartman, G. L., Rupe, J. C., Sikora, E. J., et al. (2015). *Compendium of soybean diseases and pests* (5th ed.). The American Phytopathological Society.

Huang, H., Sun, L., Bi, K., et al. (2016). The effect of phenazine-1-carboxylic acid on the morphological, physiological, and molecular characteristics of *Phellinus noxius*. *Molecules, 21*, 1–10. https://doi.org/10.3390/molecules21050613.

IAPAR. (2017). Dados Diários de Londrina. http://www.iapar.br/modules/conteudo/conteudo.php?conteudo=1828

Jung, S. C., Martinez-Medina, A., Lopez-Raez, J. A., & Pozo, M. J. (2012). Mycorrhiza-induced resistance and priming of plant defenses. *Journal of Chemical Ecology, 38*, 651–664. https://doi.org/10.1007/s10886-012-0134-6.

Lehmann, A., Leifheit, E. F., & Rillig, M. C. (2017). Mycorrhizas and soil aggregation. In N. C. Johnson & G. C. JJ (Eds.), *Mycorrhizal mediation of soil* (pp. 241–262). Amsterdam: Elsevier.

Lopes, L. P., Oliveira, A. G., Jr., Beranger, J. P. O., et al. (2012). Activity of extracellular compounds of *Pseudomonas* sp. against *Xanthomonas axonopodis in vitro* and bacterial leaf blight in eucalyptus. *Trop Plant Pathol, 37*, 233–238. https://doi.org/10.1590/S1982-56762012000400001.

Ma, Z., Hua, G. K. H., Ongena, M., et al. (2016). Role of phenazines and cyclic lipopeptides producedby pseudomonas sp. CMR12a in induced systemic resistance on rice and bean. *Environmental Microbiology Reports, 8*, 896–904. https://doi.org/10.1111/1758-2229.12454.

Mallmann, G. C., Sousa, J. P., Sundh, I., et al. (2018). Placing arbuscular mycorrhizal fungi on the risk assessment test battery of plant protection products (PPPs). *Ecotoxicology, 27*, 1–10. https://doi.org/10.1007/s10646-018-1946-0.

Munhoz, L. D., Fonteque, J. P., Santos, I. M. O., et al. (2017). Control of bacterial stem rot on tomato by extracellular bioactive compounds produced by *Pseudomonas aeruginosa* LV strain. *Cogent Food & Agriculture, 3*, 1–16. https://doi.org/10.1080/23311932.2017.1282592.

Murphy, J., & Riley, J. P. (1962). A modified single solution method for the determination of phosphate in natural waters. *Analytica Chimica Acta, 27*, 31–36. https://doi.org/10.1016/S0003-2670(00)88444-5.

Navarro, M. O. P., Simionato, A. S., Barazetti, A. R., et al. (2017). *Disease-induced resistance and plant immunization using microbes*. Singapore: Springer.

Nelson, L. (1989). A statistical editor's viewpoint of statistical usage in horticultural science publications. *Hortscience, 24*, 53–57.

Phillips, J. M., & Hayman, D. S. (1970). Improved procedures for clearing roots and staining parasitic and vesicular-arbuscular mycorrhizal fungi for rapid assessment of infection. *Transactions of the British Mycological Society, 55*(1), 158–IN18.

Querejeta, J. I. (2017). Soil water retention and availability as influenced by mycorrhizal symbiosis. In *Mycorrhizal mediation of soil* (pp. 299–317). Amsterdam: Elsevier.

Sarruge, J., & Haag, H. (1974). *Análises químicas em plantas*. Piracicaba: ESALQ.

Simionato, A. S., Navarro, M. O. P., de Jesus, M. L. A., et al. (2017). The effect of phenazine-1-carboxylic acid on mycelial growth of *Botrytis cinerea* produced by *Pseudomonas aeruginosa* LV strain. *Frontiers in Microbiology, 8*, 1–9. https://doi.org/10.3389/fmicb.2017.01102.

Smith and Read. (2008). Mycorrhizal symbiosis. *Soil Science, 137*, 204. https://doi.org/10.1097/00010694-198403000-00011.

Tapia-Torres, Y., Rodríguez-Torres, M. D., Elser, J. J., et al. (2016). How to live with phosphorus scarcity in soil and sediment: Lessons from bacteria. *Applied and Environmental Microbiology, 82*, 4652–4662. https://doi.org/10.1128/AEM.00160-16.

Zhang, Y., Wang, C., Su, P., & Liao, X. (2015). Control effect and possible mechanism of the natural compound phenazine-1-carboxamide against *Botrytis cinerea*. *PLoS One, 10*, 1–17. https://doi.org/10.1371/journal.pone.0140380.

Chapter 14
Effect of Inoculation with Glomeromycota Fungi and Fertilization on Maize Yield in Acid Soils

Eliecer Cabrales, Danilo Lopez-Hernández, and Marcia Toro

Abstract The use of arbuscular mycorrhizae (AM) to improve crop productions in acid soils has been implemented through the production of inoculants based on native Glomeromycota fungi.The inoculants were isolated by wet sieving and decanting, sucrose gradient followed by reproducetion in trap pots with *Brachiaria decumbens*. Predominant species were: *Cetraspora pellucida*, *Scutellospora calospora*, *Ambispora leptoticha* and *Acaulospora mellea*. Each native species was reproduced individually followed by mixing the fungi together as a consortium. Inocula were applied to hybrid maize HIMECA 3003 and evaluated with 0, 27, 54 and 80 kg P_2O_5 ha^{-1} as diammonium phosphate (DAP). Seeds were sowed in 4.5 m × 3.6 m plots, in a randomized block design with factorial arrangement, for a total of 72 plots. Eight g of inoculum per sowing site were added (with 150 spores /100 g soil). The seeds were placed directly over the inoculum. Maize was harvested after 3 months to determine N and P in stem. The most efficient combination of P for N nutrition was 54 kg $P_2O_5.ha^{-1}$ and mixed inoculums. The highest P content in plant was obtained with 54 kg P_2O_5 ha^{-1} and *C. pellucida.* Fertilization combined with AM allowed 25% reduction of P fertilization and increased up to 100% in maize yield (from 2 to 4 t ha^{-1}), compared to yield data of the zone.

Keywords Arbuscular mycorrhizae (AM) · Acid soils · Maize · Fertilization

E. Cabrales
Facultad de Ciencias Agrícolas, Universidad de Córdoba, Montería, Colombia

D. Lopez-Hernández · M. Toro (✉)
Laboratorio de Estudios Ambientales, Instituto de Zoología y Ecología Tropical, Facultad de Ciencias, Universidad Central de Venezuela, Caracas, Venezuela
e-mail: marcia.toro@ciens.ucv.ve

D. Zúñiga-Dávila et al. (eds.), *Microbial Probiotics for Agricultural Systems*, Sustainability in Plant and Crop Protection,
https://doi.org/10.1007/978-3-030-17597-9_14

14.1 Introduction

Tropical savannas are characterized by the presence of low fertility soils and predominance of the herbaceous component (Sarmiento et al. 2004). Nearly 270 million hectares of tropical savannas cover South America. Venezuela is the second country for extension of this type of ecosystems (Gómez and Paolini 2011). A significant portion of savanna soils is used for agricultural purposes, especially for maize without irrigation, with yields of approximately 2000 kg ha^{-1} (Vielma et al. 2005). However, agricultural activity can lead to the use of huge dose of fertilizers, producing an environmental contamination. The use of beneficial microorganisms (growth promoting bacteria and/or arbuscular mycorrhizae, AM) with complementary fertilization, may help to improve crops productivity. Glomeromycota fungi associate with 80% of vascular plants (Schüßler et al. 2001; Smith and Read 2008), thus forming arbuscular mycorrhizae. Maize is one of the crops that are frequently inoculated with AM (Harrier and Watson 2003; Rakshit and Bhadoria 2010). We focused on its study due to its importance as an important food source in the country.

Low P content of tropical soils favors the presence of AM (Cardoso and Kuyper 2006; Carrenho et al. 2007). Previous research have shown that reducing P fertilization (applied as rock phosphate) and inoculating AM, promoted maize yields of up to 3.1 Mg ha^{-1} (Hernández et al. 2012). Toro (2007) and Toro et al. (2008), report that several genera of Glomeromycota such as *Acaulospora*, *Glomus* and *Scutellospora* are associated to maize in Venezuela savannas soils. The aim of this study was to evaluate the efficiency of four species of native *Glomeromycota*, testing an inoculum composed of all native species as consortia, on maize growth and productivity in a field experiment carrid out using acid savanna soils.

14.2 Materials and Methods

This study was carried out at Estación Experimental La Iguana, in the southeastern area of Venezuela, Santa María de Ipire (8° 25' N;65° 25′ E). Yellow maize HIMECA 3005 seeds were sowed at a 0.9 m distance between rows and 0.4 m between plants. Two seeds per site were sowed, with density of 55.000 plants ha^{-1}. Four species of Glomeromycota reported as native of these soils by Toro (2007) were used: *Cetraspora pellucida, Scutellospora calospora, Ambispora leptoticha, Acaulospora mellea* alone or as a consortium (all species). Inoculation was performed adding 8 g of each inoculum per site· The inoculum consisted of a mixture of spores, rootlets and mycelium of the corresponding fungi alone or as a consortium. Maize seeds were placed above the inoculum, to allow fungal colonization of roots. All inocula contained 150 spores · 100 g^{-1} of soil. Inoculation treatments were as follows: M0 = No AM inoculation; M1 = *C. pellucida*; M2 = *S. calospora*; M3 = *A. leptoticha*; M4 = *A. mellea* and M5 = consortium of all species.

Table 14.1 Chemical analysis of soil

	pH	O.M.	S	P	Ca	Mg	Na	K	Al	CEC	Cu	Fe	Mn	Zn	Mo
Depth	1:1	%	($mg.kg^{-1}$)		($cmol.kg^{-1}$)						($mg.kg^{-1}$)				
0–20 cm	5.05	0.51	26.8	10.4	0.5	1.0	0.13	0.05	0.4	2.08	0.4	15.2	8.4	0.8	0.04

Fertilization was based on soil chemical analysis (Table 14.1) and crop requirements. The N dose was fractionated after 18-28-35 days of plant emergence by applying 45-53-52 kg/ha of N, as urea. Phosphorus and potassium doses were fully applied when sowing. P doses were: 0-27-54-80 kg/ha of P_2O_5 as DAP. 80 kg K ha^{-1} were applied as KCI.

P and N content in leaves were evaluated in the leaf opposite to the maize cob. Dried and ground foliar tissues were digested with binary mixture (sulfuric acid: perchloric acid, 4:1). P was determined as described by Murphy and Riley (1962) and N was determined by distillation with Kjeldahl (IGAC 2006).

We evaluated the effect of a combination of 4 doses of P and 6 inoculation treatments with the fungi species in foliar N and P content and grain yields. A randomized block design with factorial arrangement was used with the following factors: factor A, dose of P applied (4 levels or fertilization doses with P: 0-27-54-80 kg P_2O_5/ha); factor B, inoculation with fungi, with 6 levels (M0, M1, M2, M3, M4 and M5). Each experimental plot was integrated by 5 rows of maize. Sampling was carried out in three sites per treatment. Analysis of variance (ANOVA) was applied to assess the effect of the interaction. Descriptive statistics was carried out with SAS ver. 9.0, after checking ANOVA assumptions. Tukey's HSD (Honestly-Significant-Difference) Test was applied with a 95% confidence level.

14.3 Results and Discussion

14.3.1 Nitrogen Content in Leaves

Doses of P and fungi species, as well as their interaction (P dose · fungal species), had significant influence ($p < 0.001$) in N absorption. Foliar N content tended to increase with P doses. In Table 14.2, treatment with 0 kg P_2O_5 ha^{-1} was not included because there was no leaf opposite to the corncob at sampling.

In treatments with P fertilization, the highest level of nitrogen absorption was reached when 54 kg P_2O_5 ha^{-1} was applied (Table 14.2). Several authors have underlined the importance of P, not only in production of fruits and several crops, but also in growth and development of maize (Miller 2000; Harrier and Watson 2003; Smith et al. 2003), therefore, soils with low P contents tend to develop less vigorous plants.

When comparing the mean N absorption due to fungal inoculation, significant differences were observed ($p < 0.001$) among plants with or without AM inoculation. However, this difference in N plant uptake was not observed among mycorrhized treatments. Nevertheless, best absorption was achieved with the native

Table 14.2 Nitrogen content (and %) in the cob opposite leaf of maize inoculated with Glomeromycota fungi grown with different phosphorus doses and harvested in the flowering season

			Treatments				
P_2O_5 (kg ha^{-1})	M0	M1	M2	M3	M4	M5	Mean
27	1.60 b	2.37 a	2.59 a	2.26 a	2.36 a	2.61 a	2.30 A
	(0.33)	(0.39)	(0.08)	(0.16)	(0.24)	(0.22)	
54	1.54 b	2.55 a	2.81 a	2.39 a	2.25 a	2.88 a	2.40 A
	(0.49)	(0.07)	(0.28)	(0.11)	(0.14)	(0.13)	
80	1.55 b	2.39 a	2.25 a	2.59 a	2.49 a	2.57 a	2.31 A
	(0.31)	(0.14)	(0.02)	(0.01)	(0.07)	(0.03)	
Mean	1.56 C	2.43 AB	2.55 AB	2.41 AB	2.36 B	2.68 A	

M0 = No inoculation; M1 = *C. pellucida*; M2 = *S. calospora*; M3 = *A. leptotichum*; M4 = *A. mellea*; M5 = Consortium of native species. Standard deviation in parentheses. Mean comparison test by Tukey ($p < 0.05$). Different lowercase letters indicate significant statistical differences ($p < 0.05$) in the interaction P dose and Glomeromycota inoculum. Different capital letters indicate significant statistical differences ($p < 0.05$) in the interaction P dose (right column) and Glomeromycota inoculum factor (lower row) (coefficient of variation = 9.00%)

species consortium (M5) (Table 14.2). The higher N concentration levels were reached with AM colonized plants, which favor the uptake of N from the soil. Results agrees with previous research (Smith et al. 2003; Barea et al. 2008), who showed that Glomeromycota favor plant nutrients uptake from soil. According to Epstein and Bloom (2005), foliar N content found in this study (1.5–2.8%) indicated no N deficiency, since critical N levels for maize in leaves are below 1.46%. Furthermore, literature reports a diversity of N content in maize leaves. For instance, Reta et al. (2007) found N concentrations of maize stems sowed in soils with low fertility ranging around 1.06–1.14%, far below the concentrations reported in this study. The latter suggests that the effect of the fungi favored and increased the uptake of nutrients from soil, as stated by Smith et al. (2003) and Carrenho et al. (2007) among others.

14.3.2 Phosphorus Absorption

P concentration in the corncob opposite leaf ranged from 0.12% — in the plot with 54 kg P_2O_5 ha^{-1} inoculated with M5 — to 0.16% in the plot with 27 kg P_2O_5 ha^{-1} inoculated with M1. At 27 kg P_20_5 ha^{-1} a better P absorption was achieved (Table 14.3). Data (not shown in this paper) indicated that with a lower dose of phosphorus the AM colonization of maize improved. Although no statistic differences were appreciated in the P dose · fungal species interaction, a slight increase was detected in plots where 27 kg P_2O_5 ha^{-1} were applied. They always showed more absorption of this element, particularly inoculation of M1 which proved to be more efficient. In plots where 54 and 80 kg P_2O_5 ha^{-1} was applied, the mycorrhizae

Table 14.3 Phosphate content (%) in the cob opposite leaf of maize inoculated with Glomeromycota fungi grown with different phosphorus doses, harvested in the flowering season

			Treatments				
P_2O_5 kg.ha^{-1}	M0	M1	M2	M3	M4	M5	Mean
27	0.13 a	0.16 a	0.15 a	0.16 a	0.16 a	0.16 a	0.15 a
	(0.03)	(0.03)	(0.00)	(0.02)	(0.00)	(0.01)	
54	0.13 a	0.13 a	0.13 a	0.13 a	0.13 a	0.12 a	0.13 b
	(0.01)	(0.01)	(0.00)	(0.02)	(0.00)	(0.02)	
80	0.13 a	0.14 a	0.14 a	0.14 a	0.14 a	0.14 a	0.14 ab
	(0.02)	(0.02)	(0.01)	(0.04)	(0.04)	(0.01)	
Mean	0.10 a	0.11 a	0.10 a	0.11 a	0.11 a	0.11 a	0.10

M0 = no inoculation; M1 = C. pellucida; M2 = *S. calospora*; M3 = *A. leptotichum*; M4 = *A.mellea*; M5 = Consortium of native species. Standard deviation in parenthesis. Different lowercase letters indicate significant statistical differences (Tukey, $p < 0.05$) in the interaction P dose and Glomeromycota inoculum (right columm). No significant statistical differences was found among Glomeromycota inocula ($p < 0.05$), according to the comparison test of means by Scheffe (coefficient of variation = 21.12%)

proved to be less efficient in P uptake. It is known that soil must show low P content for the AM functionality, otherwise symbiosis may fail, even if the fungus is able to colonize the root or inoculation is applied (Miller 2000; Harrier and Watson 2003; Sylvia 2013).

In all cases, the values observed in this study are below 0.28 – 0.5% of P as reported by Faggioli and Freytes (2008) in leaf samples of AM-colonized maize, grown in soils with moderate fertility and adequate P contents. As plants absorb P with through the AM symbiotic interaction, while soils show poor supply of this element, very little P can be absorbed without the AM help.

14.3.3 Yield

The analysis of the P dose factor showed significant differences between yield levels ($p < 0.05$), being 54 kg P_2O_5 ha^{-1} the best dose with an average of 3325 kg ha^{-1}. However, no difference was appreciated between the latter and the 80 kg ha^{-1}dose. At 80 kg P_2O_5 ha^{-1} application, 3241 kg ha^{-1} were achieved. However, the dose of 27 kg P_2O_5 ha^{-1} showed the lowest average yield (2100 kgha^{-1}), close that commonly achieved in the zone of Guarico savannas (2000 kg ha^{-1}, Fig. 14.1) (Vielma et al. 2005). This difference shows that applying P increased maize yield and that an intermediate dose, together with AM inoculation, can sustain maize production in these soils (Cardoso and Kuyper 2006; Hernandez et al. 2012).

When considering the fungal factor, statistically significant differences were found ($p < 0.05$). The largest one was observed between inoculated and non-inoculated plants, the latter showing the lowest yield (1458 kg.ha^{-1}). Moreover, M4 (*A. mellea*) showed the best average with a value of 2650 kg ha^{-1}, followed by M5

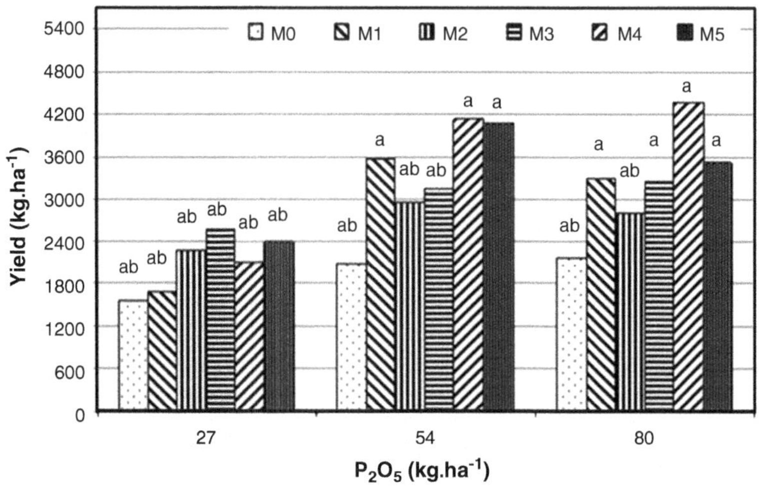

Fig. 14.1 Yield of maize inoculated with arbuscular mycorrhizae and different doses of phosphorus. M0 = no inoculation; M1 = *C. pellucida*; M2 = *S. calospora*; M3 = *A. leptotichum*; M4 = *A. mellea*; M5 = Consortium of native species. Different letters show significant differences ($p < 0.05$), according to Scheffe Test (CV = 27.95%)

(consortium of native species) with 2500 kg ha^{-1}. This suggests that in soils with low phosphoros content, the average production can be enhanced by 30–80%, by applying AM, according to the fungal species and the amount of P in the soil (Carrenho et al. 2007), Sánchez and Velásquez 2008; Barea et al. 2008).

The lowest yields were obtained when no AM inoculation was carried out and/or very low P doses were used (Fig. 14.1). At 27 kg P_2O_5 ha^{-1} with M1 (*C. pellucida)* or M0 treatments, the lowest yields were obtained (1574 and 1691 kg ha^{-1} respectively). These are below the levels reported by Vielma et al. (2005) in acid soils of Venezuela. As the P dose was raised yields improved, even more when AM inoculation was carried out. For instance, with a dose of 54 kg P_2O_5 ha^{-1} and no inoculation, yields of 2076 kg ha^{-1} were reached, but when inoculated with M4 (*A. mellea*) and M5 (consortium of native species), these values increased til 4121 and 4088 kg ha^{-1}, respectively. However, maximum yields were obtained with 80 kg P_2O_5 ha^{-1} and M4 (*A. mellea*, average yield 4384 kg ha^{-1}). These differences are important because when reducing the P application by 33% (27 kg ha^{-1}) in presence of AM, yields drop by 6% only, which means in this case 262.2 kg.

Similar soils showed a yield average of 2670 kg.ha^{-1} (Vielma et al. 2005), an amount surpassed in this study with an intermediate dose of 54 kgP_2O_5 ha^{-1} and AM inoculation. By inoculating with any of the Glomeromycota species essayed, yield averages exceeded. Nevertheless, without inoculation and at the same P doses, this average would be below the value reported by Vielma et al. (2005). Consequently, our results suggest that, in order to obtain an adequate maize yield in these low fertility soils, the use of AM is an option as long as low P doses are used. This agrees with Miller (2000), Harrier and Watson (2003) and Sanchez and Velásquez (2008), which state that P doses in soil can be reduced by using AM. In order to obtain high maize yields, enough amounts of nutrients, particularly P must be present in soil (Epstein and Bloom 2005), as this is one of the most important yield limiting factors, in the studied soils.

Thanks to the efficiency of the Glomeromycota fungi used in this study, the level of 2441 kg ha^{-1} of maize yield obtained by Uribe et al. (2007) was exceeded by inoculations with AM. In this respect, use of arbuscular mycorrhiza represents an opportunity for a sustainable program facing the scarce phosphorus available, a condition typical of soils in the tropics (Cardoso and Kuyper 2006).

14.4 Conclusions

- Using native Glomeromycota fungi, phosphorus fertilization can be reduced by up to 33% without affecting grain production and yields.
- In plantations of AM colonized maize, the raise of P in fertilization had no influence in the capacity of the plant to better absorb this element.
- *Acaulospora mellea* and the consortium of native fungi species have a potential for production of bio-fertilizers based upon Glomeromycota, suitable for maize production in acid soils with low P contents.
- In low fertility soils under savannas weather conditions, maize production can be increased by introducing AM in management plans. *Acaulospora mellea* could be inoculated alone or with other native Glomeromycota, as a consortium.

Acknowledgements The authors express their gratitude to Mrs. Maria Auxi Toro Garcia for manuscript corrections.

References

Barea, J., Ferrol, N., & Azcón-Aguilar, R. (2008) Mycorrhizal symbioses. In *The ecophysiology of plant-phosphorus interactions* (Series Plant Ecophysiology), (Vol. 7, pp. 143–163).

Cardoso, I., & Kuyper, T. (2006). Mycorrhizas and tropical soil fertility. *Agriculture, Ecosystems and Environment, 116*, 72–84.

Carrenho, R., Botelho Trufem, S. F., Ramos Bononi, V., & Schunk Silva, E. (2007). The effect of different soil properties on arbuscular mycorrhizal colonization of peanuts, sorghum and maize. *Acta Botânica Brasílica, 21*(3), 723–730.

Epstein, E., & Bloom, A. (2005). *Mineral nutrition of plant: Principles and perspectives.* Sunderland: Sinaver Associates Inc.

Faggioli, V., & Freytes, G. (2008). *Micorrización natural de maíz y su relación con la absorción de fósforo del suelo en diferentes sistemas de labranzas y fertilización.* Paper presented at the XX1st Congreso Argentino de la Ciencia del Suelo, Bahía Blanca, 5–9 May 2008.

Gómez, Y., & Paolini, J. (2011). Variación en la actividad microbiana por cambio de uso en suelos en sabanas, Llanos Orientales, Venezuela. *Revista de Biología Tropical, 59*(1), 1–15.

Harrier, L. A., & Watson, C. A. (2003). The role of arbuscular mycorrhizal fungi in sustainable cropping systems. *Advances in Agronomy, XX*, 185–225.

Hernández, R. M., Lozano, Z., Bravo, C., Morales, J., Toro, M., Ramírez, E., Castro, I., Rivero, C., & Ojeda, A. (2012). *Manejo agroecológico de suelos de sabanas bien drenadas con unidades de producción cereal ganado*. Paper presented at the XIXth Congreso Latinoamericano Ciencias del Suelo, Mar del Plata, 16–20 April 2012.

IGAC. (2006). *Métodos analíticos del laboratorio de suelos*. Bogotá: Instituto Geográfico Agustín Codazzi, IGAC.

Miller, M. (2000). Arbuscular mycorrhizae and the phosphorus nutrition of maize: A review of Guelph studies. *Canadian Journal of Plant Science, 80*, 47–52.

Murphy, J., & Riley, J. P. (1962). A modified single solution method for the determination of phosphate in natural waters. *Analytica Chimica Acta, 27*, 31–36.

Rakshit, A., & Bhadoria, P. (2010). Role of VAM on growth and phosphorus nutrition of maize with low soluble phosphate fertilization. *Acta Agrobotanica, 59*(1), 119–123.

Reta, D., Cueto, J., Mascorro, A., & Santamaria, J. (2007). Rendimiento y extracción de nitrógeno, fósforo y potasio de maíz forrajero en surcos estrechos. *Agricultura Técnica México, 33*(2), 145–151.

Sánchez, M., & Velásquez, D. (2008). *Las micorrizas: el micelio externo de los hongos Formadores de micorriza arbuscular (HMA). Cuadernos Ambientales Numero 12*. Palmira – Valle: Universidad Nacional de Colombia.

Sarmiento, G., Pinillos, M., Pereira, M., & Acevedo, D. (2004). Effects of soil water and grazing on vegetation diversity and production in a hyperseasonal savanna in the Apure Llanos, Venezuela. *Journal of Tropical Ecology, 20*, 209–220.

Schüßler, A., Schwarzott, D., & Walker, A. (2001). New fungal phylum, the Glomeromycota: Phylogeny and evolution. *Mycological Research, 105*, 1413–1421.

Smith, E., & Read, J. (2008). *Mycorrhizal symbiosis*. New York: Elsevier/Academic.

Smith, S., Smith, F., & Jakobsen, I. (2003). Mycorrhizal fungi can dominate phosphate supply to plants irrespective of growth responses. *Plant Physiology, 133*, 16–20.

Sylvia, M. (2013). *Mycorrhizal Symbioses*. In: www.ifas.ufl.edu. Accessed Feb 2017.

Toro, M. (2007). Micorrizas arbusculares en ecosistemas de sabana venezolanos. In N. Montaño, S. Camargo-Ricalde, R. García-Sánchez, & A. Monroy-Ata (Eds.), *Micorrizas arbusculares en ecosistemas áridos y semiáridos* (pp. 243–267). México: Mundi-Prensa.

Toro, M., Bazo, I., & López, M. (2008). Micorrizas arbusculares y bacterias promotoras de crecimiento vegetal, biofertilizantes nativos de sistemas agrícolas bajo manejo conservacionista. *Agronomia Tropical, 58*(3), 215–221.

Uribe, G., Petit, J., & Dzib, R. (2007). Respuesta del cultivo de maíz a la aplicación de biofertilizantes en el sistema Rosa, Tuma y Quema de suelos Alfisol (Chac lu-um, nomenclatura Maya) en Yucatán, México. *Agricultura Andina, 13*, 3–18.

Vielma, M., Cerovich, M., Miranda, F., & Marín, C. (2005). Influencia de la semilla certificada de maíz en la productividad de los sistemas de producción de maíz en grano de los estados Portuguesa y Guárico. *Agronomia Tropical, 55*(3), 343–361.

Chapter 15
Sustainability of Potato Farms and Use of Microbial Inoculants in the Central Coast of Peru

Sergio Eduardo Contreras-Liza, Huberto Noriega-Córdova, Alberto Valenzuela-Muñoz, Sady García-Bendezú, and Doris Zúñiga-Dávila

Abstract Potato is an Andean native crop with broad genetic diversity that develops in small farms in Latin America, generating rural employment. It is a priority therefore to evaluate the sustainability of the production system according to environment-friendly agronomic practices. The objective of this research was to evaluate the sustainability of potato farms and to propose management alternatives based on microbial inoculation, to optimize the potato production system in the central coast of Peru. Social, economic and environmental sustainability levels of potato farms were evaluated through the Multicriteria Analysis of Sarandon, using as a research tool a survey carried out in 127 farms, located in the provinces of Barranca, Huaral and Cañete, in the Lima Region. The effect of microbial inoculations on the sustainability of the potato production system were afterwards evaluated. It was determined that the potato producing farms presented a low level of sustainability on the weighted scale of indicators used. Hence, the work showed the need of implementing actions to reduce the vulnerability of the potato crop with respect to the conservation, among other factors, of soil life and management of agrobiodiversity. Through the strategy of inoculation of tuber propagules with microbial strains, our research demonstrated, in greenhouse and on-farm experiments, that it is possible to improve the crop performance in comparison to non-inoculated plots. Results suggest that the use of growth promoting microorganisms in potato crops is an innovative management approach, increasing the level of sustainability in the farms of the Central Coast of Peru.

S. E. Contreras-Liza (✉) · H. Noriega-Córdova · A. Valenzuela-Muñoz
Universidad Nacional José Faustino Sánchez Carrión, Huacho, Peru
e-mail: scontreras@unjfsc.edu.pe

S. García-Bendezú
Laboratorio de Análisis de Suelos y Aguas, Universidad Nacional Agraria La Molina (UNALM), Lima, Peru

D. Zúñiga-Dávila
Laboratorio de Ecología Microbiana y Biotecnología, Departamento de Biología-Facultad de Ciencias, Universidad Nacional Agraria La Molina, Lima, Peru

D. Zúñiga-Dávila et al. (eds.), *Microbial Probiotics for Agricultural Systems*, Sustainability in Plant and Crop Protection,
https://doi.org/10.1007/978-3-030-17597-9_15

Keywords Agro-ecological alternatives · Environmental sustainability · Microbial inoculation · Multicriteria analysis · Sustainability indexes

15.1 Introduction

Southern Peru and western Bolivia are the center of origin of potato and of many wild relatives. It is convenient therein to develop new technologies of agronomic management with less impact on the ecosystem. On the other hand, the use of chemical pesticides in potato is increasing in developing countries, as farmers intensify productions (FAO 2008). Sustainable agricultural practices are a response to the multifaceted problems that have arisen from the prolonged and indiscriminate use of chemical products to improve crop production, for many decades. Consequently, the search for eco-friendly options to replace chemical fertilizers and pesticides has been accentuated during the last years (Prashar et al. 2014).

The last updated statistics of the Peruvian Office of Economic and Statistical Studies of the Ministry of Agriculture and Irrigation (MINAGRI 2015) indicate that 6211 hectares were cultivated with potato during the 2014 agricultural season in the Lima region, with an average yield of 23.7 tonnes ha^{-1}. In this potato production system, intensive agricultural technology is used as agricultural inputs (particularly fertilizers, agrochemicals and crop protection products), with mechanization and irrigation. Technical assistance is also provided, but there is a need to evaluate the use of technologies as well as other socioeconomic factors from the viewpoint of environmental, economic and social sustainability.

Sarandon and Flores (2014) proposed methodological guidelines for evaluating the sustainability of agroecosystems and to determine if the crops production models are based on agro ecological principles and processes. In this way, the authors describe a methodology for the construction and implementation of sustainability indicators. The so-called "farm systems research" methodologies emphasize the understanding of traditional agricultural systems as a starting point, evaluate the historical conditions of the area, carry out on-site examinations that include interviews with farmers regarding the characteristics of the farm, and analyze why farmers apply particular methods of production (Altieri et al. 2017; Altieri 1997).

It is known that microorganisms in agricultural soil exert a profound influence on soil fertility status, in particular influencing the availability of nutrients, as well as the suppression of plant diseases and the regulation of plant tolerance to stress factors (Yang et al. 2018; Kennedy and Smith 1995). There is evidence that soil biodiversity confers stability to stress and disturbance conditions (Brussaard et al. 2007), but the mechanisms are still not fully understood. Oswald et al. (2010) demonstrated, in trials under controlled conditions with *Bacillus* sp., that the mechanisms that cause the best growth of potato plants are among others early tuberization, rapid development of leaf area, and higher rates of photosynthesis. Calvo et al. (2010) conducted a prospection of bacteria extracted from the rhizosphere of native potato plants from their natural habitat in the Andean region. Results suggested that they are a rich source of antagonists for phytopathogenic fungi, mainly isolates belonging to the genus *Bacillus* sp. Thus they concluded that the potato rhizosphere is a source of bacteria to be used in the future as inoculants to improve the agronomic performance of potato crops. Due to the large losses caused by fungal diseases, biocontrol methods using beneficial species, such as mycorrhizal fungi as protection agents, should be developed in the short term. These microorganisms can compete with taxonomically related phytopathogenic fungi, acting in synergy with growth promoting bacteria (Diallo et al. 2011).

According to Arcos and Zúñiga (2015), strains of *Bacillus subtilis* and *B. amyloliquefaciens* native in the altiplane region of Peru and Bolivia, which were inoculated on seedlings of two potato varieties (Compis and Andina), inhibited infection by *Rhizoctonia solani*, possibly through some mechanism of antagonistic action or induced resistance. Oswald et al. (2010) proposed that the use of soil microorganisms could reduce the application of fertilizers in potato cultivation and partially control soil-borne diseases that affect the production and quality of tubers.

The main objective of this research was to evaluate the sustainability of potato farms and propose alternatives to optimize the production system in the conditions of the central coast of Peru. A holistic study was developed based on a systems approach perspective, which allows to integrate the analysis of the natural environment, the productive processes and the historical performance of the potato producers. In a second step, we evaluated the potential impact of the use of inoculants as an agronomic management strategy, and its effect to improve crops sustainability.

15.2 Materials and Methods

15.2.1 Determination of the Degree of Sustainability in the Potato Farms

The area selected for the investigation included three provinces in the Lima Region: Barranca, Huaral (both north of Lima) and Cañete (south of Lima). The agro ecosystem of this coastal region is characterized by sandy loamy alluvial soils, with 0.8–2.0% of organic matter and suitable for growing potato in the winter season. The methodology to determine the sustainability indicators was based on the proposal by Sarandon (2002). It consisted in a series of weighted indicators to evaluate the critical points in the sustainability of the agricultural ecosystem (Table 15.1). The indicators evaluated were determined for their sociocultural, economic and environmental dimensions, according to the methodology proposed by Sarandon and Flores (2014).

A survey of potato producers was used as a research tool to obtain valid informations on variables and sustainability indicators. The survey consisted of a questionnaire of 70 items referring to each of the sustainability dimensions to be evaluated, interviewing a total of 127 potato producers from the provinces of

Table 15.1 Dimensions of sustainability and assessment of indicators (IS: sociocultural, IK: economical and IE: environmental) in potato farms

Sustainability dimensions	Indicators	Assessment and weighting of sustainability indicators
Sociocultural	Meeting the needs of the producer	IS = 0.25 [(A1) + (A2) + (A3) + (A4)/4] + 0.25 [(B1)] + 0.25 [(C1)] + 0.25 (D1)]
	Quality of life	(A1) Housing
	Degree of social integration	(A2) Access to education
		(A3) Access to health and health coverage
		(A4) Access to basic services
	Level of consciousness and ecological knowledge	(B1) Acceptability of the production system
		(C1) Social integration
		(D1) Ecological knowledge and consciousness
Economical	Food sufficiency	IK = [0.15(A1) +0.10 (A2)] + [0.10(B1) + 0.10(B2)] + [0.10(C1) + 0.10(C2) + 0.10 (C3) + 0.20(C4)]
	Monthly net income	
	Economic risk	(A1) Diversification of production
		(A2) Area for self-consumption (family garden)
		(B1) Monthly net income per family
		(B2) Technological level of potato producer
		(C1) Diversification for sale
		(C2) Number of marketing channels
		(C3) Dependence on external inputs
		(C4) Variability in potato planting dates
Environmental	Conservation of soil life	IE = [0.15 (A1) + 0.15 (A2)] + [0.05 (B1) + 0.10 (B2) + 0.15 (B3)] + [0.10 (C1) + 0.10 (C2) + 0.10 (C3) + 0.10 (C4)]
	Erosion risk	
	Management of agrobiodiversity	(A1) Crop rotation
		(A2) Crop diversification
		(B1) Vegetation coverage
		(B2) Type of soil
		(B3) Organic matter
		(C1) Temporal biodiversity
		(C2) Spatial biodiversity
		(C3) Biodiversity of potato varieties
		(C4) Incidence of abiotic stresses
General sustainability index		1/3 IS +1/3 IK + 1/3 IE

Barranca (67), Huaral (44) and Cañete (16). We used rating scales ranging from 0 to 4, with 0 being the least sustainable category and 4 being the most sustainable one. Regardless of the units in which they were originally obtained, the values of each indicator were referred to this scale. In this quantitative scale, the threshold value of 2 was considered as an acceptable level of sustainability (Sarandon and Flores 2014).

15.2.2 Impact of the Use of Inoculants as a Strategy of Agronomic Management in Potatoes

In order to improve the sustainability of potato agrosystems, we tested some eco-friendly technologies in the field and in greenhouse, namely potato inoculation with microorganisms, to optimize the use of fertilizers as well as to induce systemic host-plant resistance and to improve the agronomic performance of the crop. The field and greenhouse experiments were carried out during the potato growing seasons 2013 to 2017 (Table 15.2). Experimental designs were a randomized complete block design, in the field trial, and a completely randomized design, in the greenhouse. The obtained results were subjected to ANOVA analysis. The statistical post-hoc tests of Tukey (field experiments) and Scott-Knott (greenhouse experiments) were used to compare the mean values of the treatments ($p \leq 0.05$). The test proposed by Scott Knott is a procedure of means grouping, considered an effective option to perform multiple comparisons without ambiguity in controlled conditions (Bhering et al. 2008). In addition, some percent indexes were measured between inoculant treatments versus control plots.

15.3 Results and Discussion

15.3.1 Determination of the Degree of Sustainability in Potato Farms

Out of 25 sustainability sub-indicators considered in the multicriteria analysis, 14 were lower than the established threshold value (Table 15.3),. It is important to note that in the environmental dimension most had sustainability values below the threshold of 2, except for vulnerability to environmental stresses in which the value of the general sustainability index of the potato producing farms in the Lima region was 2.63. Likewise we identified, both in the socio-cultural dimension and in the economical one, up to three sub-indicators below the threshold. In the case of the socio-cultural dimension, it can be mentioned that the sub indicators access to health (medical care in regional hospitals or social security), the degree of social integration and the level of environmental consciousness, were important constraints to achieve sustainability.

In the Table 15.3, the aspects related to environmental management in the potato farms, were the most critical factors with the lowest score. Thus the environmental dimension affects notoriously the sustainability of the potato production system in the central coast. Our results showed that it is necessary to improve the environmental sustainability in the potato farms, mainly with respect to conservation of life in soil, risk of erosion and management of agro-biodiversity. In this regard, Kroschel et al. (2012) proposed some strategies of biological control of pests in potato fields and the use of inoculants to recover extinct species due to the intensive use of

Table 15.2 Agronomic experiments to determine the effect of soil microorganisms on the performance of potato cultivars (2013–2017)

No. Exp.	Treatments/ inoculants[a]	Potato cultivars	Sites and experimental conditions	Soil characteristics
I	Azo16M2 Bac15Mb B13 DZ22 Control	Única	San Vicente, Cañete 70 masl Farmer fields Tuber-seeds RCBD-Tukey	pH 8.1, EC 1.96 dS / m, OM 2.07%, P 8.8 ppm, K 104 ppm, CEC 14.7 meq 100 g^{-1}, Texture: Sandy loam
II	Azo16M2 Bac15Mb B13 DZ22 Control	Única Perricholi Canchan Inia	Quilmaná, Cañete 140 masl Farmer fields Tuber-seeds RCBD-Tukey	pH 7.8, EC 3.4 dS/m, OM 0.8%, P 16.7 ppm, K 129 ppm, CIC 7.8 meq 100 g^{-1}, texture: Sandy loam
III	Azo16M2 Bac15Mb B13 DZ22 Control	Canchan Inia Única	Lunahuana, Cañete 470 masl Farmer fields Tuber-seeds RCBD-Tukey	pH 7.7, EC 4.46 dS/m, OM 2.0%, P 18.8 ppm, K 154 ppm, CEC 13.5 meq 100 g^{-1}, texture: Loam
IV	Azo16M2 Bac15Mb B13 DZ22 Control	Canchan Inia	Huacho, Huaura Greenhouse Tuber-seed CRD-Scott-knot	Substrate: 45% Vermicompost + 45% washed sand + 10% rice husk
V	NPK + *Glomus* *Glomus* + *bacillus* Control *Glomus* NPK	Amarilis	Huacho, Huaura Greenhouse Tuber-seed CRD-Scott-knot	Substrate: 45% Vermicompost + 45% washed sand + 10% rice husk
VI	*Glomus* + *bacillus* *Glomus* NPK Control	Única Perricholi Canchan-Inia' Faustina 'Yasmine'	Huacho, Huaura Greenhouse *In vitro* plantlets CRD-Scott-knot	Substrate: 45% Vermicompost + 45% washed sand + 10% rice husk

[a]Azo16M2 = *Azotobacter* sp., Bac15 Mb = *Bacillus amyloliquefaciens*, B13 = *B. simplex*, DZ22 = *Pantoea* sp. Strains provided by Laboratorio de Ecología Microbiana y Biotecnología, UNALM, Lima. RCBD, randomized complete block design; CRD, Completely randomized design

pesticides. Mäder et al. (2002) argued that agroecological management can make agricultural ecosystems less dependent on external inputs, increasing their resilience. The findings of Oswald et al. (2010) suggest that diazotrophic strains isolated from the potato rhizosphere in the Andes have an enormous potential as biofertilizers, due to their plant growth promoting characteristic. Likewise, another action related to the use of agrobiodiversity, to improve the environmental performance of potato crops, is the use of local potato varieties. In this sense, the central and southern regions of Peru have been recognized as a center of genetic diversity of potatoes (de Haan et al. 2010), and thus as sources of local potato germplasm.

Table 15.3 Indicators of sustainability in potato farms in Lima Region according to Sarandon and Flores (2014)

Dimension	Sub-indicator	Criteria	Value weighted
Socio-cultural	A1	Type of housing	**2.99**
IS = 1.98	A2	Access to education	**2.76**
	A3	Access to health and health coverage	1.74
	A4	Basic services	**2.44**
	B1	Acceptability of the production system	**2.77**
	C1	Degree of social integration	1.48
	D1	Ecological knowledge and consciousness	1.07
Economical	A1	Production diversification	**2.01**
IK = 1.85	A2	Self-consumption production area	0.49
	B1	Monthly average income	**2.34**
	B2	Technological level	**2.52**
	C1	Diversification for sale	0.52
	C2	Number of marketing channels	0.33
	C3	Dependence on external inputs	**2.52**
	C4	Variability in potato planting dates	**2.79**
Environ-mental	A1	Crop rotation	1.65
IE = 1.52	A2	Crop diversification	1.26
	B1	Vegetation coverage	1.34
	B2	Type of soil	1.92
	B3	Organic matter applied to the soil	1.09
	C1	Temporal biodiversity	1.15
	C2	Spatial biodiversity	1.45
	C3	Biodiversity in potato varieties	1.41
	C4	Vulnerability to environmental stresses	**2.63**
		General sustainability index	1.78

Index values in bold are above the threshold value established (2.0)

15.3.2 *Impact of the Use of Inoculants as a Strategy of Agronomic Management in Potatoes*

To improve the environmental sustainability index (IE), an agro ecological management strategy was evaluated in potato production agroecosystems under coastal conditions, consisting in the inoculation of the potato tuber propagules with microbial inoculants. The starting hypothesis was that by means of this strategy, it would be feasible to increase the environmental sustainability indicators in the potato farms. Thus farms wouls achieve a reduction in dependence on high energy consumption resources and in the risk associated with potato farming in Lima region. Potato "seed" certification systems are indeed still precarious in this context. The results obtained with the inoculation strategy are presented below.

In the experiment I (San Vicente) the inoculation of potato tubers cv. Unica with strain Az16M2 (*Azotobacter* sp.), showed significant differences with respect to the control for tuber yield and fresh weight of biomass per hectare (Table 15.4).

Table 15.4 Effect of bacterial strains on the agronomic performance of potato varieties in farms in the Lima region (2014–2015)

Experiment	Potato cultivars tested	Treatments	Tuber yield t ha^{-1}		% of control	Biomass t ha^{-1}		% of control
I. San Vicente	Unica	Azo16M2	22.07	a	153.6	24.90	a	136.8
		Bac15Mb	19.57	ab	136.2	22.70	ab	124.7
		B13	16.43	ab	114.4	18.80	ab	103.3
		DZ22	16.37	b	113.9	19.20	ab	105.5
		Control	14.37	b	100.0	18.20	b	100.0
II. Quilmaná	Unica	DZ22	18.48	a	103.2	57.60	b	101.0
	Perricholi	B13	18.48	a	103.2	62.90	a	110.4
	Canchan-Inia	Azo16M2	17.78	a	99.3	76.00	a	133.3
		Bac15Mb	16.37	a	91.5	65.70	ab	115.3
		Control	17.90	a	100.0	57.00	b	100.0
III. Lunahuana	Unica	DZ22	23.00	a	125.0	34.9	ab	114.8
	Canchan-Inia	B13	22.90	a	124.5	34.4	ab	113.2
		Azo16M2	22.60	a	122.8	34.3	ab	112.8
		Bac15Mb	21.40	a	116.3	37.2	a	122.2
		Control	18.40	a	100.0	30.4	b	100.0

The mean values followed by the same letters did not statistically differ following Tukey test ($p < 0.05$). When several cultivars were tested, the results for each treatment are shown as mean values for all cultivars

The experiment II (Quilmaná), was performed with three potato cultivars: Canchán-Inia, Perricholi and Unica (Table 15.4). The mean values for the three cultivars indicate that the inoculation produced significant differences with respect to the control for weight of total biomass per hectare, with a stimulating effect of several strains, especially *Azotobacter* sp. Azo16M2. No statistical difference was observed for tuber yields due to the inoculation with bacterial strains, but some of them showed a tendency towards an increased weight of tubers with respect to the control (Table 15.4).

In the experiment III (Lunahuana), the bacterial strains were inoculated in potato cultivars Unica and Canchan-Inia, and in this location the strain Bac15Mb (*Bacillus amyloliquefaciens*) significantly increased the fresh weight of the total biomass compared to the control. Although no significant difference was shown for tuber yields, there was also a tendency towards higher productions, as a result of inoculation (Table 15.4).

In the greenhouse experiment V (Table 15.5), the inoculation of potato cv. Amarilis with *Glomus intraradices* + *Bacillus subtilis* and the treatment of NPK + *Glomus intraradices* led to a significant increase vis-à-vis the control for foliar area ($\sqrt{FA}$) and dry weight of the foliage ($\sqrt{FDW}$). For the experiment VI a significantly higher tuber weight per plant (TWP) was obtained for the inoculation with the consortia of microorganisms *Glomus intraradices* + *Bacillus subtilis*, compared to the inoculation with *Glomus intraradices* alone. Both treatments produced also significantly higher values than the uninoculated control.

Table 15.5 Effect of microbial strains on the agronomic performance of potato genotypes in greenhouse conditions

Experiment	Potato genotypes	Treatments⁺	TWP, g		% of control	FDW, g		% of control
IV. Greenhouse Huaura	Canchán-Inia	B13	104.0	a	130.8	12.7	a	115.2
		Azo16M2	102.5	a	128.9	13.3	a	121.2
		DZ22	102.0	a	128.3	14.0	a	127.3
		Bac15Mb	88.0	a	110.7	12.0	a	109.1
		Control	79.5	a	100.0	11.0	a	100.0
			$\sqrt{}$ FA, cm^2			$\sqrt{}$ FDW, g		
V. Greenhouse Huaura	Amarilis	NPK + *Glomus*	15.71	a	198.4	5.78	a	191.4
		Glomus + *Bacillus*	13.28	a	167.7	4.54	b	150.3
		Glomus	8.52	b	107.6	2.99	c	99.0
		NPK	7.17	b	90.5	2.47	c	81.8
		Control	7.92	b	100.0	3.02	c	100.0
			TWP, g			FDW, g		
VI. Greenhouse Huaura,	Única	*Glomus* + *Bacillus*	9.34	a	244.5	26.71	a	114.1
	Canchán-Inia	*Glomus*	6.49	b	169.9	26.71	a	114.1
	Perricholi	NPK	4.00	c	104.7	25.24	a	107.8
	Faustina	Control	3.82	c	100.0	23.41	a	100.0
	Yasmine							

TWP, tuber weight per plant; FDW, foliage dry weight; FA, foliar area. The data corresponding to the parameters preceded by the symbol $\sqrt{}$ ($\sqrt{}$ FA and $\sqrt{}$ FDW in the experiment V) were transformed to square root. Mean values followed by the same letter did not significantly differ ($p < 0.05$) following Scott-Knott test

The indicators showed that the potato agroecosystems in the Lima Region need an improvement to reach appropriate levels of sustainable production. The central coast of Peru (where the Lima region is located), accounts for only 2.5% of the surface devoted to potato crop in the Andean region of Peru and Bolivia. However, the average yield of this region is almost double than in the rest of Peru, due to a medium to high technological level, based on high energy consumption inputs. This led to a poor score in the environmental items in the Sarandon and Flores (2014) multicriteria analysis for sustainability, in the mentioned region. Agro-ecological management consisting on partial substitution of the conventional chemical inputs by microorganisms-based products (PGPRs and mycorrhiza) can lead to an improvement of the environmental performance of the crop, which will be of interest in the Lima region. To date there is sufficient scientific evidence that, through the symbiosis between plants and microorganisms, it is possible to reduce the use of high energy consumption inputs, such as fertilizers and agrochemicals, with negative environmental impacts, thus mitigating their harmful effects on the ecosystem. The services derived by the plants and microorganisms interactions include the biological fixation of atmospheric nitrogen, solubilization and absorption of phosphorus, induced systemic resistance and plant immunity, among other physiological aspects. Besides the results obtained in this work, which indicated an increase in the crop yield as a result of the use of microbial inoculants, other evidences may be considered from the literature published (Table 15.6). The identification and characterization of the specific microorganisms for potato crop has advanced markedly in the last years, together with the industrial activity of derived products, at least at a pilot scale(Oswald et al. 2010; Calvo et al. 2010; Pii et al. 2015).

There is also a tendency towards the reduction of biodiversity in all its variants which is causing genetic erosion as a consequence of the need for uniformity in crop production, This trend is especially observed in potato crops, due to the market demand. The existing microbial diversity associated with potato plants is very broad, and the inoculants based in autochthonous strains can help to the maintenance of the natural microbial populations. Moreover, this helps to reduce the loss of soil biodiversity as a consequence of monoculture, strengthening the agrosystems invested to potato productions (Sessitsch and Mitter 2015). Soil microbial communities can furthermore help to the restore degraded soils, mainly regarding their fertility, capacity for nutrient retention and availability of elements necessary for crops nutrition (Turnbull et al. 2014).

Although the inoculation of pulse legumes with rhizobial inoculants is a very common and well stablished practice, especially in Latin America (Chibeba et al. 2017), the use of other microbial inoculants for other crops such as potato, is far for being a common practice. The generalization of their use is often hampered by the lack of efficacy and reliability in field conditions (Van Loon 2007), which makes the diffusion of this innovation difficult. To overcome this issue, efforts must be focused in developing adequate formulations, in order to attain a sufficient number of viable cells to colonize the crop rhizosphere (Bashan et al. 2016).

15.4 Conclusions

This research shows that the potato producing farms in the Lima Region have a low level of sustainability (1.78), falling below the established threshold (2) on the weighted scale of indicators used (from 0 to 4). It is therefore necessary to implement measures that reduce the vulnerability of the potato crop in terms of conservation of soil life, reduction of erosion risk and improvement of the agrobiodiversity management. The use of microbial inoculants is a viable alternative for sustainable agronomic management of potato. The results obtained in this research indicate that after inoculating the potato tuber propagules with selected bacterial strains or mycorrhizal fungi, several significant differences were observed compared to the uninoculated control, namely increased tubers yield and weight of plant biomass, both in on-farms and greenhouse experiments. Hence, the use of microorganisms that promote potato growth can help to improve crops performance. In conclusion, the use of inoculants can result in a reduced need for chemicals and an improvement of soil biodiversity, thereby representing a strategy to increase the level of environmental sustainability in the potato producing farms of the central coast of Peru.

Table 15.6 Microbial taxonomic groups associated with several agronomic effects in potato cultivars

Cultivars	Microbial group	Conditions	Country	Traits[a]	References
Kennebec	*Pseudomonas* sp. PsJN	*In vitro*	Canada	N.D.	Conn et al. (1997)
Yungay, Canchan, Desiree, Unica, Mariva, Reiche	*Bacillus* sp.	Greenhouse Field	Peru	Early tuberization, tuber yield[+]	Oswald et al. (2010)
Peruanita	*Bacillus amyloliquefaciens*	*In vitro*	Peru	Biofertilizers	Calvo et al. (2010)
Ccompis, Andina	*Bacillus, Azotobacter*	Field	Peru	*Rhizoctonia*	Arcos and Zúñiga (2015)
Fripapa, Yungay, Waycha	*Pseudomonas, Paenibacillus, Bacillus*	Field	Andean region	Yield [+], VOCs	Velivelli et al. (2015)
Kuroda	*Azospirillum sp. TN10*	*In vitro*, hydroponics	Pakistan	BNF, AIA	Naqqash et al. (2016)
Shepody	N.D.	*In vitro*, field	Canada	BNF, AIA	Turnbull et al. (2014)
Innovator	*Bacillus subtilis KPS-11*	*In vitro*	Pakistan	P-sol	Hanif et al. (2015)
Desiree	*Azotobacter*	Greenhouse, field	Peru	BNF, AIA, *Rhizoctonia*	Kohashikawa (2010)
Unica	*Azotobacter, Bacillus, Pantoea*	Field	Peru	Yield [+], Biomass	Luyo (2015)
Unica, Perricholi,Canchan	*Azotobacter, Bacillus, Pantoea*	Field	Peru	Vigor, WFoliage, Biomass	Custodio (2016)
N.D.	*Azotobacter*	Greenhouse	Peru	P-sol, AIA, *Rhizoctonia*	Rico (2009)
Serranita	Actinomicetos, *Pseudomonas*	Greenhouse	Peru	Tuber	Camacho and La Torre (2015)
Waycha	*Bacillus subtilis, Glomus sp.*	Field	Bolivia	*Rhizoctonia*, Yield [+]	Main and Franco (2016)
CIP germplasm	*Gigaspora, Glomus, Scutellosporas*	Field	Peru	Formonetina, tuber	Davies et al. (2005)
Desireé	*Azospirillum*	Field	Chile	BNF, Yield [+]	Castillo et al. (2016)
Golden Wonder	*Glomus intraradices*	*In vitro*, greenhouse	Ireland	Yield [+], tuber	Duffy and Cassells (2000)
Superior	*Glomus intraradices*	Field	United Kingdom	Yield [+], tuber	Douds et al. (2007)

[a]Vigor, vegetative vigor; Tuber, induction of tuberization; *Rhizoctonia*, tolerance to *R. solani*; Biomass, biomass production; Yield [+], high yield of tubers; BNF, Biological nitrogen fixation; AIA, production of indole acetic acid, VOCs, production of volatile organic compounds; DM, dry matter of the tubers; WFoliage, fresh weight of foliage; P-sol, Phosphorus solubilization

Acknowledgements The "Universidad Nacional José Faustino Sánchez Carrión" (Perú) funded the project as a part of the program of Socioeconomic Development Fund of Camisea – FOCAM; the "Laboratorio de Ecología Microbiana y Biotecnología (UNALM)" provided the bacterial strains used in the research and contributed to the design of the experiments; the "Laboratorio de Análisis de Suelos y Aguas (UNALM)" provided soil characterization analysis and designed the field surveys for farmers. We want to thank Luis Luyo Sánchez, Juan Carlos Custodio, Juan Carlos Bravo and Herson Torres for the technical support in the field trials, and Dr. Rodomiro Ortiz (SLU) for the critical reading of this manuscript.

References

Altieri, M. A. (1997). *Enfoque Agroecológico para el Desarrollo de Sistemas de Producción Sostenibles en los Andes*. Ed. CIED. Lima-Perú. 92 p.

Altieri, M. A., Nicholls, C. I., & Montalba, R. (2017). Technological approaches to sustainable agriculture at a crossroads: An agroecological perspective. *Sustainability, 9*(3), 349. https://doi.org/10.3390/su9030349.

Arcos, J., & Zúñiga, D. (2015). Efecto de rizobacterias en el control de *Rhizoctonia solani* en el cultivo de papa. *Ecología Aplicada, 14*(2), 95–101.

Bashan, Y., de-Bashan, L. E., & Prabhu, S. R. (2016). Superior polymeric formulations and emerging innovative products of bacterial inoculants for sustainable agriculture and the environment. In H. B. Singh, B. K. Sarma, & C. Keswani (Eds.), *Agriculturally important microorganisms: Commercialization and regulatory requirements in Asia* (pp. 15–46). Singapore: Springer.

Bhering, L. L., Cruz, C. D., Vasconcelos, E. D., Ferreira, A., & Resende Junior, M. D. (2008). Alternative methodology for Scott-Knott test. *Crop Breeding and Applied Biotechnology, 8*, 9–16.

Brussaard, L., De Ruiter, P. C., & Brown, G. G. (2007). Soil biodiversity for agricultural sustainability. *Agriculture, Ecosystems & Environment, 121*(3), 233–244.

Calvo, P., Ormeño-Orrillo, E., Martínez-Romero, E., & Zúñiga, D. (2010). Characterization of Bacillus isolates of potato rhizosphere from andean soils of Peru and their potential PGPR characteristics. *Brazilian Journal of Microbiology, 41*, 899–906.

Camacho, M., & La Torre, M. I. (2015). Efecto promotor de bacterias PGPR sobre cultivo de papa bajo diferentes sustratos a nivel de invernadero. *The Biologist, 13*(1), 75–90.

Castillo, C., Huenchuleo, M. J., Michaud, A., & Solano, J. (2016). Micorrización en un cultivo de papa adicionado del biofertilizante Twin-N establecido en un Andisol de la Región de La Araucanía. *Idesia (Arica), 34*(1), 39–45.

Chibeba, A., Kyei-Boahen, S., De Fátima, G. M., Nogueira, M., & Hungria, M. (2017). Isolation, characterization and selection of indigenous Bradyrhizobium strains with outstanding symbiotic performance to increase soybean yields in Mozambique. *Agriculture, Ecosystems and Environment, 246*, 291–305.

Conn, K. L., Lazarovits, G., & Nowak, J. (1997). A gnotobiotic bioassay for studying interactions between potatoes and plant growth-promoting rhizobacteria. *Canadian Journal of Microbiology, 43*(9), 801–808.

Custodio, J. C. (2016). Efecto de inoculantes bacterianos sobre el comportamiento agronómico de tres genotipos de *Solanum tuberosum* L. "papa" en el distrito de Quilmaná, Cañete. Tesis para Ingeniero Agrónomo. Facultad de Ingeniería Agraria, Industrias Alimentarias y Ambiental. UNJFSC, Huacho. 86 p.

Davies, F. T., Calderón, C. M., Huaman, Z., & Gómez, R. (2005). Influence of a flavonoid (formononetin) on mycorrhizal activity and potato crop productivity in the highlands of Peru. *Scientia Horticulturae, 106*(3), 318–329.

de Haan, S., Núñez, J., Bonierbale, M., & Ghislain, M. (2010). Multilevel agrobiodiversity and conservation of Andean potatoes in central Peru species, morphological, genetic, and spatial diversity. *Mountain Research and Development, 30*, 222–231.

Diallo, S., Crépin, A., Barbey, C., Orange, N., Burini, J. F., & Latour, X. (2011). Mechanisms and recent advances in biological control mediated through the potato rhizosphere. *FEMS Microbiology Ecology, 75*, 351–364.

Douds, D. D., Nagahashi, G., Reider, C., & Hepperly, P. R. (2007). Inoculation with arbuscular mycorrhizal fungi increases the yield of potatoes in a high P soil. *Biological Agriculture and Horticulture, 25*(1), 67–78.

Duffy, E. M., & Cassells, A. C. (2000). The effect of inoculation of potato (*Solanum tuberosum* L.) microplants with arbuscular mycorrhizal fungi on tuber yield and tuber size distribution. *Applied Soil Ecology, 15*(2), 137–144.

FAO. (2008). *Gestión de las plagas y enfermedades. Secretaria del año internacional de la Papa*. International Year of the Potato celebrated in 2008. FAO. Roma.

Hanif, M. K., Hameed, S., Imran, A., Naqqash, T., Shahid, M., & Van Elsas, J. D. (2015). Isolation and characterization of a β-propeller gene containing phosphobacterium *Bacillus subtilis* strain KPS-11 for growth promotion of potato (*Solanum tuberosum* L.). *Frontiers in Microbiology, 6*, 583. https://doi.org/10.3389/fmicb.2015.00583.

Kennedy, A. C., & Smith, K. L. (1995). Soil microbial diversity and the sustainability of agricultural soils. In H. P. Collins, G. P. Robertson, & M. J. Klug (Eds.), *The significance and regulation of soil biodiversity* (pp. 75–86). Dordrecht: Springer.

Kohashikawa, T. N. (2010). Comportamiento de diferentes bacterias PGPR sobre el crecimiento del cultivo de papa (*Solanum tuberosum*). Tesis Biólogo. UNALM. Lima.

Kroschel, J., Mujica, N., Alcazar, J., Canedo, V., & Zegarra, O. (2012). Developing integrated pest management for potato: Experiences and lessons from two distinct potato production systems of Peru. In Z. He, R. Larkin, & W. Honeycutt (Eds.), *Sustainable potato production: Global case studies* (pp. 419–450). Dordrecht: Springer.

Luyo, L. (2015). Efecto de rizobacterias y ácido acetil salicílico en el comportamiento agronómico de la papa (*Solanum tuberosum* L.) cv. Unica. Tesis para Ingeniero Agrónomo. Facultad de Ingeniería Agraria, Industrias Alimentarias y Ambiental. UNJFSC. Huacho. 82 p.

Mäder, P., Fliessbach, A., Dubois, D., Gunst, L., Fried, P., & Niggli, U. (2002). Soil fertility and biodiversity in organic farming. *Science, 296*(5573), 1694–1697.

Main, G., & Franco, J. (2016). Efecto de la bacteria Bacillus subtilis y el hongo micorrízico arbuscular *Glomus fasciculatum* en la fertilización fosfórica en el cultivo de la papa (*Solanum tuberosum* ssp. andigena). *Revista Latinoamericana de la Papa, 16*(2), 250–269.

MINAGRI. (2015). *Intenciones de siembra 2015–2016*. DGESEP. Ministerio de Agricultura y Riego. Lima. Julio 2015. 36 p.

Naqqash, T., Hameed, S., Imran, A., Hanif, M. K., Majeed, A., & van Elsas, J. D. (2016). Differential response of potato toward inoculation with taxonomically diverse plant growth promoting Rhizobacteria. *Frontiers in Plant Science, 7*, 144. https://doi.org/10.3389/fpls.2016.00144.

Oswald, A., Calvo, P., Zúñiga, D., & Arcos, J. (2010). Evaluating soil rhizobacteria for their ability to enhance plant growth and tuber yield in potato. *The Annals of Applied Biology, 157*, 259–271.

Pii, Y., Mimmo, T., Tomasi, N., Terzano, R., Cesco, S., & Crecchio, C. (2015). Microbial interactions in the rhizosphere: Beneficial influences of plant growth-promoting rhizobacteria on nutrient acquisition process. A review. *Biology and Fertility of Soils, 51*(4), 403–415.

Prashar, P., Kapoor, N., & Sachdeva, S. (2014). Rhizosphere: Its structure, bacterial diversity and significance. *Reviews in Environmental Science and Bio/Technology, 13*(1), 63–77.

Rico, M. A. (2009). Capacidad promotora del crecimiento vegetal por bacterias del género Azotobacter y actinomicetos aislados de cultivos de *Solanum tuberosum* Linnaeus, 1753 (papa) cultivados en zonas alto andinas del Perú. Tesis para de Biólogo. Universidad Nacional Mayor de San Marcos. Lima.

Sarandon, S. J. (2002). El desarrollo y uso de indicadores para evaluar la sustentabilidad de los agroecosistemas. En: Santiago J. Sarandón (Ed.), *Agroecología. El camino hacia una Agricultura sustentable*. Ediciones Científicas Americanas (Vol. 20, pp. 393–414).

Sarandon, S. J., & Flores, C. C. (2014). *Agroecología: Bases teóricas para el diseño y manejo de agroecosistemas sustentables* (1a ed., p. 467). La Plata: Universidad Nacional de La Plata.

Sessitsch, A., & Mitter, B. (2015). 21st century agriculture: Integration of plant microbiomes for improved crop production and food security. *Microbial Biotechnology, 8*(1), 32–33.

Turnbull, A. L., Campbell, I., & Lazarovits, G. (2014). Resistance of bacterial communities in the potato rhizosphere to disturbance and its application to agroecology. *Soil Biology and Biochemistry, 79*, 125–131.

Van Loon, L. C. (2007). Plant responses to plant growth-promoting rhizobacteria. *European Journal of Plant Pathology, 119*, 243–254.

Velivelli, S. L., Kromann, P., Lojan, P., Rojas, M., Franco, J., Suarez, J. P., & Prestwich, B. D. (2015). Identification of mVOCs from Andean rhizobacteria and field evaluation of bacterial and mycorrhizal inoculants on growth of potato in its center of origin. *Microbial Ecology, 69*(3), 652–667.

Yang, G., Wagg, C., Veresoglou, S. D., Hempel, S., & Rillig, M. C. (2018). How soil biota drive ecosystem stability. *Trends in Plant Science, 23*, 1057–1067. https://doi.org/10.1016/j.tplants.2018.09.007.

Chapter 16
Characterization of Plant Growth-Promoting Bacteria and *In Vitro* Antagonistic Activity on Root-Knot Nematodes (*Meloidogyne* spp.)

Nora Vera-Obando, Katty Ogata-Gutiérrez, and Doris Zúñiga-Dávila

Abstract Root-knot nematodes (*Meloidogyne* spp.) are the most important plant-parasitic nematodes affecting a wide variety of crops. Thus, recent research has been directed at finding microorganisms, including Plant Growth Promoting Rhizobacteria (PGPR), with activity against root-knot nematodes. In this work five potential PGPR were selected from the LEMYB culture collection, which had been isolated from potato and coffee plants in previous work. Five *in vitro* PGPR features, indol-3-acetic acid (IAA) production, phosphate solubilization, cellulase, protease, and chitinase activity, were tested. Antagonism assays against phytopathogenic fungi and *Meloidogyne* spp. were performed as well. All studied isolates demonstrated at least one *in vitro* PGPR ability. Among the five isolates, AZO16M2 was found to be the best producer of IAA and could solubilize bi- and tri-calcium phosphate at all tested temperatures. Additionally, Bac15Mb showed the highest antagonistic activity against *Fusarium* and *Rhizotocnia* spp. Root-knot nematode antagonism assays showed that AZO16M2 exhibited the highest nemastatic and nematicidal effect. Data results suggest that these strains could be used in future studies as PGPR and biocontrol agents against *Meloidogyne* spp. and some phytopathogenic fungi.

Keywords *Meloidogyne* · Root-knot nematode · *Rhizoctonia* · *Fusarium* · PGPR

N. Vera-Obando · K. Ogata-Gutiérrez · D. Zúñiga-Dávila (✉)
Laboratorio de Ecología Microbiana y Biotecnología, Departamento de Biología-Facultad de Ciencias, Universidad Nacional Agraria La Molina, Lima, Peru
e-mail: dzuniga@lamolina.edu.pe

D. Zúñiga-Dávila et al. (eds.), *Microbial Probiotics for Agricultural Systems*, Sustainability in Plant and Crop Protection,
https://doi.org/10.1007/978-3-030-17597-9_16

16.1 Introduction

Plant parasitic nematodes cause yield losses estimated to range between 8.8 and 14.6% of total crop production (Mokrini et al. 2018). However, in most severe cases, losses may reach up to 75% (Wei et al. 2014). Worldwide losses caused by nematodes are estimated to reach 157 billion USD annually (Abad et al. 2008). However losses have not been clearly estimated for all crops, since the damage caused by nematodes may vary depending on many factors, such as soil type, cultivar, climate, and crop management (Aballay et al. 2013).

The most important plant parasitic nematodes are root-knot nematodes (RKNs) (*Meloidogyne* spp.) (Park et al. 2014). They are sedentary, endoparasitic and polyphagous species that attack a wide variety of crops, and are considered the most damaging nematodes in the world (Coyne et al. 2018). Parasitism begins with root penetration of second-stage juveniles (J2), which moult and develop in roots inducing formation of galls, which impede the normal uptake of water and nutrients. Also, RKNs facilitate infection by soil-borne phytopathogenic fungi (Huang et al. 2015).

RKNs are often controlled by chemicals applications. However total control cannot be achieved, due to the lack of persistence of soil-applied nematicides which have to penetrate a large volume of soil to be effective. In addition, chemical methods can cause soil and water pollution and are very toxic to humans, animals, beneficial microorganisms, and sometimes even plants. For that reason, new nematicide production techniques are required, being the use of biological agents an alternative method of control (Park et al. 2014).

Biological control has attracted interest as an alternative option to chemical methods of controlling plant pathogens. Some related strategies are being used to suppress RKNs, such as the synthesis of secondary plant metabolites with inhibitory properties (Asif et al. 2017; Wang et al. 2018). Antagonistic soil microorganisms with nematicidal activity with rapid multiplication rates are non-toxic and environmentally compatible. Thus, they might be used to achieve sustainable control over a long period, in open-field conditions (Park et al. 2014).

In recent years, nematophagous bacteria have been used extensively as biocontrol agents against nematodes, because of their fast multiplication and easy production. In addition, a number of rhizobacterial mechanisms for biocontrol of RKNs have been studied (Niu et al. 2007). The mode of action of antagonistic bacteria against sedentary and migratory endoparasitic nematodes includes parasitism, reduction in juvenile penetration, hatching inhibition due to competition for nutrients, and antibiosis associated with bioactive metabolites (Xiang et al. 2018). The aim of this work was to identify the most effective potential PGPR bacteria with activity against *Meloidogyne* spp., under *in vitro* conditions.

16.2 Materials and Methods

16.2.1 In vitro *PGPR Activity*

Five rhizosphere isolates obtained from the LEMYB culture collection at Universidad Nacional Agraria La Molina were evaluated: AZO16M2 (*Rhanella* sp.) (Ogata-Gutiérrez et al. 2016), Bac15Mb (*Bacillus* sp.) (Calvo et al. 2010) from potato plants and the actinobacteria ACC17, ACC20, and ACC40 from coffee plants. Indole-3-acetic acid (IAA) production (μg ml^{-1}) was detected using the Gordon and Weber (1951) methodology with a pink tone reaction considered positive. Phosphate solubilization assays were perfomed as described by Nautiyal (1999), observing formation of a solubilization halo up to 20 dpi. For these assays, bi and tri-calcium phosphate were used as the phosphate (P) source, considering incubation temperatures of 20 °C, 28 °C and 35 °C. The phosphate-solubilization index (SI) was determined measuring the colony and halo zone diameters, by applying the formula proposed by Edi-Premono et al. (1996). Assays were performed with three replicates for statistical analysis.

16.2.2 Qualitative enzymatic activity of bacteria

Rhizobacteria ability to produce cellulase was measured in agar plates containing 1-carboxymethylcellulose according to Prasad et al. (2013) and incubated at 28 °C for 4 days. Plates were revealed by staining with 0.1% Congo Red solution and counterstaining with 1 M NaCl for 15 min at room temperature. Formation of clear zones around colonies was considered positive (Mohanta 2014). Protease production was detected inoculating each strain in Casein Skim Milk Agar according to Venant et al. (2013). Plates were incubated at 28 °C for 7 days The presence of a transparent halo around colonies indicated protease production. Chitinase activity was detected according to Jha et al. (2016). Rhizobacteria were incubated at 28 °C for 5 days. Colloidal chitin was prepared from prawn shell powder following the method of Agrawal and Kotasthane (2012) using 2% colloidal chitin agar plates (CCA). Plates were revealed by staining with 0.1% Congo Red solution and counterstaining with 1 M NaCl for 15 min at room temperature. Clear zones around the colonies indicated chitinase activity. Experiments were performed using the five isolates with three replicates each.

16.2.3 In vitro *Antifungal Activity Assay*

Antagonistic activity against *Rhizoctonia* sp. and *Fusarium* sp. was tested through the dual culture technique using Potato Dextrose Agar (PDA). Cultures were incubated at 25 °C in the dark for 3–7 days. Micelial growth inhibition was determined

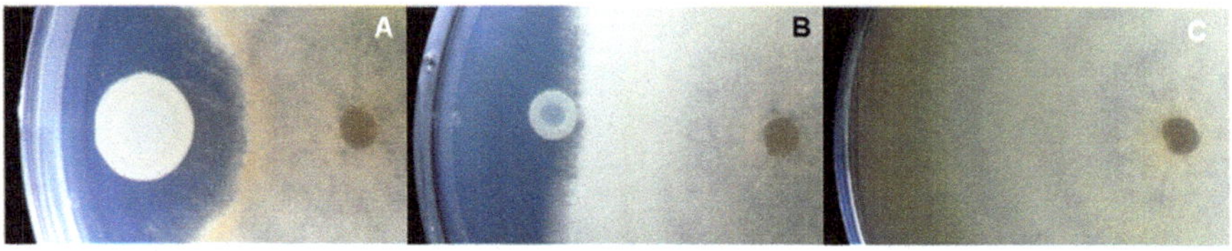

Fig. 16.1 Mycelial growth inhibition of *Rhizoctonia* sp. by isolates Bac15Mb (**a**), AZO16M2 (**b**), and control (**c**)

using the methodology of Prashar et al. (2013) (Fig. 16.1). Assays were performed using three replicated fungi-bacteria pairs.

16.2.4 In vitro *Assay for Nematostatic and Nematicidal Activity Against* Meloidogyne *spp.*

Bacteria isolates tested for *in vitro* nematostatic and nematicidal activity were grown at 150 rpm in broth media (TSB for isolates Bac15Mb and AZO16M2, and ISP2 for actinomycetes). The incubation temperature was 28 °C for 1 and 7 days for TSB and ISP2 plates, respectively. Bacteria cells were removed by centrifugation at 10000 g for 15 min. The supernatant was used to evaluate 20 surface-sterilized J2 (Ruanpanun and Chamswarng 2016). Nematode inoculum was obtained by extracting egg masses from a tomato greenhouse stock culture. Assays were performed at 25 °C for 48 and 72 h in well plates with three replicates. Mobile juveniles were counted using an Olympus SZ61 stereoscope with 90x magnification. The mean percentage of immobile juveniles was determined for this assay. Nematodes were considered dead if they were straight, stiffened, and did not move when probed with a fine needle (Park et al. 2014). Nematostatic and nematicidal effects of the supernatant was determined by testing the recovery capacity of stiff and non-moving specimens in sterile water. Results were reported considering culture media effect on nematodes. Experiments were carried out in three replicates for statistical analysis.

16.2.5 Statistical Analysis

Data was analyzed with analysis of variance for a completely randomized design using Statgraphics Centurion package, followed by a LSD ($p \leq 0.05$) post hoc test.

16.3 Results

16.3.1 Plant Growth Promoting Traits and In Vitro *Antifungal Activity*

Production of IAA was observed in all isolates. AZO16M2 showed the highest production (37.57 μg ml^{-1}). This isolate proved to be the most efficient phosphate solubilizer on NBRIP agar with both bi and tri-phosphate calcium (Table 16.1). All tested isolates were able to inhibit both *Rhizoctonia* sp. and *Fusarium* sp. growth. Percentages of fungal growth inhibition were higher than 15% for all tested isolates (Table 16.1) (Fig. 16.1).

16.3.2 Qualitative Enzymatic Activity

Results of enzymatic activity determinations are shown in Table 16.2. Isolates ACC 17 and ACC 20 showed activity for all the tested enzymes. Isolate Bac15Mb showed positive results for cellulase and protease activities (Fig. 16.2b) while isolate ACC40

Table 16.1 IAA production, Solubilization Index (SI), and Fungal Growth Inhibition (%) of the tested bacterial isolates

	IAA (ppm)	SI/bicalcium phosphate			SI/tricalcium phosphate			Fungal growth inhibition (%)	
Isolates	28 °C	20 °C	28 °C	35 °C	20 °C	28 °C	35 °C	*Rhizoctonia*	*Fusarium*
Bac15Mb	3.69 c	0.0 c	0.0 c	0.0 c	0.0 b	0.0 b	0.0 b	55.0 a	60.18 a
AZO16M2	37.57a	3.67a	5.58 a	4.91a	1.44 a	6.06a	4.42 a	39.0 bc	30.09 c
ACC 17	6.97 b	1.35 b	1.34 b	1.48 b	0.0 b	0.0 b	0.0 b	45.35 b	36.58 c
ACC 20	5.64 b	0.0 c	0.0 c	0.0 c	0.0 b	0.0 b	0.0 b	30.2 c	45.53 b
ACC 40	2.36 c	0.0 c	0.0 c	0.0 c	0.0 b	0.0 b	0.0 b	15.5 d	21.14 d

Means with the same letters in each column denote no significant difference at $P < 0.05$, based on LSD test

Table 16.2 Enzymatic activity of bacterial isolates

	Enzymatic activity (28 °C) Halo (mm)[a]		
Isolates	Cellulase	Protease	Chitinase
Bac15Mb	27.08 ± 1 d	18.11 ± 0.95 c	0.0 a
AZO16M2	0.0 a	0.0 a	0.0 a
ACC 17	6.08 ± 0.25 b	1.15 ± 1.06 a	5.14 ± 1.94 b
ACC 20	14.99 ± 0.79 c	7.34 ± 0.24 b	6.52 ± 0.32 b
ACC 40	33.63 ± 0.85 e	0.0 a	12.04 ± 0.37 c

Mean values ± standard deviation (SD)
[a]Means with the same letters in column have no significant difference at, based on LSD test ($p \leq 0.05$)

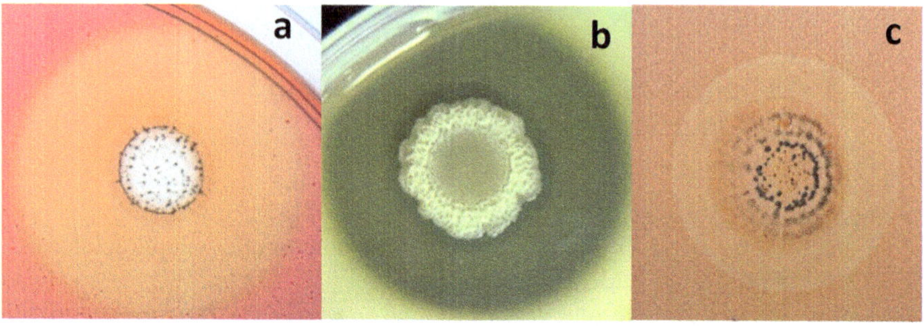

Fig. 16.2 Hydrolysis halo for cellulase (**a**), protease (**b**) and chitinase (**c**) activities

Table 16.3 Nematostatic and nematicidal activities of different bacterial isolates against *Meloidogyne* spp.

	Effect (%)	
Isolate	Nematostatic[a]	Nematicidal[a]
Bac15Mb	41.00 c	29.37 c
AZO16M2	92.34 e	92.34 d
ACC17	68.48 d	12.33 b
ACC20	29.68 bc	5.21 b
ACC40	75.19 d	14.09 b
TS broth	43.33 c	10.56 b
ISP2 broth	23.72 b	0.00 a
Control	5.25 a	1.30 a

[a]Means with same letters in column have no significant difference, based on LSD test ($p \leq 0.05$)

for cellulase (Fig. 16.2a) and chitinase (Fig. 16.2c). Isolate AZO16M2 did not produce any of these enzymes under the evaluated conditions. These assays indicated potential of the isolates against fungi and nematodes.

16.3.3 In vitro *Assay for Activity Against* Meloidogyne *spp.*

Results obtained from the *in vitro* nematicidal activity assays (Table 16.3) indicated that isolate AZO16M2 had significantly higher nematicidal and nematostatic effects (92.34%) (Fig. 16.3a and b respectively). Furthermore, isolate Bac15Mb showed a significantly stronger nematicidal effect (29.37%) against *Meloidogyne* spp., compared to the control.

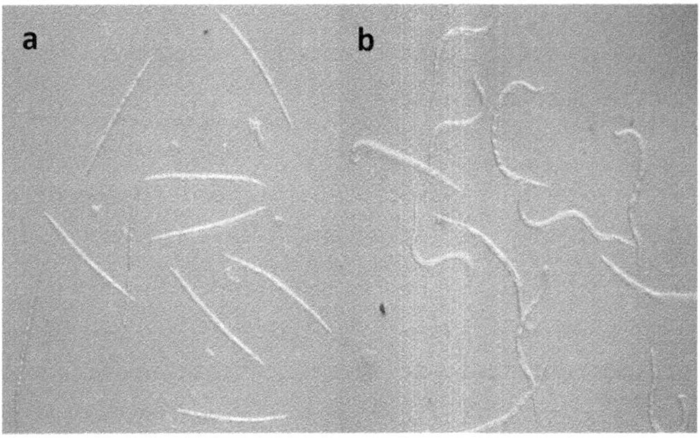

Fig. 16.3 Second-stage juveniles (J2) of *Meloidogyne* spp. exposed to AZO16M2 (**a**) and sterile water (**b**, control)

16.4 Discussion

Some rhizobacteria are known to promote plant growth through several mechanisms, including atmospheric nitrogen fixation, phytohormones synthesis (e.g. IAA), phosphate solubilization, and antibiotic production. In the present study, all tested rhizobacteria isolates produced IAA, and two solubilized phosphate (AZO16M2 and ACC17). These results are similar to other studies that reported PGPR properties by rhizobacteria, such as *Bacillus*, *Pseudomonas* and many diazotrophs (Angulo et al. 2014; Bakhshandeh et al. 2015; Liu et al. 2017; Gouda et al. 2018). Isolate AZO16M2 had high rates of both bi and tri-calcium phosphate solubilization, whereas ACC17 only showed solubilization of bi-calcium phosphate. The effectiveness of phosphate solubilization by rhizospheric bacteria depends on the solubility of P in the medium (Panhwar et al. 2009) and on the organic acids produced.

Antagonistic activity against *Rhizoctonia* sp. and *Fusarium* sp. ranged from 15.5 to 55% and 21.14 to 60.18%, respectively. *In vitro* antagonism towards fungi has been previously reported in rhizobacteria isolates (Calvo et al. 2010; Ogata-Gutiérrez et al. 2016). Several mechanisms have been proposed to explain a wide range of activity against many phytopathogenic fungi, including the secretion of hydrolytic enzymes, competition, parasitism, and the production of various antifungal metabolites (Jadhav et al. 2017). Lytic enzymes, such as cellulase, are also associated with PGPR features in rhizobacteria, enabling them to limit fungal pathogens growth (El-Sayed et al. 2014). In this study we determined the production of cell wall-degrading enzyme cellulase resulting in four positive isolates (Bac15Mb, ACC17, ACC20 and ACC40). Strain Bac15Mb showed high cellulolytic activity and the highest inhibition effect, towards both phytopathogenic fungi (Table 16.2).

Nematophagous bacteria use different mechanisms to kill nematodes: some of them include microbial toxins, insecticidal crystal proteins, hydrolytic enzymes, and secondary metabolites. Chitin plays an important role in the formation of nematode egg shells and cuticle, by acting as an important protective barrier. Several bacteria could use chitin as a carbon source and infect hosts including nematodes, using chitinases. Also, some microbial proteases can destroy the nematode cuticle or intestine tissues (Chen et al. 2015; Geng et al. 2016). In our study, strains Bac15Mb and ACC40 produced the greatest chitinase and protease activities, respectively (Table 16.2), showing also nematicidal activity (Table 16.3). It is worth to note that AZO16M2 had the best nematicidal effect, even if protease and chitinase production was not observed in this isolate. Therefore, nematicidal activity of AZO16M2 can be attributed to other mechanisms that could play an important role against nematodes (Gao et al. 2016).

Rhizobacteria have been reported to be good candidates for biological nematode control. Results from this work agree with other studies tested under similar assay conditions, which showed nematicidal effects of rhizobacteria against *Meloidogyne* sp. (Huang et al. 2015). Similar results have been reported by Turatto et al. (2017), who evaluated rhizobacteria ability for hatching reduction of *Meloidogyne javanica* eggs and decreased motility of *Ditylenchus* spp. Xiang et al. (2017) also reported *in vitro* nematicidal effect of potential PGPR strains against *Heterodera glycines* J2 as well as the ability of bacteria to reduce nematode population densities on soybean rhizospheres.

The maximum temporal nematostatic and permanent nematicidal *in vitro* effects against *Meloidogyne* spp. were recorded for the diazotrophic isolate AZO16M2 (92.34%). Previous research showed that the application of diazotrophic bacteria (*Rahnella aquatilis*) resulted in disease suppression of *Erwinia amylovora* and crown gall induction by *Agrobacterium vitis* (Guo et al. 2014). Additionally, several microbial metabolites have been reported to have antibiosis effects against nematodes, such as volatile fatty acids produced by *Gluconacetobacter diazotrophicus* (Pankaj et al. 2010; Chawla et al. 2014).

Isolate Bac15Mb showed an *in vitro* nematicidal effect of 29.37% (Table 16.2). *Bacillus* species are known to produce different exotoxins and antibiotics, some of which are reported to be nematicidal (Siddiqui et al. 2009; Abbasi et al. 2014). For example, a gene of *Bacillus amyloliquefaciens* FZB42 involved in biosynthesis of plantazolicin has also been reported to be involved in antagonistic effects against nematodes (Liu et al. 2013). Analysis of the whole FZB42 genome revealed a strong potential to produce a diverse spectrum of secondary metabolites aimed at suppressing harmful microbes and nematodes within the plant rhizosphere. This activity may be due to synthesis of metabolites such as bacillomycin D, fengycin, polyketides, bacilysin, and bacteriocins (Chowdhury et al. 2015).

16.5 Conclusion

The results obtained in this study showed the potential of bacterial isolates such as PGPR with an antagonistic activity against phytopathogenic fungi and *Meloidogyne* spp., under *in vitro* conditions. This study could be used as a basis for future greenhouse and field trials.

Acknowledgements This work was supported by project 177-2015-FONDECYT and project 007-2016-INIA-PNIA/UPMSI/IE.

References

Abad, P., Gouzy, J., Aury, J. M., Castagnone-Sereno, P., Danchin, E. G., Deleury, E., Perfus-Barbeoch, L., Anthourd, V., Artigueve, F., & Blok, V. C. (2008). Genome sequence of the metazoan plant parasitic nematode *Meloidogyne incognita*. *Nature Biotechnology, 26*, 909–915.

Aballay, E., Ordenes, P., Martensson, A., & Persson, P. (2013). Effects of rhizobacteria on parasitism by *Meloidogyne ethiopica* on grapevines. *European Journal of Plant Pathology, 135*, 137–145.

Abbasi, M. W., Ahmed, N., Zaki, M. J., Shuakat, S., & Khan, D. (2014). Potential of *Bacillus* species against *Meloidogyne javanica* parasitizing eggplant (*Solanum melongena* L.) and induced biochemical changes. *Plant and Soil, 375*, 159–173.

Agrawal, T., & Kotasthane, A. S. (2012). Chitinolytic assay of indigenous *Trichoderma* isolates collected from different geographical locations of Chhattisgarh in Central India. *Springerplus, 1*(1), 73. https://doi.org/10.1186/2193-1801-1-73.

Angulo, V. C., Sanfuentes, E. A., Rodríguez, F., & Sossa, K. E. (2014). Characterization of growth-promoting rhizobacteria in *Eucalyptus nitens* seedlings. *Revista Argentina de Microbiología, 46*(4), 338–347

Asif, M., Tariq, T., Khan, A., & Siddiqui, M. A. (2017). Biocidal and antinemic properties of aqueous extracts of Ageratum and Coccinia against root-knot nematode, *Meloidogyne incognita in vitro*. *The Journal of Agricultural Sciences, 12*(2), 108–122. https://doi.org/10.4038/jas.v12i2.8229.

Bakhshandeh, E., Rahimian, H., Pirdashti, H., & Nematzadeh, G. A. (2015). Evaluation of phosphate-solubilizing bacteria on the growth and grain yield of rice (*Oryza sativa* L.) cropped in northern Iran. *Journal of Applied Microbiology, 119*(5), 1371–1382.

Calvo, P., Ormeño-Orrillo, E., Martínez-Romero, E., & Zúñiga, D. (2010). Characterization of *Bacillus* isolates of potato rhizosphere from Andean soils of Peru and their potential PGPR characteristics. *Brazilian Journal of Microbiology, 41*, 899–906.

Chawla, N., Phour, M., Suneja, S., Sangwaan, S., & Goyal, S. (2014). *Gluconacetobacter diazotrophicus*: An overview. *Research in Environment and Life Sciences, 7*(1), 1–10.

Chen, L., Huang, J., Cheng, Q., Chen, J., Wu, G., Kumar, A., Sun, M., & Liu, Z. (2015). Enhanced nematicidal potential of the chitinase pachi from *Pseudomonas aeruginosa* in association with Cry21Aa. *Scientific Reports, 5*, 14395.

Chowdhury, S. P., Hartmann, A., Gao, X., & Borriss, R. (2015). Biocontrol mechanism by root-associated *Bacillus amyloliquefaciens* FZB42 – a review. *Frontiers in Microbiology, 6*, 780. https://doi.org/10.3389/fmicb.2015.00780.

Coyne, D. L., Cortada, L., Dalzell, J. J., Claudius-Cole, A. O., Haukeland, S., Luambano, N., & Talwana, H. (2018). Plant-parasitic nematodes and food security in Sub-Saharan Africa. *Annual Review of Phytopathology, 56*, 381–403. https://doi.org/10.1146/annurev-phyto-080417-045833.
Edi-Premono, M., Moawad, A. M., & Vlek, P. L. G. (1996). Effect of phosphate- solubilizing *Pseudomonas putida* on the growth of maize and its survival in the rhizosphere. *Indonesian Journal of Crop Science, 11*, 13–23.
El-Sayed, W. S., Akhkha, A., El-Naggar, M. Y., & Elbadry, M. (2014). *In vitro* antagonistic activity, plant growth promoting traits and phylogenetic affiliation of rhizobacteria associated with wild plants grown in arid soil. *Frontiers in Microbiology, 5*, 651. https://doi.org/10.3389/fmicb.2014.00651.
Gao, H., Qi, G., Yin, R., Zhang, H., Li, C., & Zhao, X. (2016). Bacillus cereus strain S2 shows high nematicidal activity against Meloidogyne incognita by producing sphingosine. *Scientific Reports, 6*, 28756.
Geng, C., Nie, X., Tang, Z., Zhang, Y., Lin, J., Sun, M., & Peng, D. (2016). A novel serine protease, Sep1, from *Bacillus firmus* DS-1 has nematicidal activity and degrades multiple intestinal-associated nematode proteins. *Scientific Reports, 6*, 25012.
Gordon, S. A., & Weber, R. P. (1951). Colorimetric estimation of indolacetic acid. *Plant Physiology, 26*, 192–195.
Gouda, S., Kerry, R. G., Das, G., Paramithiotis, S., Shin, H. S., & Kumar, P. J. (2018). Revitalization of plant growth promoting rhizobacteria for sustainable development in agriculture. *Microbiological Research, 206*, 131–140.
Guo, Y., Jiao, Z., Li, L., Wu, D., Crowley, D. E., Wang, Y., & Wua, W. (2014). Draft genome sequence of *Rahnella aquatilis* strain HX2, a plant growth-promoting Rhizobacterium isolated from vineyard soil in Beijing, China. *Journal of Bacteriology, 194*(23), 6646–6647.
Huang, Y., Ma, L., Fang, D. H., Xi, J. Q., Zhu, M. L., Mo, M. H., Zhang, K. Q., & Jie, Y. P. (2015). Isolation and characterisation of rhizosphere bacteria active against *Meloidogyne incognita*, *Phytophthora nicotianae* and the root knot-black shank complex in tobacco. *Pest Management Science, 71*, 415–422.
Jadhav, H. P., Shaikh, S. S., & Sayyed, R. Z. (2017). Role of hydrolytic enzymes of rhizoflora in biocontrol of fungal phytopathogens: an overview. In S. Mehnaaz (Ed.), *Rhizotrophs: Plant growth promotion to bioremediation* (pp. 183–203). Singapore: Springer.
Jha, S., Modi, H. A., & Jha, C. K. (2016). Characterization of extracellular chitinase produced from *Streptomyces rubiginosus* isolated from rhizosphere of *Gossypium sp*. *Cogent Food & Agriculture, 2*, 1198225. https://doi.org/10.1080/23311932.2016.1198225.
Liu, Z., Budiharjo, A., Wang, P., Shi, H., Fang, J., Borriss, R., Zhang, K., & Huang, X. (2013). The highly modified microcin peptide plantazolicin is associated with nematicidal activity of *Bacillus amyloliquefaciens*. *Applied Microbiology and Biotechnology, 97*, 10081–10090.
Liu, K., Newman, M., McInroy, J. A., Hu, C. H., & Kloepper, J. W. (2017). Selection and assessment of plant growth-promoting Rhizobacteria for biological control of multiple plant diseases. *Phytopathology, 107*(8), 928–936.
Mohanta, Y. K. (2014). Isolation of cellulose-degrading actinomycetes and evaluation of their cellulolytic potential. *Bioengineering and Bioscience, 2*(1), 1–5.
Mokrini, F., Janati, S., Houari, A., Essarioui, A., Bouharroud, R., & Mimouni, A. (2018). Management of plant parasitic nematodes by means of organic amendment. *Review of Marketing Science. Agron Vét, 6*(3), 337–344.
Nautiyal, C. S. (1999). An efficient microbiological growth medium for screening phosphate solubilizing microorganisms. *FEMS Microbiology Letters, 170*, 265–270.
Niu, Q. H., Huang, X. W., Zhang, L., Lian, L. H., Li, Y. X., Li, J., Yang, J. K., & Zhang, K. Q. (2007). Functional identification of the gene bace16 from nematophagous bacterium *Bacillus nematocida*. *Applied Microbiology and Biotechnology, 75*, 141–148.
Ogata-Gutiérrez, K., Alvarado, D., Chumpitaz-Segovia, C., & Zúñiga-Dávila, D. (2016). Characterization of plant growth promoting Rhizobacteria isolated from the Rhizosphere of Peruvian Highlands native crops. *International Journal of Plant & Soil Science, 11*(1), 1–8.

Panhwar, Q. A., Radziah, O., Sariah, M., & Ismail, M. R. (2009). Solubilization of different phosphate forms by phosphate solubilizing bacteria isolated from Aerobic Rice. *International Journal of Agriculture and Biology, 11*(6), 667–673.

Pankaj, K., Bansal, R., & Nandal, S. (2010). Biocontrol of *Meloidogyne graminícola* using Rhizobacteria on rice seedling. *Nematology Mediterránea, 38*, 115–119.

Park, J., Seo, Y., & Ho Kim, Y. (2014). Biological control of *Meloidogyne hapla* using an antagonistic bacterium. *Plant Pathology Journal, 30*(3), 288–298.

Prasad, P., Singh, T., & Bedi, S. (2013). Characterization of the cellulolytic enzyme produced by *Streptomyces griseorubens* (Accession No.AB184139) isolated from Indian soil. *Journal of King Saud University – Science, 25*(3), 245–250.

Prashar, P., Kapoor, N., & Sachdeva, S. (2013). Isolation and characterization of *Bacillus sp.* with in-vitro antagonistic activity against *Fusarium oxysporum* from Rhizosphere of Tomato. *Journal of Agricultural Science and Technology, 15*, 1501–1512.

Ruanpanun, P., & Chamswarng, C. (2016). Potential of actinomycetes isolated from earthworm castings in controlling root-knot nematode *Meloidogyne incognita*. *Journal of General Plant Pathology, 82*(1), 43–50.

Siddiqui, Z. A., Qureshi, A., & Akhtar, M. S. (2009). Biocontrol of root-knot nematode *Meloidogyne incognita* by *Pseudomonas* and *Bacillus* isolates on *Pisum sativum*. *Archives of Phytopathology and Plant Protection, 42*(12), 1154–1164.

Turatto, M. F., Dourado, F. D. S., Zilli, J. E., & Botelho, G. R. (2017). Control potential of *Meloidogyne javanica* and *Ditylenchus* spp. using *Pseudomonas fluorescent* and *Bacillus* spp. *Brazilian Journal of Microbiology, 49*(1), 54–58.

Venant, N., Pascal, K., & Ernest, S. (2013). Isolation of *Bacillus* strains from the rhizosphere of tomato and their *in vitro* antagonistic effects against phytopathogenic fungi. *Global Advanced Research Journal of Microbiology, 2*(3), 65–71.

Wang, C., Masler, E. P., & Rogers, S. T. (2018). Responses of *Heterodera glycines* and *Meloidogyne incognita* infective juveniles to root tissues, root exudates, and root extracts from three plant species. *Plant Disease, 102*(9), 1733–1740.

Wei, L., Shao, Y., Wan, J., Feng, H., Zhu, H., & Huang, H. (2014). Isolation and characterization of a Rhizobacterial antagonist of root-knot nematodes. *PLoS One, 9*(1), e85988. https://doi.org/10.1371/journal.pone.0085988.

Xiang, N., Lawrence, K. S., Kloepper, J. W., Donald, P. A., & McInroy, J. A. (2017). Biological control of *Heterodera glycines* by spore-forming plant growth-promoting rhizobacteria (PGPR) on soybean. *PLoS One, 12*(7), e0181201.

Xiang, N., Lawrence, K. S., & Donald, P. A. (2018). Biological control potential of plant growth-promoting rhizobacteria suppression of *Meloidogyne incognita* on cotton and *Heterodera*. *Journal of Phytopathology, 166*(7–8), 449–458.

Chapter 17
Combined Application of Microbial and Non-Microbial Biostimulants to Improve Growth of Peanut Plants Exposed to Abiotic Stresses

Ana Furlan, Eliana Bianucci, Micaela Sequeira, Lucía Álvarez, Juan Manuel Peralta, Carina Valente, Valmiro Guarnieri, and Stella Castro

Abstract Peanut is a widespread legume, with an important agricultural and economic significance. It symbiotically interacts with rhizobia, increasing atmospheric nitrogen assimilation by the biological nitrogen fixation process, therefore improving yield. The presence of the environmental pollutant arsenic and the occurrence of water deficit episodes constitute severe abiotic stresses affecting this symbiosis, being biostimulants a sustainable alternative to increase crop yields. Thus, the objective of this work was to determine the effects of the joint application of a commercial seed non-microbial plant biostimulant (Nutrifer® 202) and a microbial plant biostimulant on growth, nodulation and oxidative stress indicator-levels, on peanut plants exposed to arsenic or drought. Biostimulant addition reduced As translocation to leaves and improved plant growth and nodulation in the drought stress condition, in association with proline accumulation, with a protective function on the cellular redox balance. Therefore the application of the biostimulant combination Nutrifer® 202 and *Bradyrhizobium* sp. C-145 is promising for peanut crops growing in regions susceptible to water deficit or arsenic exposure.

Keywords Abiotic stress · Arsenic · Biological nitrogen fixation · Biostimulant · Drought

A. Furlan (✉) · E. Bianucci · M. Sequeira · L. Álvarez · J. M. Peralta · S. Castro
Instituto de Investigaciones Agrobiotecnológicas (INIAB-CONICET), Facultad de Ciencias Exactas, Físico-Químicas y Naturales, Universidad Nacional de Río Cuarto, Córdoba, Argentina
e-mail: afurlan@exa.unrc.edu.ar

C. Valente · V. Guarnieri
NUTRIFER S.A., C.A.B.A., Argentina

D. Zúñiga-Dávila et al. (eds.), *Microbial Probiotics for Agricultural Systems*, Sustainability in Plant and Crop Protection,
https://doi.org/10.1007/978-3-030-17597-9_17

17.1 Introduction

Peanut (*Arachis hypogaea* L.) is one of the most important regional crops in the center-south of the province of Córdoba (Argentina) where more than 90% of the national primary production is concentrated (Fernandez and Giayetto 2017). The crop has a high demand for nitrogen, satisfied through symbiotic associations with microorganisms providing biological nitrogen fixation (BNF) (Pedelini 2008). However, abiotic stresses such as arsenic exposure and drought, frequently affect this symbiosis (Serraj 2003; Paudyal et al. 2007; Akerblom et al. 2007; Sobolev and Begonia 2008; Mhadhbi et al. 2015).

Arsenic (As) is a toxic metalloid, naturally found in soils and water, being groundwater the principal source of human contamination (Smedley and Kinniburgh 2002). In Argentina, groundwater containing As is widely extended, with the higher values found in the Córdoba province. In some areas As levels are up to 20 μM (Francisca et al. 2006; Cabrera et al. 2005), while the Argentinian Alimentary Codex (AAC) limited As concentration in drinking water to 0.1 μM. The problem arises because crop plants could directly absorb the metalloid from groundwater or incorporate it by artificial irrigation, thus acting as first stage of As distribution in the trophic chain (Smedley and Kinniburgh 2002; Bustingorri and Lavado 2014). Some of the consequences induced by this metalloid in plants are related to root darkening, growth inhibition, chlorosis (Bianucci et al. 2018). In the symbiotic interaction legume-rhizobia, metalloid addition originates redox imbalances, leading to oxidative stress, reduced nitrogen assimilation and impaired nodule organogenesis (Mandal et al. 2008; Lafuente et al. 2010, 2015; Bianucci et al. 2018).

In the central-southern region of the province of Córdoba, the occurrence of periods of drought stress in the summer is common, coinciding with the reproductive development of peanuts (R1) (Boote 1992; Fernandez and Giayetto 2017). Although peanut is a relatively drought-tolerant crop, a correct level of humidity during the stage of reproductive development is required to achieve maximum performances. Some of the morpho-physiological and biochemical changes that this stress produces are: the decrease of photosynthesis and growth; the accumulation of compatible osmolytes such as proline, abscisic acid (ABA) and reactive oxygen species (ROS); the activation of the antioxidant system and BNF decrease (Furlan et al. 2013, 2014, 2016). In this scenario, proline has become an important molecule being an osmoprotective molecule, a molecular chaperone and a non-enzymatic antioxidant (Szabados and Savouré 2009) by its ability to detoxify the OH· radical (Signorelli et al. 2014).

Biostimulants are becoming increasingly integrated into agricultural production systems in order to optimize crop yields. Their heterogeneous and indefinite–composition, and the possibility that their activity can not be explained by the presence of any individual constituent, make difficult their definition. In the present work we will refer to biostimulants as "*formulated products of biological origin that improve plant productivity as a result of the novel or emerging properties of their complex of constituents, and not as a single consequence of the presence of well-known essential plant nutrients, plant growth regulators or protective compounds*" (Yakhin et al. 2017).

Plant biostimulants are divided into two major classes: microbial plant biostimulant (MPB) and non-microbial plant biostimulant (NMPB) (Calvo et al. 2014). Applications of NMPBs and MPBs have been described as recommended management practices in peanut crops. They stimulate seedling establishment and initial growth (Kearney et al. 2011); increase yield (Cerioni et al. 2014; Morla et al. 2013); and provide favorable conditions of nitrogen nutrition (Valetti et al. 2008; Baliña et al. 2013). However, they have been evaluated separately and under non-stressful conditions. The joint use of these technologies to manage exposition to abiotic stresses appears as an interesting approach. Selection of the best NMPB and MPB combination is not easy to achieve due to several challenges that crops must overcome such as edaphic conditions, climatic factors and the occurrence of different abiotic or biotic stresses. Besides, the knowledge of the strategies performed by plants to mitigate these adverse effects have not yet been fully elucidated, even more when exposed to different stresses. Therefore, the aim of the present work was to elucidate the effects of the joint application of a commercial seed NMPB (Nutrifer® 202) and MPBs on growth, nodulation and oxidative stress indicators on peanut plants exposed to arsenic or drought.

17.2 Materials and Methods

17.2.1 Plant material and treatments

The inoculant strains (MPB) used were *Bradyrhizobium* sp. SEMIA6144 and *Bradyrhizobium* sp. C-145, from MIRCEN (Porto Alegre, Brazil) and INTA (Castelar, Buenos Aires, Argentina), respectively. The peanut cultivar was Granoleico, widely used in the peanut growing area in Córdoba (Argentina) obtained from "Criadero El Carmen" (General Cabrera, Córdoba, Argentina). The NMPB used was Nutrifer® 202 (Nutrifer S.A. 2017) which includes in its composition an algal extract which acts as a carrier of other elements, regulating the osmotic pressure, conferring resistance to extreme temperatures and drought, and stimulating growth due to its phytohormone content. In addition, this product can correct deficiencies of molybdenum and cobalt, two very specific micronutrients of the rhizobium-legume symbiotic relationship.

The peanut seeds were sterilized (Vincent 1970) and Nutrifer® 202 was added according to the supplier's recommended dose (100 g 100 k^{-1} seeds). Subsequently, a subgroup of seeds were inoculated with 4 ml of each MPB grown in yeast extract–mannitol medium containing $1 \cdot 10^9$ CFU ml^{-1} and equal volume of sterile arabic gum (0.6% w/v) and the number of viable cells attached to seeds was determined (Somasegaran and Hoben 1994). The treatments were: i) uninoculated; ii) uninoculated and subsequent fertilization in pots by irrigation with 5 mM KNO_3; iii) NMPB addition (Nutrifer® 202); iv) addition of the MPB *Bradyrhizobium* sp. SEMIA6144 ($3.1 \cdot 10^7$ CFU $seed^{-1}$); v) addition of the NMPB and MPB *Bradyrhizobium* sp. SEMIA6144 ($1.2 \cdot 10^7$ CFU $seed^{-1}$); vi) addition of the MPB *Bradyrhizobium* sp.

C-145 (4.2 · 10^8 CFU seed^{-1}); vii) addition of the NMPB and the MPB *Bradyrhizobium* sp. C- 145 (5.5 · 10^8 CFU seed^{-1}). Pots containing perlite: sterile sand (1:2) were placed in a growth chamber with controlled light and temperature conditions (light intensity: 200 μmol/m^2/s, photoperiod: 16-h day/8-h darkness, temperature: 28 ° C, relative humidity: 50%).

Plants were irrigated twice a week alternatively with distilled water and nitrogen-free Hoagland nutrient solution (Hoagland and Arnon 1950), in order to keep the field capacity (13%) determined as Burk (1996). Arsenic exposure was performed at the beginning of the experiment. Plants were irrigated twice a week with Hoagland medium either in the absence or presence of 3 μM As as $NaHAsO_2{\cdot}7H_2O$ (the As concentration found in groundwater of some producing areas of Córdoba). The drought stress treatment was imposed 30 days after sowing, in plants in the phenological stage R1 (flowering) according to Boote (1992), separated at random into the experimental groups: a) the control, where plants were kept under normal irrigation conditions; b) drought stress, where the irrigation was suspended for 10 days. At harvest, the fully expanded second nodal leaves were used to measure the relative water content (RWC) (Barr and Weatherley 1962). Nodules and leaves were frozen into liquid nitrogen and stored at −80 °C until use.

17.2.2 Analytical procedures

Growth and biological nitrogen fixation variables

Shoots and roots were dried at 70 °C until constant weight to determine the dry weight (DW). Nodules were counted and the dry weight was recorded. The nitrogen content per plant was determined by Kjeldahl's modified method (Nelson y Sommers 1973).

Arsenic content

Metalloid concentration was determined in peanut shoots, roots and nodules digested with an acid oxidizing mixture $H_2O{:}HNO_3{:}H_2O_2$ (5:3:2) as described by Sobrino-Plata et al. (2009). Arsenic concentration was measured using inductively coupled plasma mass spectrometry (ICP-MS).

ROS production and oxidative stress markers

In order to detect superoxide anion (O_2^-) production, fully expanded leaves from the second node were detached and vacuum-infiltrated with nitroblue tetrazolium (NBT) (Frahry and Schopfer 2001). Freshly cut nodules were incubated with a 0.1% (w/v) NBT solution and visualized as described by Doke (1983).

Hydrogen peroxide (H_2O_2) was measured spectrophotometrically after reaction with KI (Alexieva et al. 2001). The amount of hydrogen peroxide was calculated using a standard curve prepared with known concentrations of H_2O_2.

The level of lipid peroxides was determined as thiobarbituric acid reactive-substances (TBARs) content where thiobarbituric acid (TBA) reacts with the aldehyde group of MDA (final product of lipoperoxidation) and other TBARs giving a compound of pink color measured at 532 nm (Heath and Packer 1968). The final concentration was calculated based on a curve made with 1 mM tetraethoxypropane.

The amount of total chlorophyll was determined by the method described by Arnon (1949) in tissues grinded with 80% (v/v) acetone and spectrophotometer reading at 652 nm. The final amount was calculated by:

$$\text{Total chlorophyll} = \left[\left(\text{OD}_{652} \times 1000\right) / 3.45\right] \times 10 \ \text{ml} / 1000 \times 0.1 \ \text{g}$$

The concentration of proline in leaves and nodules was determined according to Bates et al. (1973) by the colorimetric quantification of the amino acid after the reaction with acid ninhydrine in toluene. The calibration curve was performed using 1 mM L-proline (Sigma).

17.2.3 *Statistical analysis*

Experiments were conducted in a completely randomized design and repeated three times. The data were analysed using ANOVA and Duncan's test at $p \leq 0.05$. Prior to the significance test, the normality and homogeneity of variance were verified using the modified Shapiro–Wilk and Levene tests, respectively. Data were transformed using normal scores when homogeneity of variance was not given. Principal component analysis (PCA) was performed in the software R v. 2.25.2.

17.3 Results and Discussion

17.3.1 *NMPB and MPB impact on growth and nodulation of peanut plants exposed to environmental stresses*

17.3.1.1 Arsenic stress

The uninoculated plants were the only group that showed a negatively affected plant growth when exposed to a realistic As dose (3 μM). However, As addition reduced nodulation variables compared to non-contaminated plants in a distinctive manner. A significant decrease of nodule number was observed in peanut plants, regardless of the treatment applied. Interestingly, this reduction was accompanied with a significant reduction of nodule dry weight only in plants inoculated with the MPB *Bradyrhizobium* sp. SEMIA6144 and supplied with the NMPB (Table 17.1). The results obtained for nitrogen content of peanut plants exposed to As showed that application of the NMPB, regardless of the used strain, increased in a significant manner the variable compared to plants devoid of the metalloid. Nitrogen content of inoculated peanut plants exposed to the contaminant and treated with NMPB

Table 17.1 Effect of NMPB and MPB addition on nodulation of peanut plants exposed to arsenic

	Nodule number		Nodule dry weight (g plant^{-1})		N content (mg g^{-1} dry weight)	
Treatments	Control	Arsenic	Control	Arsenic	Control	Arsenic
Uninoculated	nd	nd	nd	nd	27.2 ± 3.2 abc1	27.2 ± 1.7 a1
Fertilized	nd	nd	nd	nd	29.8 ± 1.0 bc1	29.7 ± 3.3 a1
NMPB	nd	nd	nd	nd	21.2 ± 2.3 a1	33.2 ± 4.07 ab1
Bradyrhizobium sp. SEMIA6144	33.5 ± 3.2 a1	24.0 ± 1.4 ab2	0.03 ± 0.01 a1	0.03 ± 0.002 a1	36.4 ± 0.8 d1	46.5 ± 4.0 c1
NMPB + *Bradyrhizobium* sp. SEMIA6144	29.4 ± 1.3 a1	21.5 ± 1.2 a2	0.04 ± 0.01 a1	0.02 ± 0.003 a2	36.0 ± 2.6 d1	47.7 ± 2.7 c2
Bradyrhizobium sp. C- 145	46.5 ± 3.4 b1	29.8 ± 1.8 c2	0.03 ± 0.01 a1	0.03 ± 0.003 a1	34.9 ± 1.3 cd1	42.2 ± 5.8 bc1
NMPB + *Bradyrhizobium* sp. C-145	39.0 ± 5.2 ab1	26.5 ± 1.3 bc2	0.03 ± 0.01 a1	0.03 ± 0.003 a1	26.4 ± 3.7 ab1	42.6 ± 3.3 bc2

Data represent the mean ± SE (n = 6). Different numbers per row indicate significant differences between growth condition (devoid of As or not) of peanut plants for each treatment according to the Duncan's multiple range test ($p \leq 0.05$). Different letters per column indicate significant differences between treatments for each growth condition according to the Duncan's multiple range test ($p \leq 0.05$)

significantly increased compared to plants growing without its addition. In plants devoid of As, this variable showed similar results between treatments. However, plants treated with Nutrifer® 202 showed the higher values. Thus, although the NMPB addition did not improve plant growth in plants treated with As or not, joint application with bradyrhizobia improved atmospheric nitrogen assimilation.

NMPBs effects on plants have been extensively studied. However, knowledge about its effects on plants growing on adverse conditions, as metalloid exposure, have not yet been elucidated. It is known that As has a negative impact on plant growth and in the symbiotic interaction established between legumes and rhizobia. Bianucci et al. (2018) and Peralta et al. (2018) also reported that a low As concentration did not affect growth variables in soybean and peanut plants, although nodulation was significantly affected. Moreover, it was shown that As reduced the number of nodules in *Medicago sativa* and *Glyicine max* plants (Reichman 2007; Vázquez et al. 2009) as observed in this work. Lafuente et al. (2010) explained that the negative impact of As on the establishment of the symbiotic interaction could be due to the interference of the metalloid with the expression of genes involved in plant defense and in early nodule development. Nevertheless, it was shown that once the interaction was established, nodule development continued normally (Pajuelo et al. 2008). Although reduction of nodule number could be associated to a reduction in the total nitrogen content of the plant (Bianucci et al. 2018), in this study, peanut plants exposed to As reduced nodulation but nitrogen content remained unchanged, similar to control condition. Moreover, NMPB application increased the variable in a significant way in plant devoid or not of As, regardless of the treatment tested. In addition, most nodules showed a brownish red deposit in the infection zone (representing active nodule by leghemoglobin presence), indicating that all of them were effective (data not shown). Thus, although plants exposed to a realistic As dose did not modify their growth; NMPB addition increased N fixation in the peanut plants exposed to arsenic, showing that micronutrients present in this NMPB make the BNF effective under metalloid exposure.

The joint application of NMPB and bradyrhizobia showed the lower levels of As content in peanut leaves compared to the other treatments tested. In roots, NMPB and inoculation reduced As concentration on roots compared to uninoculated or fertilized plants. However, inoculated plants (without NMPB addition) showed higher metalloid concentration in leaves in comparison with plants inoculated and treated with the NMPB.

Arsenic accumulation in nodules did not show differences, regardless of bradyrhizobia strain inoculated and treatment employed. Total As concentration revealed that uninoculated plants treated with the NMPB showed similar metalloid content than control plants (uninoculated or fertilized) (Table 17.2).

It is known that non-hyperaccumultor plants such as lupin, soybean, *Vigna*, alfalfa and peanut present similar metalloid distribution pattern (Pajuelo et al. 2008; Mandal et al. 2008; Panigrahi and Randhawa 2010; Panigrahi et al. 2013; Bianucci et al. 2017, 2018; Peralta et al. 2018) that is highest in roots, followed by nodules and finally by leaves. Moreover, it is known that inoculation with a specific symbiotic partner significantly reduced As content in shoot (Bianucci et al. 2018).

Table 17.2 Arsenic content in peanut plants treated with biostimulants

	As content (μg As g^{-1} dry weight)[a]			Total (%)
Treatments	Leaves	Roots	Nodules	
Uninoculated	3.4 ± 0.4 b	17.5 ± 1.1 de	Nd	20.9 ± 0.8 b
Fertilized	3.7 ± 0,2 b	24.7 ± 2.5 f	Nd	28.4 ± 2.3 c
NMPB	3.6 ± 0.4 b	18.9 ± 0.4 e	Nd	22.6 ± 0.6 b
Bradyrhizobium sp. SEMIA6144	3.1 ± 0.2 b	9.2 ± 0.6 ab	7.3 ± 0.7 a	13.2 ± 1.2 a
NMPB + *Bradyrhizobium* sp. SEMIA6144	1.9 ± 0.1 a	10.3 ± 0.2 bc	8.9 ± 1.5 a	14.2 ± 0.3 a
Bradyrhizobium sp. C- 145	3.5 ± 0.3 b	8.6 ± 0.3 a	7.7 ± 0.9 a	13.3 ± 1.1 a
NMPB + *Bradyrhizobium* sp. C-145	1.5 ± 0.2 a	14.3 ± 0.67 cd	5.9 ± 0.5 a	16.5 ± 1.1 a

[a]Data represent mean ± SE (n = 6). Different letters in columns indicate significant differences among treatments, according to the Duncan's multiple range test ($p \leq 0.05$)

Remarkably, the results presented in this work revealed that peanut plants inoculated and treated with the NMPB showed both lower total As concentration and minor content in leaves. These results suggest that joint application of rhizobia and NMPB limit metalloid translocation to aerial part.

17.3.1.2 Drought stress

The plant water status was assessed by relative water content (RWC) determination, which decreased significantly in plants exposed to drought stress with respect to well-irrigated plants, in all the treatments analyzed. Upon exposure to drought stress, combined application of biostimulants, regardless of the used strain, improved root/shoot length-ratio in comparison with plants treated only with each MPB. Plants treated solely with each of the MPBs had the lowest root/shoot dry weight-ratios in the well-irrigated condition. Besides, plants inoculated with the MPB *Bradyrhizobium* sp. C-145, regardless of the NMPB addition, had a lower root/shoot dry weight-ratio than noninoculated plants (Table 17.3).

The RWC integrates the water balance of the plants being, in comparison with the water potential, more stable and sensitive (Clavel et al. 2006). It is also a satisfactory indicator of the water status in peanut plants (Wright and Nageswara Rao 1994). Thus, it allowed measuring water deficit magnitude in our experimental system to ensure that plants were effectively exposed to drought stress. In this condition, Puangbut et al. (2009) and Jongrungklang et al. (2011) found that peanut genotypes that increased yield had an increased root-weight and root-length in the deeper layers of the soil. Extension of a deep root system would then allow soil exploration in search for water. In a similar manner, NMPB application in the interaction peanut-*Bradyrhizobium* sp. C-145 or SEMIA6144 promoted the development of a radical system extended in depth, that would be beneficial in the stress situation increasing the exploration surface for water sources.

Table 17.3 Effect of biostimulants addition on the relative water content (RWC), Root/Shoot length-ratio and Root/Shoot dry weight (DW)-ratio of peanut plants exposed to drought stress

	RWC (%)[a]		Root/Shoot length-ratio[a]		Root/Shoot DW-ratio[a]	
Treatments	Control	Drought stress	Control	Drought stress	Control	Drought stress
Uninoculated	96.77 ± 1.28 bc1	89.43 ± 2.13 c2	0.62 ± 0.05 ab1	0.59 ± 0.13 a1	0.30 ± 0.03 d1	0.26 ± 0.04 cd1
Fertilized	98.16 ± 1.03 c1	65.35 ± 4.59 b2	0.56 ± 0.05 a1	0.74 ± 0.06 ab1	0.25 ± 0.01 cd1	0.30 ± 0.03 d1
NMPB	87.22 ± 2.47 ab1	57.15 ± 1.32 ab2	0.67 ± 0.04 bc1	0.67 ± 0.12 a1	0.22 ± 0.02 bcd1	0.21 ± 0.01 bc1
Bradyrhizobium sp. SEMIA6144	86.38 ± 2.88 ab1	42.53 ± 3.89 a2	0.82 ± 0.05 cd1	0.68 ± 0.03 a2	0.18 ± 0.02 ab1	0.21 ± 0.04 bc1
NMPB + *Bradyrhizobium* sp. SEMIA6144	83.17 ± 1.88 a1	62.76 ± 7.43 a2	0.89 ± 0.07 cd1	0.91 ± 0.05 b1	0.19 ± 0.01 abc1	0.19 ± 0.01 bc1
Bradyrhizobium sp. C- 145	77.69 ± 4.91 a1	45.63 ± 6.32 a2	0.90 ± 0.12 cd1	0.80 ± 0.05 ab1	0.16 ± 0.01 a1	0.18 ± 0.02 ab1
NMPB + *Bradyrhizobium* sp. C-145	84.52 ± 2.53 a1	41.88 ± 7.35 a2	1.12 ± 0.08 d1	0.91 ± 0.11 b1	0.21 ± 0.02 bcd1	0.16 ± 0.01 a1

[a]Data represent the mean ± E.S. (n = 6). Different numbers per row indicate significant differences among the water status of peanut plants for each treatment, according to the Duncan's multiple range test ($p \leq 0.05$). Different letters per column indicate significant differences among treatments, for each water state according to Duncan's multiple range test ($p \leq 0.05$).

Upon exposition to drought stress, plants inoculated with *Bradyrhizobium* sp. SEMIA6144 and supplemented with NMPB, had an increased N content when compared to non-inoculated plants. On the other hand, *Bradyrhizobium* sp. C-145 promoted higher nodule number, nodule dry weight and normalized-nodule weight (the ratio among nodule dry weight and shoot dry weight). Besides, addition of the NMPB to plants inoculated with *Bradyrhizobium* sp. C-145 improved the mentioned variables and also the nitrogen content per plant (Table 17.4).

Drought stress adversely affects nodulation, nodule growth and BNF in different legumes, including peanuts (Furlan et al. 2014; Pimratch et al. 2008). The responses of nodule functioning to drought are related to closure of the oxygen diffusion barrier that reduces oxygen exposure of bacteroids causing a lack of energy to support the highly energy-dependent BNF (Serraj et al. 1999). Besides, Furlan et al. (2014) found that the peanut- *Bradyrhizobium* sp. SEMIA6144 interaction exposed to drought had a lower BNF, as a consequence of a decrease in nitrogenase activity, which was inhibited by 90%. In the present work, BNF, estimated on the nitrogen content basis, showed that NMPB addition was beneficial for the BNF process in the peanut–*Bradyrhizobium* sp. C-145 interaction, as shown by higher nitrogen content per plant, with respect to only inoculation (Table 17.4), suggesting that peanut BNF in drought stress conditions can be improved by the application of NMPB and the adequate inoculant.

17.3.2 Biostimulants effects on ROS production and oxidative stress-indicators

Drought stress showed the most contrasting responses among treatments and the more pronounced benefits by biostimulant utilization. Consequently, its redox state was evaluated. $O_2^{\cdot-}$ production was visualized by formation of a blue precipitant (formazan) resulting from the reaction of NBT with $O_2^{\cdot-}$ in drought-stressed leaves and nodules occupied by *Bradyrhizobium* sp. SEMIA6144. Interestingly, the interaction NMPB – MPB had a synergistic effect with respect to only MPB, limiting the production of the $O_2^{\cdot-}$ in leaves and nodules (Fig. 17.1a). The H_2O_2 content revealed that drought stress did not induce changes in the production of H_2O_2 in leaves and nodules when comparing distinct water regimes (Fig. 17.1b, c).

Abiotic stresses induce oxidative stress as a result of a ROS burst that exceeds the capability of the antioxidant defense system. In peanut, ABA triggers H_2O_2 production within 12 h of drought stress, which is then exacerbated by ABA-independent mechanisms (Furlan et al. 2013). In the present study, after 10 d of water withholding, H_2O_2 was not detected but evidences of its precursor ($O_2^{\cdot-}$) were found. Thus, we quantified the levels of several oxidative stress indicators.

Total chlorophyll content in plants inoculated with each bradyrhizobial strain and/or treated with NMPB, and exposed to drought stress, remained the same as that of the normally irrigated control plants. However, uninoculated or fertilized plants showed the lowest chlorophyll content upon exposition to drought. The use of NMPB together with the MPB *Bradyrhizobium* sp. SEMIA6144 turned out to be the treatment that increased the most the chlorophyll content (Fig. 17.2a).

Table 17.4 Effect of biostimulant addition on variables related to the nodulation process of peanut plants exposed to drought stress

	Nodule number[a]		Nodule dry weight (g)[a]		Normalized nodule weight[a]		Nitrogen content (mg $plant^{-1}$)[a]	
Treatments	Control	Drought stress	Control	Drought stress	Control	Drought stress	Control	Drought stress
Uninoculated	ND	ND	ND	ND	ND	ND	27.2 ± 3.2 ab1	13.1 ± 3.2 a2
Fertilized	ND	ND	ND	ND	ND	ND	23.6 ± 7.9 ab1	25.3 ± 1.2 ab1
NMPB	ND	ND	ND	ND	ND	ND	21.2 ± 2.7 a1	25.8 ± 3.7 ab1
Bradyrhizobium sp. SEMIA6144	35.9 ± 3.5 a1	33.5 ± 3.9 ab1	0.03 ± 0.00 a1	0.03 ± 0.00 a1	0.025 ± 0.004 a1	0.022 ± 0.003 a1	36.0 ± 2.6 c1	31.4 ± 3.9 b1
NMPB + *Bradyrhizobium* sp. SEMIA6144	41.2 ± 2.9 a1	26.4 ± 1.9 a2	0.04 ± 0.01 a1	0.03 ± 0.00 a1	0.026 ± 0.002 a1	0.029 ± 0.002 b1	36.4 ± 0.9 c1	32.7 ± 1.8 b1
Bradyrhizobium sp. C- 145	59.3 ± 5.2 b1	42.5 ± 2.3 bc2	0.05 ± 0.01 a1	0.04 ± 0.01 ab1	0.026 ± 0.005 a1	0.034 ± 0.002 b1	34.9 ± 1.3 bc1	19.1 ± 0.9 a2
NMPB + *Bradyrhizobium* sp. C-145	30.5 ± 6.1 a1	46.7 ± 4.8 c1	0.03 ± 0.01 a1	0.05 ± 0.01 b1	0.040 ± 0.010 a1	0.034 ± 0.005 b1	27.3 ± 6.3 abc1	38.9 ± 1.9 c1

[a]Data represent the mean ± SE. (n = 6). Different numbers per row indicate significant differences between the water status of peanut plants for each treatment according to the Duncan's multiple range test ($p \leq 0.05$). Different letters in columns indicate significant differences among treatments for each water state, according to the Duncan's multiple range test ($p \leq 0.05$)

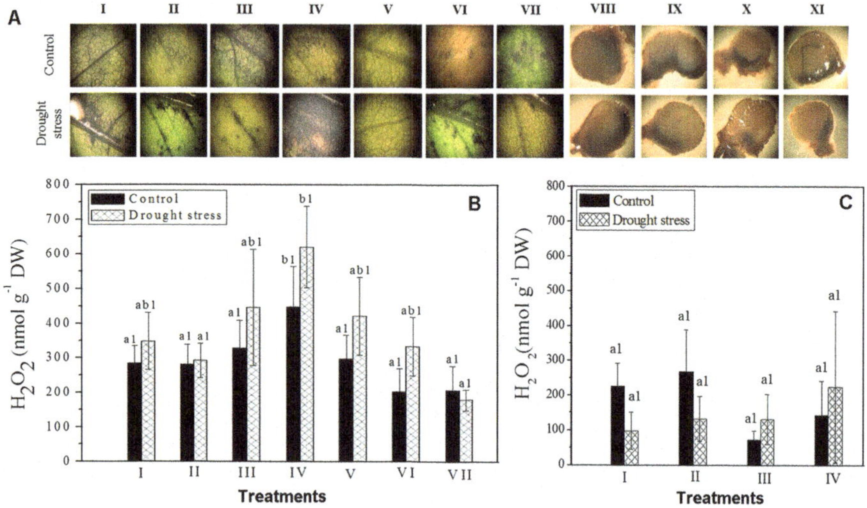

Fig. 17.1 Biostimulants addition effects on ROS production in peanut leaves and nodules exposed to drought stress. (**a**) $O_2^{\cdot -}$ production. (I) uninoculated; (II) fertilized by irrigation with 5 mM KNO_3; (III) NMPB; (IV) MPB *Bradyrhizobium* sp. SEMIA6144; (V) NMPB and MPB *Bradyrhizobium* sp. SEMIA6144; (VI) MPB *Bradyrhizobium* sp. C-145; (VII) NMPB and MPB *Bradyrhizobium* sp. C- 145; (VIII) MPB *Bradyrhizobium* sp. SEMIA6144; (IX) NMPB and MPB *Bradyrhizobium* sp. SEMIA6144; (X) MPB *Bradyrhizobium* sp. C-145; (XI) NMPB and MPB *Bradyrhizobium* sp. C- 145. (**b**) H_2O_2 content in peanut leaves (the numbers correspond to the treatments mentioned in panel A). (**c**) H_2O_2 content in peanut nodules. (I) MPB *Bradyrhizobium* sp. SEMIA6144; (II) NMPB and MPB *Bradyrhizobium* sp. SEMIA6144; (III) MPB *Bradyrhizobium* sp. C-145; (IV) NMPB and MPB *Bradyrhizobium* sp. C- 145. Data represent the mean ± SE ($n = 6$). Different numbers per column indicate significant differences between the water status for each treatment and different letters per column indicate significant differences between treatments for each water status according to the Duncan's multiple range test ($p \leq 0.05$)

TBARs content of plants exposed to drought stress increased only in plants inoculated with the MPB *Bradyrhizobium* sp. SEMIA6144 and in those treated with NMPB and inoculated with this strain. In nodules of peanut plants, TBARs were not detected (Fig. 17.2b, c).

Proline quantification showed that drought stress induced accumulation of proline, with the exception of plants treated with *Bradyrhizobium* sp. SEMIA6144 alone or together with NMPB and in those plants inoculated with *Bradyrhizobium* sp. C-145 without NMPB addition. In nodules occupied by *Bradyrhizobium* sp. SEMIA6144, proline accumulation was induced in response to drought stress (Fig. 17.2d, e).

A hyper-accumulation of ROS can affect the cell-membrane integrity by reacting with membrane lipids through a process known as lipid peroxidation, which can be estimated by TBARs content (Esterbauer and Cheeseman 1990). In this work, the symbiotic association peanut-*Bradyrhizobium* sp. SEMIA6144 showed oxidative stress in the aerial part of drought-stressed plants which can be associated with the

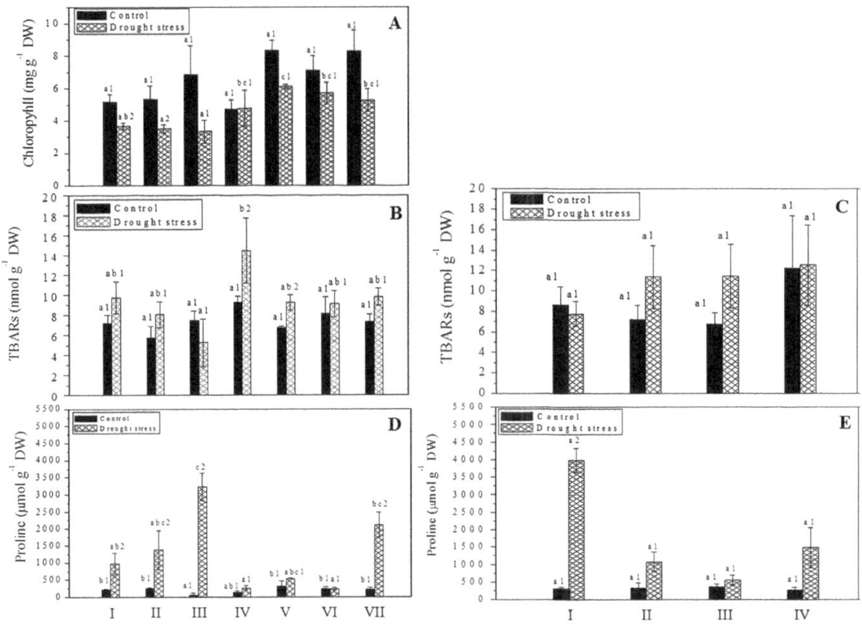

Fig. 17.2 Biostimulants addition effects on oxidative stress markers-levels. (**a**) Chlorophyll content; (**b**, **c**) TBARs content; (**d**, **e**) proline content. Panels A, B and D present quantifications made in leaves in the following treatments: (I) uninoculated; (II) fertilized by irrigation with 5 mM KNO_3; (III) NMPB addition; (IV) MPB *Bradyrhizobium* sp. SEMIA6144; (V) NMPB and MPB *Bradyrhizobium* sp. SEMIA6144 addition; (VI) MPB *Bradyrhizobium* sp. C-145; (VII) NMPB and MPB *Bradyrhizobium* sp. C- 145 addition. Panels C and E present quantifications made in nodules in the following treatments: (I) MPB *Bradyrhizobium* sp. SEMIA6144; (II) NMPB and MPB *Bradyrhizobium* sp. SEMIA6144 addition; (III) MPB *Bradyrhizobium* sp. C-145 addition; (IV) NMPB and MPB *Bradyrhizobium* sp. C- 145 addition. Data represent the mean ± SE (n = 6). Different numbers per column indicate significant differences between the water status for each treatment and different letters per column indicate significant differences between treatments for each water status according to the Duncan's multiple range test ($p < 0.05$)

$O_2^{\cdot -}$ accumulation. On the other hand, the low content of TBARs in nodules suggests that this organ would present an active antioxidant defense system that would allow maintainance of the cellular redox balance.

The negative effects of drought stress on plant growth are triggered by the reduction of leaf area and stomatal conductivity and the over-excitation of photosystems with consequent photo-oxidative damage in photosystem II, within which chlorophylls are the main components (Demmig-Adams and Adams 1992; Reddy et al. 2003). In our work, NMPB and MPB addition enhanced total chlorophyll content under stress conditions compared to uninoculated plants, associated with low oxidative stress symptoms. Nitrogen is a structural element of chlorophylls and proteins (Amaliotis et al. 2004), and this very close relationship was observed in the inoculated plants which had the highest nitrogen and total chlorophyll contents, under drought stress conditions.

In general, drought-tolerant plants show some physiological characteristics such as high RWC, high proline content and low peroxidation of membrane lipids under water deficit (Naser et al. 2010; Liu et al. 2013). Similarly, in this work, plants treated with NMPB, alone or together with *Bradyrhizobium* sp. C-145 inoculation, accumulated the highest amount of proline in the drought stress condition, and simultaneously, recorded a low TBARs content, an improvement in growth variables in comparison to the noninoculated plants, which could explain peanut tolerance to drought.

Principal components analysis (PCA) allows ordering and representing multivariate data through combinations of the original variables simplifying its visualization. The analysis revealed that variables that contributed to the explanation of the variance of the first principal component (PC1) were nitrogen content, dry weight of roots and dry weight of aerial part. The metabolites that contributed the most to the second principal component (PC2) were chlorophyll and proline contents. The biplot analysis showed that treatment of peanut plants with NMPB jointly with the inoculation with *Bradyrhizobium* sp. C-145 was positively associated with an improvement in the growth and nitrogen content variables and the least oxidative damage and greater accumulation of proline. In addition, peanut plants inoculated with *Bradyrhizobium* sp. SEMIA6144 showed the lowest proline content and the highest oxidative damage (Fig. 17.3).

In the literature there is little information about how NMPBs and MPBs can limit the production of ROS in plants exposed to unfavorable growth conditions. The results obtained in our experimental system allow us to suggest that their combined usage is associated with low ROS production and oxidative damage improving plant growth and nodulation.

17.4 Conclusion

Inoculation of peanut plants together with addition of the tested NMPB reduced As translocation to leaves, preventing possible metalloid accumulation in peanut grains and therefore human contamination. Besides, NMPB application in the interaction peanut-*Bradyrhizobium* sp., improved plant growth in the drought stress condition promoting biomass production and development of a radical system extended in depth that would be beneficial in the stress situation increasing the exploration surface for water. A lower production of ROS and less oxidative damage was associated with proline accumulation and improvements in growth and nodulation variables in the peanut-*Bradyrhizobium* sp. C-145 interaction, in the presence of the NMPB. Thus, selection of the best combination of microbe and non-microbe plant biostimulants is advised for peanut crops growing in regions susceptible to the occurrence of water deficit or arsenic exposure.

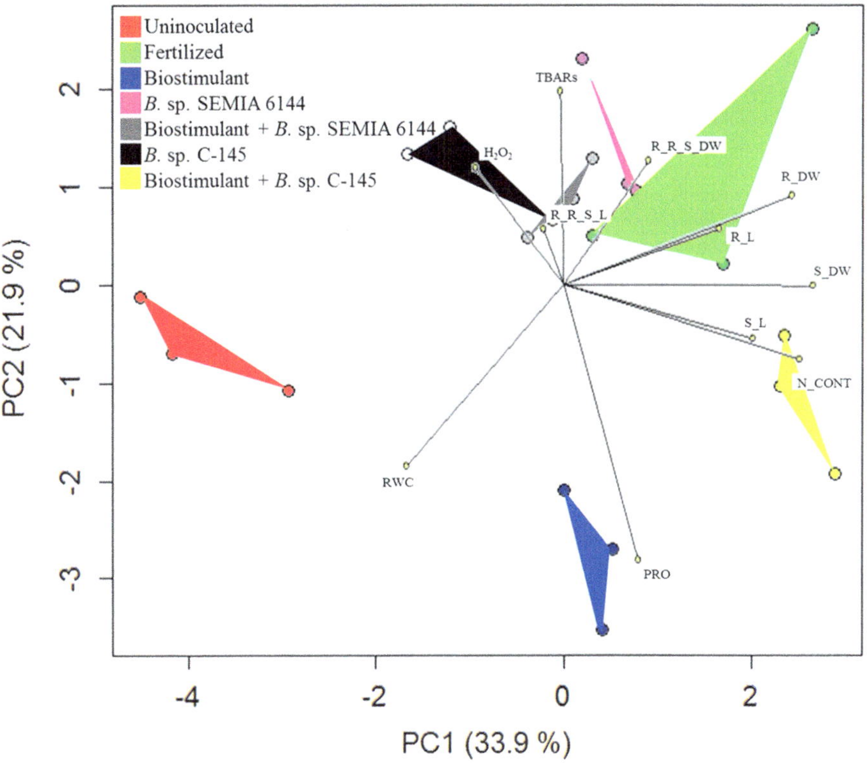

Fig. 17.3 Principal component analysis (PCA) of the 11 variables evaluated in peanut shoots of plants exposed to the different treatments. The values are expressed as the log_2 ratio of stressed/control values. The lines represent the convex hull for each treatment. H_2O_2: hydrogen peroxide; N_CONT: nitrogen content; PRO: proline; S_DW: shoot dry weight; S_L: shoot length; R_DW: root dry weight; R_L: root length; R_R_S_DW: root/shoot dry weight ratio; R_R_S_L: root/shoot length ratio; RWC: relative water content; TBARs: thiobarbituric acid-reactive substances

Acknowledgements The authors thank Secretaría de Ciencia y Técnica (Universidad Nacional de Río Cuarto); FONCYT PICT 2015-2104; FONCYT PICT 2014-0956; NUTRIFER S.A. for providing financial assistance for this research. AF and EB are members of the research career from CONICET. JMP has a scholarship from CONICET.

References

Akerblom, S., Baath, E., Bringmark, L., & Bringmark, E. (2007). Experimentally induced effects of heavy metal on microbial activity and community structure of forest more layers. *Biology and Fertility of Soils, 44*, 79–91.

Alexieva, V., Sergiev, I., Mapelli, S., & Karanov, E. (2001). The effect of drought and ultraviolet radiation on growth and stress markers in pea and wheat. *Plant, Cell & Environment, 24*, 1337–1344.

Amaliotis, D., Therios, I., & Karatissiou, M. (2004). Effect of nitrogen fertilization on growth, leaf nutrient concentration and photosynthesis in three peach cultivars. *ISHS, Acta Horticulture, 449*, 36–42.

Arnon, D. (1949). Copper enzymes in chloroplasts polyphenoloxydases in *Beta vulgaris*. *Journal of Plant Physiology, 24*, 1–15.

Baliña, R. M., Díaz-Zorita, M., & Kearney, M. I. T. et al. (2013). *Combinación de microorganismos y rendimiento de maní*. XXVIII Jornada Nacional del Maní. INTA – CIA Cabrera. General Cabrera, Córdoba, Argentina.

Barr, H., & Weatherley, P. (1962). A re-examination of the relative turgidity technique for estimating water deficit in leaves. *Australian Journal of Biological Sciences, 15*, 413–428.

Bates, L. S., Waldren, R. P., & Teare, I. D. (1973). Rapid determination of free proline for water stress study. *Plant and Soil, 39*, 205–207.

Bianucci, E., Furlan, A., Tordable, M. D. C., et al. (2017). Antioxidant responses of peanut roots exposed to realistic groundwater doses of arsenate: Identification of glutathione S-transferase as a suitable biomarker for metalloid toxicity. *Chemosphere, 181*, 551–561.

Bianucci, E., Godoy, A., Furlan, A., Peralta, J. M., Hernández, L. E., Carpena-Ruiz, R. O., & Castro, S. (2018). Arsenic toxicity in soybean alleviated by a symbiotic species of *Bradyrhizobium*. *Symbiosis, 74*(3), 167–176.

Boote, K. (1992). Growth stages of peanut (*Arachis hypogaea* L.). *Peanut Science, 9*, 35–40.

Burk, R. (1996). *Soil survey laboratory methods manual*. Soil Survey Investigations Report 42, Ver. 3.0. (National Soil Survey Center: Lincoln, NE).

Bustingorri, C., & Lavado, R. S. (2014). Soybean as affected by high concentrations of arsenic and fluoride in irrigation water in controlled conditions. *Agricultural Water Management, 144*, 134–139.

Cabrera, A., Blarasin, M., Matteoda, E., Villalba, G., Gomez, M. L. (2005). *Composición química del agua subterránea en el sur de córdoba: línea de base hidroquímica o fondo natural en referencia a arsénico y flúor*. http://www.produccion-animal.com.ar. Accessed 1 March 2018.

Calvo, P., Nelson, L., & Kloepper, J. W. (2014). Agricultural uses of plant NMPBs. *Plant and Soil, 383*, 3–41.

Cerioni, G. A., Morla, F. D., Kearney, M. I. T., Mattana, F., Bassino, S., Pironello, A., Giayetto, O., Fernandez, E. M., Righi, D., & Stefani, R. (2014). Efecto de bioestimulantes e inoculante sobre el crecimiento y rendimiento en el cultivo de maní. XXIX Jornada Nacional de Maní. INTA - CIA Cabrera. General Cabrera, Córdoba, Argentina.

Clavel, D., Diouf, O., Khalfaoui, J. L., et al. (2006). Genotypes variations in fluorescent parameters along closely related groundnut (*Arachis hypogaea* L.) lines and their potential for drought screening programs. *Field Crops Research, 96*, 296–306.

Demmig-Adams, B., & Adams, W. W. (1992). Photoprotection and other responses of plants to high light stress. *Annual Review of Plant Physiology and Plant Molecular Biology, 43*, 599–626.

Doke, N. (1983). Involvement of superoxide anion generation in the hypersensitive response of potato tuber tissues to infection with an incompatible race of *Phytophora infestans* and to the hyphall walls components. *Physiological Plant Pathology, 23*, 345–357.

Esterbauer, H., & Cheeseman, K. H. (1990). Determination of aldehydic lipid peroxidation products: malonaldehyde and 4-hydroxynonenal. *Methods in Enzymology, 186*, 407–421.

Fernandez, E. M., & Giayetto, O. (Eds). (2017). *El cultivo del maní en Córdoba*. Uni Rio. 144 pp.

Frahry, G., & Schopfer, P. (2001). NADH-stimulated, cyanide-resistant superoxide production in maize coleoptiles analyzed with a tetrazolium-based assay. *Planta, 212*(2), 175–183.

Francisca, F. M., Celollada-Verdaguer, M. P., & Carro-Pérez, M. E. (2006) *Presented in part at Conference VIII Congreso Latinoamericano de hidrología subterránea*. Distribución espacial del arsénico en las aguas subterráneas de la provincia de Córdoba, Argentina. Asunción.

Furlan, A., Llanes, A., Luna, V., et al. (2013). Abscisic acid mediates hydrogen peroxide production in peanut induced by water stress. *Biologia Plantarum, 57*(3), 555–558.

Furlan, A. L., Bianucci, E., Tordable, M., et al. (2014). Antioxidant enzyme activities and gene expression patterns in peanut nodules during a drought and rehydration cycle. *Functional Plant Biology, 2014*(41), 704–713.

Furlan, A., Bianucci, E., Tordable, M., et al. (2016). Dynamic responses of photosynthesis and antioxidant system during a drought and rehydration cycle in peanut plants. *Functional Plant Biology, 43*(4), 337–345.

Heath, R. L., & Packer, L. (1968). Photoperoxidation in isolated chloroplasts. I. Kinetics and stoichiometry of fatty acid peroxidation. *Archives of Biochemistry and Biophysics, 25*, 189–198.

Hoagland, D., & Arnon, D. (1950). The water culture method for growing plants without soil. *California Agriculture (Berkeley), 347*, 1–39.

Jongrungklang, N., Toomsan, B., Vorasoot, N., et al. (2011). Rooting traits of peanut genotypes with different yield responses to pre-flowering drought stress. *Field Crops Research, 120*, 262–270.

Kearney, M. I. T., Cerioni, G. A., Stefani, R. (2011). *Bioestimulante aplicado a la semilla de maní sobre la emergencia, el rendimiento y la calidad*. Paper presented at the 26° Jornada Nacional de Maní. 15 de septiembre de 2011. General Cabrera, Córdoba (AR): INTA – CIA. pp. 90–92.

Lafuente, A., Pajuelo, E., Caviedes, M. A., & Rodríguez-Llorente, I. D. (2010). Reduced nodulation in alfalfa induced by arsenic correlates with altered expression of early nodulins. *Journal of Plant Physiology, 167*(4), 286–291.

Lafuente, A., Pérez-Palacios, P., Doukkali, B., Molina-Sánchez, M. D., Jiménez-Zurdo, J. I., Caviedes, M. A., Rodríguez-Llorente, I. D., & Pajuelo, E. (2015). Unraveling the effect of arsenic on the model *Medicago-Ensifer* interaction: A transcriptomic meta-analysis. *The New Phytologist, 205*, 255–272. https://doi.org/10.1111/nph.13009.

Liu, H., Sultan, M. A. R. F., & Zhao, H. X. (2013). The screening of water stress tolerant wheat cultivars with physiological indices. *GJESM, 3*, 211–218.

Mandal, S. M., Pati, B., Das, R., Amit, K., & Ghosh, K. A. (2008). Characterization of a symbiotically effective *Rhizobium* resistant to arsenic: Isolated from root nodules of *Vigna mungo* (L.) Hepper grown in arsenic contaminated field. *The Journal of General and Applied Microbiology, 54*, 93–99.

Mhadhbi, H., Mylona, P. V., & Polidoros, A. N. (2015). Legume-rhizobia symbiotic performance under abiotic stresses.In M. M. Azooz & P. Ahmad (Eds.), *Legumes under Environmental Stress*. https://doi.org/10.1002/9781118917091.ch8.

Morla, F. D., Kearney, M. I. T., & Cerioni, G. A. (2013). *Bioestimulantes en cultivos*. II maní Paper presented at the XIX Jornadas científicas – SBC – La Falda, Córdoba, Argentina 8–10 agosto 2013.

Naser, L., Kourosh, V., & Bahman, K. (2010). Soluble sugars and proline accumulation play a role as effective indices for drought tolerance screening in Persian walnut (*Juglans regia* L.) during germination. *Fruits, 65*, 97–112.

Nelson, D., & Sommers, L. (1973). Determination of total nitrogen in plant material. *Agronomy Journal, 65*, 109–112.

Nutrifer S.A. (2017) *Propuesta de nutrición en maní*. In: http://www.nutriter.com.ar/productos_cultivo_otros_cultivos_4/

Pajuelo, E., Rodríguez-Llorente, I. D., Dary, M., & Palomares, A. J. (2008). Toxic effects of arsenic on *Sinorhizobium-Medicago sativa* symbiotic interaction. *Environmental Pollution, 154*, 203–211.

Panigrahi, D. P., & Randhawa, G. S. (2010). A novel method to alleviate arsenic toxicity in alfalfa plants using a deletion mutant strain of *Sinorhizobium meliloti*. *Plant and Soil, 336*, 459–467.

Panigrahi, D. P., Sagar, A., Dalal, S., & Randhawa, G. S. (2013). Arsenic resistence and symbiotic efficiencies of alfalfa and cowpea rhizobial strain isolated from arsenic agricultural fields. *The Journal of Experimental Biology, 3*, 322–333.

Paudyal, S. P., Aryal, R. R., Chauhan, S. V. S., & Maheshwari, D. K. (2007). Effect of heavy metals on growth of *Rhizobium*strains and symbiotic efficiency of two species of tropical legumes. *Science World, 5*, 27–32.

Pedelini, R. (2008). *Maní: Guía práctica de su cultivo*. General Cabrera, Córdoba: OT General Cabrera- INTA EEA Manfredi. Boletín de divulgación técnica N° 2.

Peralta, J. M., Travaglia, C. N., Gil, R. A., Furlan, A., Castro, S., Bianucci, E. C. (2018). *An effective rhizoinoculation restraints arsenic translocation in peanut and maize plants exposed to a realistic groundwater metalloid dose*. International Congress and Exhibition: Arsenic In The Environment, Beijing, P.R.. Environmental Arsenic in a Changing World As 2018. China 2018.

Pimratch, S., Jogloy, S., & Vorasoot, B. (2008). Effect of drought stress on traits related to N_2 fixation in eleven peanut (*Arachis hypogaea* L.) genotypes differing in degrees of resistance to drought. *Asian Journal of Plant Sciences, 7*, 334–342.

Puangbut, D., Jogloy, S., & Vorasoot, N. (2009). Association of root dry weight and transpiration efficiency of peanut genotypes under early season drought. *Agricultural Water Management, 96*(10), 1460–1466.

Reddy, T. Y., Reddy, V. R., & Anbumozhi, V. (2003). Physiological responses of groundnut (*Arachis hypogaea* L.) to drought stress and its amelioration: A review. *Acta Agronomica Hungarica, 51*, 205–227.

Reichman, S. M. (2007). The potential use of the legume-rhizobium symbiosis for the remediation of arsenic contaminated sites. *Soil Biology and Biochemistry, 39*, 2587–2593.

Serraj, R. (2003). Effects of drought stress on legume symbiotic nitrogen fixation: Physiological mechanisms. *Indian Journal of Experimental Biology, 41*(10), 1136–1141.

Serraj, R., Sinclair, T. R., & Purcell, L. C. (1999). Symbiotic N_2 fixation response to drought. *Journal of Experimental Botany, 50*, 143–155.

Signorelli, S., Coitiño, E., & Borsani, O. (2014). Molecular mechanisms for the reaction between •OH radicals and proline: Insights on the role as reactive oxygen species scavenger in plant stress. *The Journal of Physical Chemistry. B, 118*, 37–47.

Smedley, P. L., & Kinniburgh, D. G. (2002). A review of the source, behaviour, and distribution of arsenic in natural waters. *Applied Geochemistry, 17*, 517–568.

Sobolev, D., & Begonia, M. (2008). Effects of heavy metal contamination upon soil microbes: Lead induced changes in general and denitrifying microbial communities as evidenced by molecular markers. *International Journal of Environmental Research and Public Health, 5*, 450–456.

Sobrino-Plata, J., Ortega-Villasante, C., Flores-Cáceres, M. L., et al. (2009). Differential alterations of antioxidant defenses as bioindicators of mercury and cadmium toxicity in alfalfa. *Chemosphere, 77*, 946–954.

Somasegaran, P., & Hoben, H. (1994). Quantifying the growth of rhizobia. In *Handbook for rhizobia* (pp. 165–169). New York: Springer-Verlag Inc..

Szabados, L., & Savouré, A. (2009). Proline: A multifunctional amino acid. *Trends in Plant Science, 15*, 89–97.

Valetti, L., Angelini, J., Cerioni, G., et al. (2008). *Desarrollo y evaluación a campo de un inoculante para maní elaborado a partir de aislamientos rizobianos nativos de la zona mansera de la provincia de Córdoba*. Paper presented at the XXIII Jornada Nacional de Maní. Gral. Cabrera – Córdoba. (pp. 36–38).

Vázquez, S., Goldsbrough, P., & Carpena, R. O. (2009). Comparative analysis of the contribution of phytochelatins to cadmium and arsenic tolerance in soybean and white lupin. *Plant Physiology and Biochemistry, 47*, 63–67.

Vincent, J. (1970). *A manual for the practical study of root nodule bacteria*. En: IBP Handbook N° 15. Oxford: Blackwell Scientific Publication.

Wright, G. C., Nageswara Rao, R. C. (1994). Groundnut water relations. In: Smartt, J (Ed). The groundnut crop, a scientific basic for improvement. Chapman & Hall. Londres. Cap. 9. p. 281–335.

Yakhin, O. I., Lubyanov, A. A., & Yakhin, I. A. (2017). Biostimulants in plant science: A global perspective. *Frontiers in Plant Science, 7*, 2049. https://doi.org/10.3389/fpls.2016.02049.